Bel-Tib New Non-Fiction
577.0911 Sale 2017
The Arctic  : the complete story
31111041584945

P9-DYE-880

Richard Sale and Per Michelsen

# The Arctic

Whittles Publishing

Published by
**Whittles Publishing Ltd.,**
Dunbeath,
Caithness, KW6 6EG,
Scotland, UK

**www.whittlespublishing.com**

ISBN 978-184995-342-9

© 2018 Richard Sale and Per Michelsen

*All rights reserved.*
*No part of this publication may be reproduced,*
*stored in a retrieval system, or transmitted,*
*in any form or by any means, electronic,*
*mechanical, recording or otherwise*
*without prior permission of the publishers.*

*Printed in Malta by Melita Press*

# Contents

# Preface

In 1829 John Ross sailed to the Arctic. He had been before, commanding one of the first attempts by the British Royal Navy to find a North–West Passage. The expedition had not been a success, but for Ross himself it had been a disaster. Vilified for his apparent lack of gumption, he had been ignored as a succession of further expeditions set out to find a passage. In 1829 Ross was in command of a private expedition seeking the elusive route. He overwintered in the ice, something he had avoided on his first trip. It must have been a traumatic experience as, writing in the spring of 1830 of the Inuit he had encountered, he noted that it was 'for philosophers to interest themselves in speculating on a horde so small, and so secluded, occupying so apparently hopeless a country, so barren, so wild, and so repulsive; and yet enjoying the most perfect vigour, the most well-fed health, and all else that here constitutes, not merely wealth, but the opulence of luxury; since they were as amply furnished with provisions, as with every other thing that could be necessary to their wants'.

In that one, succinct, sentence Ross encapsulated the lure of the Arctic for travellers from temperate regions to the south. Here was a wilderness populated by a people and animals that not only survived its harshness but seemed to thrive, to the fascination of scientists and laymen alike. The country was, as Ross contended, wild, but its wildness was also its beauty. Travellers discovered a land that could not only be harsh and unforgiving, but seemingly of rare beauty. In winter, of crystal and silent cold, at times filled by the ghostly pale, trembling light of the aurora. In summer diffused with a light of breathtaking purity. The light often illuminated a monochromatic landscape, white geese and swans on a black tundra, white ice on a dark sea, but occasionally finding patches of glowing colour. A land where the sun, when it appeared after the Arctic night, could be cold and red and dishevelled, not the sun they knew. A land that seemed empty, with the people and animals thinly spread so the loneliness could be awesome.

Early travellers brought back tales of amazing creatures and of the endurance required of visitors, the Arctic becoming a land of inspiration and imagination. When Mary Shelley wrote her tale of Dr Frankenstein and the creature he created, she ends with the creature heading towards the North Pole.

The Arctic still inspires. Adventurers test themselves against it. Its wildlife still amazes – when film and television show Earth's natural wonders it is always the polar regions that draw the biggest audiences.

But today the Arctic is in retreat. Humanity's relentless exploitation of the Earth's resources in the pursuit of progress has, it seems, altered the climate and threatens the ice and ice-living organisms. It is a cliché that the loss of a species diminishes us, but it is true nonetheless. Even to people who have never seen a Polar Bear its loss will be immeasurable as the bear is iconic, both defining and reflecting the Arctic.

This book celebrates the Arctic, exploring the nature and scenery, the history and the natural history that has so inspired generations. It ends with an assessment of the Arctic's future: it is a bleak one, but while there is chance to save this wonderful place we should strive to do so.

Please note that within the text the reader may encounter terms with which they are not familiar. To enable us to concentrate on the extensive and comprehensive coverage of the subject a glossary has not been included, but definitions for such terms will be readily found online.

Midnight, 21 June. Looking west from the sea ice of Baffin Bay. To the left is Baffin Island, to the right Bylot Island.

# 1 Defining the Arctic

Though the fact that stars moved around the night sky and the Sun rose and set at different times throughout the year would doubtless have been known to the first human observers, this knowledge was only written down, by the Chaldeans of what is now southern Iraq, around 5,000 years ago. Later, the Greeks systematised the knowledge, naming individual star patterns and noting how they moved. The Greeks realised the Earth was spherical and they knew that at noon on any day of the year the Sun would be directly overhead at a given latitude on Earth. From this they reasoned that if they travelled far enough north they would reach a point where the Sun would be visible all day at midsummer, and absent all day at midwinter. Noting the rising and setting of star formation allowed the Greeks to construct a celestial sphere, on which they drew lines that corresponded to the extreme latitudes at which the Sun was directly overhead on at least one day. These lines passed through the constellations of Cancer, to the north, and Capricorn, to the south. Projected onto the Earth these are the Tropics. And when they projected the latitude of where the Sun would be continuously present or absent for six months at a time the Greeks found it went through a constellation they called *Arktikos*, the Great Bear. They therefore called the latitude at which this phenomenon would occur the Bear's Circle – the Arctic Circle.

Earth and the other planets of the Solar System orbit the Sun on, or close to, a plane known as the plane of the ecliptic. As well as orbiting along this plane, the planets also rotate, the Earth's rotation being about an axis through the North and South Poles. If this axis was at a right angle to the plane of the ecliptic then for all points at the same latitude, both north and south of the Equator, the lengths of night and day would be the same throughout the year. But the axis is at an angle. Known as the equatorial inclination, the angle is 23.44°, and the Arctic Circle lies at a latitude

> ### The Great Bear
>
> Astronomers still use the names of stars and constellations set down by the Greek astronomer Ptolemy in the 2nd century AD, and so have retained the Greek name for the Great Bear, though it has been converted to its Latin form – Ursa Major. To casual observers it is the Plough, the stellar pattern that allows Polaris, the Pole Star (and therefore north) to be located. As the patterns of stars can be interpreted as many different forms – though the attributions of the Greeks are now, more or less, accepted everywhere – it is fascinating that both the Greeks and the natives of northern North America had bear myths associated with the constellation of Ursa Major. In North America it was said that the first bear was lowered to earth in a golden cradle. The cradle was then returned to the sky where it became Ursa Major.

of 90-23.44° = 66.56°, a latitude more usually written as 66°34'N.

Because of equatorial inclination the Sun is visible for six months each year at the North Pole. For the other six months the Sun does not rise above the horizon. At the South Pole the reverse is true.

If we stood at the North Pole during the northern summer, we would see the Sun circling the sky. On any given day the elevation of the Sun is (more or less) constant, but the elevation changes each day. The elevation reaches a maximum (of 23.44° = 23°26') at Midsummer's Day, and then falls until the autumnal equinox when the Sun skims the horizon. On that day the Sun appears at the South Pole, heralding the beginning of the austral summer. For an observer at the South Pole the Sun now rises each day, reaching a maximum elevation (again of 23°26') at the southern summer solstice, which corresponds to the northern winter solstice. This is illustrated in Figures 1.1 and 1.2.

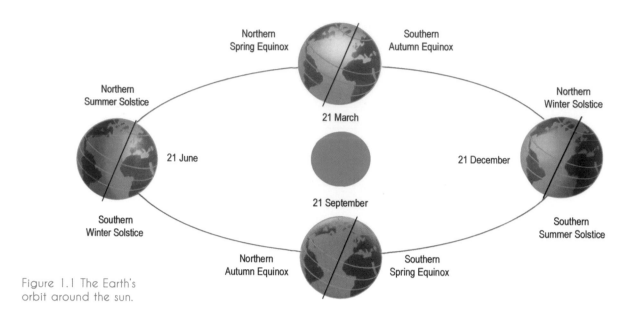

Figure 1.1 The Earth's
orbit around the sun.

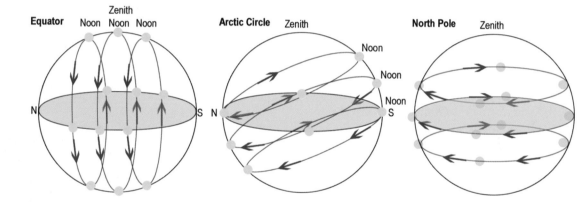

Figure 1.2 The suns path at solstice and equinox. For the
Equator the three parts are, from the left, June solstice,
Equinox and December solstice; for the Arctic Circle and
North Pole the paths are, top to bottom, June solstice,
Equinox and December solstice.

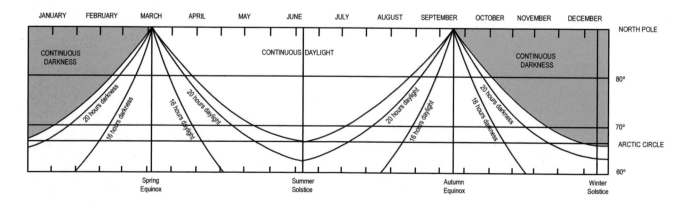

Figure 1.3 Variation of daylight hours
with lattitude throughout the year.

Figure 1.3 shows the variation of the hours of daylight with time of the year for places north of 60°.

However, while it is a neat idea that we at the North Pole would see the Sun continuously for six months and then not at all for the next six months, the phenomenon of atmospheric refraction (the deflection of light by the atmosphere) means that the Sun may appear above its true position by about 2½ times its diameter. The Midnight Sun – the romantic name given to the phenomenon of the continuously visible Sun – is therefore visible at sea level for about 150km south of the Arctic Circle. The phenomenon also means that the Arctic summer lasts longer than the Arctic winter, by an amount that increases with distance north: at the North Pole the summer is about 16 days longer than the winter. Refraction can also cause the Sun to rise after it has set for the Arctic winter, or to rise early for the Arctic summer. One of the most extreme examples of the latter occurred during the 1596–97 overwintering of the Dutch expedition of Willem Barents on Novaya Zemlya, when the Sun appeared almost two weeks before it was due to rise. Such images are usually distorted or broken.

Given that the Arctic Circle defines that part of the Earth which experiences the cold and dark northern winter, it would seem the obvious definition for the area. However, in reality the Circle has very limited climatic significance and, therefore, limited significance for either people or Arctic wildlife. To the west, an influx of cold air and cold water chills North America, while to the east the North Atlantic Drift (more usually called the Gulf Stream, though the two, while related, are not actually the same) moves vast quantities of warm water to north-west Europe. This warm water, and the warm, damp air above it, has a huge influence on the climate of the region, particularly on the British Isles and Norway. North of the Arctic Circle in Norway there are large towns, and both industry and agriculture are possible. The effect of the Drift is less pronounced in Sweden and Finland, though both benefit to an extent, and it is eventually lost altogether in western Russia. The effect is most clearly seen in Svalbard, the Arctic archipelago that lies to the north of Norway. The capital of the archipelago, Longyearbyen on Spitsbergen, the largest island, lies at 78°N. Longyearbyen is a town with hotels

Noon, 21 December, The Arctic Circle lies a few kilometres north of Rovaniemi in northern Finalnd. At the Circle a collection of buildings is devoted to Christmas, visitors arriving from across the world to post cards and letters bearing a Santa Claus postmark.

Arcitc Circle indicator board on the Dalton Highway, Alaska.

One suggested definition avoids the problems created by climate and geography and considers only the incident solar energy on the Earth. In the Arctic the Sun is always at a low angle in the sky and, as a consequence, light from it must traverse more of the atmosphere to reach the ground, losing energy to absorption and scattering on the way. Because of the low angle the Sun also illuminates a larger area of the Earth than it does at the Equator, for example. Both reduce the energy input per unit surface area. A proposal from the 1960s defined the Arctic as covering an area where the incident energy was less than 15kcal/cm²/year. This definition, though scientifically sound, has the disadvantage of not being an easily recognisable unit or feature in the way that, say, temperature or the treeline (the northern limit of trees) are. The search for a definition was therefore transferred to these options.

Use of a definition based on a land feature has one distinct advantage: the large land masses surrounding the Arctic mean that the extrapolation required across water is much more limited than that required by a sea-based definition. Use of the treeline appears to smooth out climatic differences: in Europe, the influence of the North Atlantic Drift allows tree growth well north of the Arctic Circle, while in North America trees have generally faded away before 60°N is reached. Although attractive, use of the treeline brings its own problems. One is that a tree must be defined, a definition which is far from trivial. Artic Willow (*Salix polaris*) is a tree by all the usual definitions, even changing the colour of its leaves in autumn as the southerly deciduous trees do. Yet in places where the trees have to contend with severe winds and extreme cold, Arctic Willow occasionally grows to a height of just a few centimetres; travellers from temperate places are used to walking beneath a forest canopy, but in the Arctic they may be effectively walking on the canopy. Another problem is that the treeline is not as precise and easily identified boundary as might be expected. Local geology and geography, as well as local climatic effects, play a role in defining the habitability of an area. Ground elevation and aspect, drainage, and soil composition all influence plant growth so that occasionally patches of forest exist to the north of areas, sometimes significant areas, of tree-free tundra. On paper the treeline is a solid, immutable line, yet on the ground it is rather less substantial, forming a band

and shops, and it is served by an airport with scheduled flights from Norway that arrive and depart throughout the year. At the same latitude in North America the land is essentially uninhabitable (for people), as are places on the same latitude in those parts of Eurasia that do not receive any warming effect from the North Atlantic Drift.

The Antarctic Circle, which crosses continental Antarctica in several places, is not used to define the southern polar region, a more convenient boundary being the Antarctic Convergence, where cold polar waters meet warmer waters from further north. But in the north a search for an Arctic Convergence is defeated by geography. The Arctic is an ocean, a frozen ocean, largely surrounded by land, whereas Antarctica is land surrounded by ocean: an Arctic Convergence exists, but it is discontinuous and much less clearly defined. While comparing the two polar regions it is worth noting other major differences. There are no terrestrial mammals in Antarctica, no birds beyond 70°S and limited plant life beyond 80°S; there are many mammals in the Arctic, many birds beyond 70°N, and abundant plant life beyond 80°N.

Another sea-based suggestion in our search for a definition of the Arctic is the southern limit of pack ice (the frozen Arctic Ocean), but this too has problems: the limit is seasonal, there are unpredictable annual variations, and global warming, as we shall see later in the book, is causing sea-ice coverage to shrink. It would also be extremely difficult to interpolate the position of the pack-ice edge across land masses.

over which the transition from true boreal forest to true tundra occurs. In Siberia this transitional band can be as much as 300km wide. A further complication is that there are places where local conditions (e.g. shelter from wind) allow trees to grow north of the latitude at which those species can propagate by seed formation, because the summer is neither warm enough nor long enough for this process to occur. Though the ancestors of these trees all germinated from seeds, climatic changes mean their seeds cannot do the same. They instead propagate by a form of 'suckering', in which branches that touch the ground produce roots from which new trunks grow. The branch dies off but the new tree continues to grow. In some places whole stands of trees grow, each a clone of the original parent tree.

To overcome the difficulties of treeline definition, scientists turned to a temperature-based definition. The initial proposal was use of the 10°C summer isotherm, a line that links points on the Earth's surface at which the mean temperature of the warmest month of the year is 10°C. The isotherm has the advantage of being closely aligned to the treeline. It is usually assumed that the factor limiting the northerly spread of trees is the cold; that is correct, but not in the sense that is usually inferred. It is not winter cold that is the limiting factor – in Siberia trees grow at a latitude that experiences the lowest winter temperatures recorded in the northern hemisphere. The limit is summer cold. In the Arctic summer there is abundant light, but the tree can only utilise this energy source if its cell temperatures are sufficiently high for the chemical reactions of photosynthesis to occur. Thus it is summer temperature – which must be high enough for a long enough period – that is critical to tree growth. To the north of the 5°C isotherm vegetation cover is thinner, with shrubs growing only to about 20 cm in height, while north of the 2°C isotherm there are only lichens and mosses, with small flowering plants occasionally seen in sheltered hollows.

As temperature is both easily measured and understood and the 10°C summer isotherm is closely aligned with the treeline, a more-or-less tangible feature of the landscape, the use of this isotherm as a definition of the Arctic would seem ideal. Again there are drawbacks though. The isotherm is poorly (often very poorly) defined across the intercontinental waters of the Arctic fringe, and the definition makes no allowance for winter cold. The former is not too much of a problem since the isotherm is a less valuable measure in the oceans, but the latter means that the place where the lowest-ever temperature in the northern hemisphere was recorded (at Oymyakon in Siberia – on 6 February 1933 a temperature of -67.8 °C was recorded there; only in Antarctica have lower temperatures been recorded) lies south of the 10°C isotherm.

Despite these drawbacks, the isotherm has been adopted as a useful measure of the border between the Arctic and the sub-Arctic by many specialists, though there have been attempts to address the problem of winter cold. One of the most often quoted suggestions was that of Swedish scientist Otto Nordenskjöld – nephew of the first man to sail the North-East Passage – who suggested replacing 10°C with a temperature V, where V < 9°C-(K/10), with V and K being the mean temperatures of the warmest and coldest months of the year. For the Siberian forest, where the mean temperature of the coldest month might be -40°C the mean temperature of the warmest month would then be 13°C. On the Nordenskjöld formula the 10°C isotherm is applicable for sites where the mean temperature of the coldest month is -10°C. Use of the formula pushes the Arctic boundary south in Asian Russia and North America, but still excludes some areas that would be considered Arctic by the layman – Iceland, much of Alaska and northern Fennoscandia (Fennoscandia being the combination of Norway, Sweden, Finland and the Kola Peninsula, and land immediately south of the White Sea in Russia).

These exclusions seem anomalous for reasons other than common perception. For instance, although Iceland lies almost entirely south of the Arctic Circle it is north of the treeline (though whether the island's present treeless state is a man-made rather than natural phenomenon is a matter of debate). This issue, and others, were addressed by the Arctic Council – a joint initiative of the Scandinavian countries (including Iceland), Russia, and Canada – which defined a boundary for CAFF, the programme for the Conservation of Arctic Flora and Fauna, which pushed the Arctic boundary well south, including not only Iceland, but extensive areas of Fennoscandia as well as the hinterlands of Russia and Canada, and much of south-western Alaska. However, the CAFF

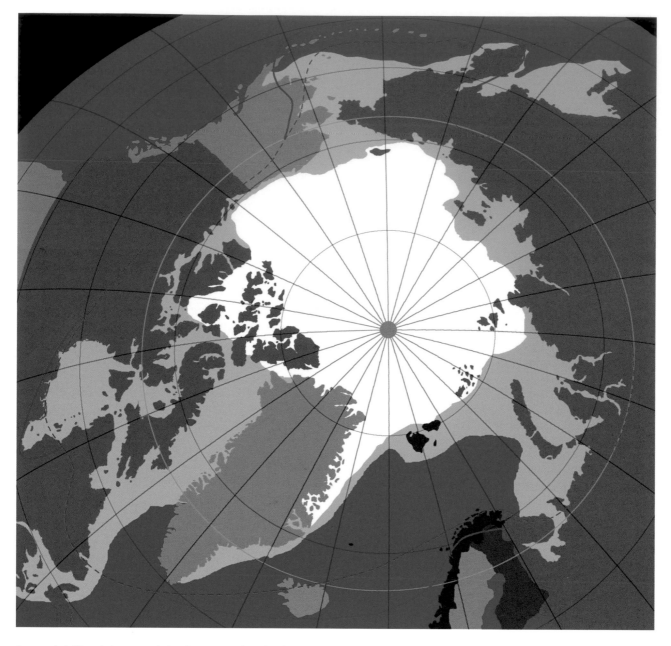

Figure 1.4 The definition of the Arctic used in this book. The green line is the Arctic Circle. The red, dotted line is the modified 10°c isotherm. The purple line is where the definition departs from the isotherm line. White is the usual boundary of permanent sea ice, pale blue is the usual boundary of winter sea ice. Clearly climate change is affecting both, but for the present the areas are still OK in most years.

definition excludes Russia's Kamchatka Peninsula and Commander Islands, while including the Aleutian chain, a decision that is surprising.

In this book, I have taken a pragmatic approach, the southern boundary of the Arctic being essentially defined by the Nordenskjöld modified 10°C isotherm, but being pushed south to take in areas whose exclusion seems inappropriate. In North America, James Bay is included. It would be excluded by the modified 10°C isotherm line; however, its importance to Polar Bears requires its inclusion. Alaska's Denali National Park is also included. A specific problem exists in the Bering Sea where the use of the modified 10°C isotherm omits much of what is usually considered Arctic. Here it is assumed that the Pribilof Islands, the Aleutian Island chain, the Commander Islands, the Kamchatka Peninsula, and the north-eastern coast of the Sea of Okhotsk lie within the Arctic. In Eurasia, northern

## Time at the poles

Before 1884, towns and cities throughout the world had their own times based on local sunrise, so it was possible for clocks in places that were relatively close to show different times. When travel between the towns was measured in hours that hardly mattered, although once the telegraph and the railway had been invented these minor differences became a nuisance. In October 1884, it was agreed that time would be standardised throughout the world. Since Greenwich, England was already recognised as sitting on the Prime Meridian (0° longitude), standard time was referred to as Greenwich Mean Time (GMT), with all other clocks being set relative to it. As the day is 24 hours long and there are 360° of longitude, a traveller following the Equator finds local time changing by one hour for every 15° of longitude (though time zones are not always so rigorously applied). But if that same traveller is intent on reaching the North Pole this time change becomes increasingly meaningless. The traveller will head out towards the pole from one of the surrounding land masses with a watch set to local time. As our traveller moves north, the crowding together of the lines of longitude means that although the time differences between them remains the same, the distance that needs to be travelled between them reduces fast. At the Equator, the circumference of the Earth is about 40,000 km, so the difference that needs to be travelled there for a time difference of 1 hour is around 1,670 km. At the Arctic Circle this distance reduces to about 663 km. At 85° it is down to 145km and by the time the traveller is within 1° of the pole it is a mere 29 km. At the pole itself, of course, the distance has shrunk to zero and the heavily clothed traveller can circle the Earth as fast as he can turn circles and for as long as he can repel giddiness. At the pole the time is all times – it is the time the traveller is facing right now, but also the time faced with a turn of the head. This seemingly strange state of affairs is illustrated in Figure 1.6.

Standardising time also required an agreed line at which the day changes – head west across it and today becomes tomorrow, head east and it becomes yesterday. The agreed line, the International Date Line, was chosen to run away from occupied lands, and takes a more-or-less north-south course from the North to the South

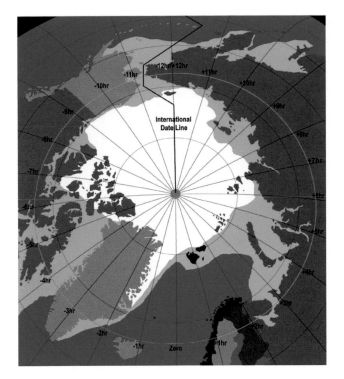

Figure 1.6 The world's time zones.

Fennoscandia and the northern coast of Russia are included. The modified 10°C isotherm and boundary of the Arctic assumed in this book are shown in Figure 1.4.

One other definition must be also addressed. Many books and reports dealing with Arctic species use the terms 'High' and 'Low' Arctic without necessarily defining what is meant by them. This is partly because such definitions are largely arbitrary. Here I have avoided use of these terms, but do differentiate between the polar desert and the tundra. In the former the average summer temperature is about 4°C and the plant growing period only 8–10 weeks. It is dry, the number of species of plants, insects, birds, and mammals is limited and their distribution is sparse (see Chapter 5 for further data on the definition of the polar desert in terms of precipitation and plant coverage). The tundra is wetter, summer temperatures average 7°C or more, the growing period is 3–4 months, and the number of species increases significantly. The transition is of course gradual, with polar desert becoming semi-desert becoming tundra, but the definition is more useful than the vague 'High' and 'Low'. The approximate transition line is show in Figure 1.5.

For all the travellers the International Date Line has comic potential, but because of the effect of decreasing distance for 15° of longitude travelled as one heads north, the Arctic traveller has the better deal. Here the cliffs of Chukotka, about 100km away, are seen through morning haze. In the foreground a Vega Herring Gull and Glaucous Gull stand at the water's edge on the western tip of St Lawrence Island. Over there in Chukotka it is already tomorrow.

Pole through the Bering Strait and across the Bering Sea, deviating to ensure that the Aleutian Islands, part of the United States, all lie to its east, while Russia's Commander Islands lie to the west. For all travellers the line has comic potential, but because of the effect of decreasing distance for 15° of longitude the further north travelled, the Arctic traveller has the better deal. In the northern Bering Sea, the islands of Big Diomede, Russian territory, is separated by only 4 km from the USA's Little Diomede. It is also separated by 21 hours, the two islands often being termed Tomorrow Island and Yesterday Island respectively. The time difference is only 21 hours as western Alaska does not accommodate the full 12 hour time difference from London. However, eastern Russia does. Arctic travellers in, for instance, Kamchatka can therefore be a full 12 hours from friends in the UK. As Kamchatka is 9 hours different from Moscow, and the flight time from Petropavlosk to Moscow is about 9 hours, travellers can also enjoy the slightly surreal experience of leaving Kamchatka as the Sun is going down in the evening and arriving in Moscow to see the same phenomenon. If the flight has made good time, arrival can even be a few minutes before departure. Take an onward flight to London and arrival can be on the same day – although that day has lasted 36 hours.

# 2 The geology of the Arctic

The similarity of the outlines of the east coast of South America and the west coast of Africa had been noticed by scholars such as Francis Bacon as early as the 1620s. Over the succeeding centuries, the similarity of species on different continents raised issues that scientists found hard to explain. Why was it that fossil plants of the genus *Glossopteris* were found in coal measures, apparently of the same age, in southern South America, southern Africa, southern India, and across Australia? The distances between these areas were so vast and the seeds of the plants so large (relatively speaking) that wind could be discounted as a dispersal mechanism. Instead it was suggested that at one time the continents had been connected by land bridges. The name Gondwanaland was coined for the huge, hypothetical southern continent across which *Glossopteris* flourished, the existence of which also explained other apparently anomalous fossil organism distributions. But while Gondwanaland solved one mystery, it created others. Since *Glossopteris* was a deciduous plant with a ringed trunk, implying a seasonal climate, why were its fossils found in tropical areas? Why was the fauna of Madagascar, an island on which *Glossopteris* had grown, so different from that of nearby Africa? And there were other questions: why, for instance, were there coal seams on Svalbard, a cold, inhospitable archipelago on which plants could barely grow at all, let alone produce the growth necessary for laying down such deposits?

## Plate Tectonics

In 1915 the German meteorologist and geophysicist Alfred Wegener published a book entitled *Die Entstehung der Kontinente und Ozeane* (first published in English as *The Origin of Continents and Oceans* in 1924). In it he presented his theory of continental drift, a theory he had developed from watching the movement of Arctic ice floes during two expeditions

*Glossopteris* fossil. The curious distribution of this fossil plant across the world puzzled scientists until the theory of plate tectonics was accepted.

to Greenland. Wegener's theory was that the continents had once been joined together in a single supercontinent he called Pangaea, from the Greek for 'all land'. Wegener's Pangaea was surrounded by a single ocean, Panthalassa ('all water'), and had broken up, the land masses drifting apart to form the world we see today. Wegener's theory helped explain the distribution of *Glossopteris* and other fossils across the continents, making the similarities of geology at continental edges more readily explainable.

However, despite the explanation it offered for observable distributions, Wegener's theory was rejected by most scientists, partly because it was based on flawed data: he had compared survey data from his 1906 expedition with data collected in 1870 and calculated that Greenland and Scotland had moved apart over that 36 year period at a rate of 18–36 metres annually (the presently accepted drift is closer to 4 cm/year, about the same rate as fingernail growth). On that basis, Wegener proposed the break-up of Pangaea occurred at the start of the Tertiary period, about 65 million years ago, which required a rapid evolution of continental biotas, for which no evidence existed.

One geologist who did not dismiss Wegener's theory in spite of this apparently overwhelming difficulty was the South African Alexander Du Toit. Du Toit found further fossil forms that fitted with the theory of continental drift,

Close to Iceland's Skaftfell National Park Svartifoss, the Black Waterfall, drops over a cliff of basalt columns.

but most telling was the evidence of striations left by Palaeozoic glaciers. The scourings indicate the direction of flow of the glacier that creates them, and what Du Toit, and others, found was that in southern Australia and southern South America the striations indicated glaciers flowing from an area that was now only ocean. There was also new data that showed that Wegener's rate of separation of the northern land masses was wrong, the measured movement being slower. Du Toit therefore concluded that Wegener's theory was correct, but that his timescale had been wrong: Pangaea had broken up not at the start of the Tertiary, but at the end of the Palaeozoic era, about 250 million years ago. Sadly by then Wegener was dead; he died of exposure in 1930 during another expedition to Greenland.

The new evidence solved the biogeographical riddle of the drift theory, but left the problem of how continents could move through a continuous crust, a problem so apparently insurmountable that scepticism remained. Not until the 1960s was the solution discovered. It came from evidence amassed by the American geophysicist Harry Hess, in part from data obtained from ship-borne echo sounders while Hess, at that time a rear-admiral, commanded a vessel during World War II. Hess realised that the depth of sediment on the oceanic floors was insufficient to account for the accumulation that must have taken place throughout the life of the Earth (the sedimentation rate was known to be about 1 cm per millennium, but that implied that the sediment depth should be some 30 times deeper than that observed). Hess also noted, as others had, that ridges rising from the beds of the ocean were often positioned mid-ocean (the Mid-Atlantic Ridge being the prime example; the ridge was discovered in the 1870s by a British naval research vessel but its significance was not understood at the time), and that there was a trench at the peak of these ridges. The uniformity and comparative youth of the oceanic beds led Hess to suggest that the ridges were the driving force behind continental drift, and his theory now forms the basis of plate tectonics.

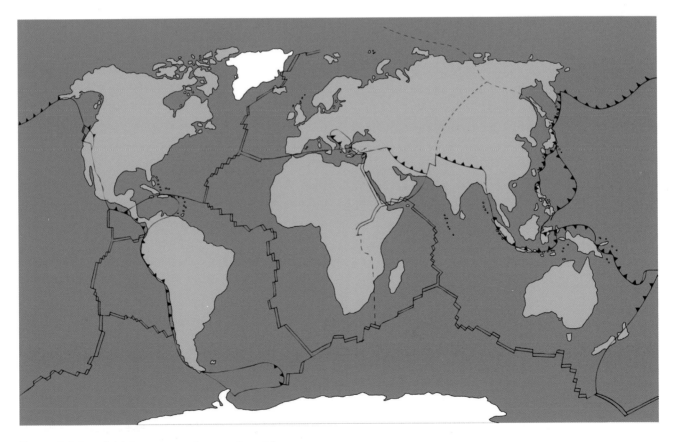

Figure 2.1 Parallel lines represent spreading ridges. Single and dashed lines represent transform boundaries. Single lines with arrowheads are subduction boundaries, the arrow indicating the direction of subduction.

The plates in question are a series of rigid structures that together make up the Earth's crust (geologists generally consider there to be seven major plates, together with some smaller ones and a number of microplates – Figure 2.1) that float on the mantle. The mantle is known to be predominantly solid (despite being much hotter than the melting point of the rocks of which it is comprised – they remain solid due to the enormous pressure on them), with zones that are demarcated by the elasticity of the material. The crustal plates are of two basic forms, continental and oceanic, the latter being denser and therefore floating lower on the mantle. The difference in density reflects a difference in composition: oceanic plates are made of mafic, mainly basaltic rocks of high iron and magnesium content (the name mafic deriving from Mg and Fe), while continental plates consist predominantly of felsic, a largely granitic rock with a higher aluminium content and less iron. The name felsic derives from feldspar, a mineral rich in silicon and aluminium that is common in continental plates. The two crustal plate forms make up the lithosphere, a 100 km-thick layer that floats on the upper, plastic portion of the mantle (known as the asthenosphere). The world's oceans lie above the oceanic plates, lapping the edges of the continental plates.

Convection currents occur in the mantle due to the temperature gradient between the Earth's hot core and the cold crustal plates. In a fluid, such currents result in motions familiar to everyone who has watched water boil: hot liquid rising from the base, cooling at the surface and falling back down again, leading to a circular motion of water and bubbles. The mantle is not fluid, but similar currents arise, the material of the mantle moving by crystalline creep, a movement similar to (but much slower than) that in hot water. The currents (Figure 2.2) circulate the material of the asthenosphere, moving the plates lying directly above. It was originally thought that the movement of the plates was due to viscous drag (essentially that the plate was embedded in the asthenosphere), although this is now thought to be just one of several components of the movement.

Although most scientists were now convinced, further observations led to Wegener's continental drift theory receiving general assent. One was the curious discovery that the Earth's North Magnetic Pole moves (as does the South Magnetic Pole). When the North Magnetic Pole

Figure 2.2 The present understanding of plate tectonics is illustrated here. Convective currents within the mantle give rise to mid-ocean ridges where upward flows are adjacent.

was first reached by James Clark Ross in 1831 it was on the western coast of Canada's Boothia Peninsula, but when it was rediscovered in 1903 it had moved north. Cooling rock is magnetised by the Earth's magnetic field, the precise orientation of the magnetism being dependent on the position of the two Magnetic Poles. Studies of the declination (the direction of magnetism) of rocks in Europe showed that the North Magnetic Pole did indeed wander over considerable distances with time, yet when the studies were repeated in North America, although the movement was observed, the path was different. There was no mechanism by which different continents could have different poles, though it was found that the paths for each continent coincided if the continents were assumed to have been much closer together millions of years ago; they must therefore have drifted apart.

A second discovery was the magnetic 'striping' of the ocean beds. Over time the polarity of the Earth's magnetic field has reversed many times. As molten rock emerging from mid-ocean ridges crystallises, it acquires a magnetic polarisation consistent with the current orientation of the Earth's field. Studies showed that the changes in polarisation formed a series of magnetic 'stripes' (of reversed polarity) and that the pattern of these stripes on each side of a ridge was identical. Clearly the sea floor had spread from mid-ocean ridges and this spread had pushed the continental plates apart. The mid-ocean ridges explained the youth of the ocean beds and the separation of continents, but since the surface area of the Earth had clearly remained constant over time, if oceanic bed was being created, somewhere plate material had to be lost. The creation of material at the spreading ridges and the loss of material elsewhere led to geological processes that are evident in many places in the Arctic.

## The formation of the Arctic – continental movements and land bridges

Pangaea formed about 350 million years ago and began to break up about 100 million years later. To the south, Antarctica and Australia separated from the other continents, with Australia later breaking away as Antarctica drifted across the South Pole. The northern movement of Africa began to close the Mediterranean, while the northern movement of India against Eurasia has created, and is still raising, the Himalayas. The Mid-Atlantic Ridge began to rotate the Americas away from Eurasia and Africa, forming the Atlantic Ocean. Other spreading ridges have formed, but unlike the Mid-Atlantic Ridge they have failed to produce oceans. Spreading ridges formed the Labrador Sea and also Lancaster Sound/Baffin Bay, but the land mass that was to become North America, Greenland, and Eurasia resisted the spreading, perhaps because they formed a circumpolar continental mass capable of doing so.

However, the spreading Atlantic lead to a rotation of the continental land masses so that Eurasia was brought into contact with North America. The Arctic Ocean formed within this pole-circling landmass.

About 55 million years ago the Arctic Ocean was essentially landlocked, with Beringia – the Bering Sea land bridge – connecting Alaska and Siberia. Greenland and the Canadian High Arctic archipelago were also connected to the north by the Thulean land bridge, south of which Baffin Bay represented a failed spreading ridge. South again the Davis Strait land bridge connected what would become Baffin Island to Greenland. The exact nature of the separation of Greenland from North America has still not been resolved. On the north-eastern side of ancestral Greenland the De Geer land bridge connected it to Fennoscandia, the land bridge including what is now the Svalbard archipelago. Further south Greenland was attached to Scotland. South of the De Geer land bridge was an embryonic sea that would become the Norwegian Sea. It was connected to the proto-North Sea, but that, too, was landlocked by the land bridge that connected what would become the countryside around Dover and Calais

Greenland separated from Europe around 35 million years ago when an extension of the Mid-Atlantic Ridge

At Þingvellir, close to Icelsnd's capital Reykjavik, the mid-Atlantic Ridge is visible. At one point a path has been laid through a rift in the ridge allowing visitors to walk with the North American plate to their left and the Eurasian plate to their right.

Evidence in support of plate tectonics include the rocks of the mountain range folded by the Caledonian Orogeny in north-east Greenland where they have been exposed at Badlanddalen after glaciation.

forced the two land masses apart. The Greenland–Scotland land bridge was severed as the ocean bed spread, but to the north the De Geer land bridge was severed by transform faulting – a shear motion – the Spitsbergen Fracture Zone opening as Greenland slid past Svalbard. The fracture zone connected the Arctic and Atlantic oceans, the mixed waters forming what is now called the Fram Strait. However, this mixing was of surface waters only, deep currents of cooler water being prevented from flowing into or out of the Arctic Ocean since the spreading ridge between Greenland and Scotland presented a deep sea barrier. Bathed in relatively warm water, the Arctic at the time bore little resemblance to the region today, but at about this time the global climate started to deteriorate. This became more marked when Antarctica, by then positioned over the South Pole, became detached from South America as the latter drifted north. Now surrounded by cool water – the Southern Ocean – which could circulate unimpeded around it, Antarctica grew colder. Glaciers formed on its high mountains and sea ice became a feature of its coastline. The cold southern waters sank and were replaced by warmer waters flowing in from the north. The interaction of the two caused clouds to form and snow to fall on the continent. The Antarctic ice sheet began to grow; since ice reflects up to 90% of incident radiation, the growing ice sheet led to a cooling of the Earth. Despite this, the Arctic remained relatively warm for several million years more, though tectonic activity eventually allowed colder deep waters to flow. Seasonal sea ice may have begun to form in the Arctic by about 3½ million years ago, but permanent ice probably did not appear until global temperatures fell – according to an analysis of ice cores – about 2½ million years ago. Only then did the Arctic as we now know it appear.

## Formation of the Arctic – rocks, seas, and mountains

The central areas of the tectonic plates that enclose the Arctic Ocean have shields of exposed Precambrian bedrock, and are surrounded by further regions that isolate them from the geologically active areas at the continental margins. Some of these Precambrian rocks are among the oldest on Earth. Close to the Acasta River, in Canada's North-West Territories, rocks have been dated as being just over 4,000 million years old. In the Isua upland close to Nuuk, the Greenlandic capital in the southwest of the country, pillow lavas and fine-grained sedimentary rocks laid down in deep waters have been dated to 3,750 million years ago. Most of the Arctic's shield rocks are 2,500–3,500 million years old. During the period from about 1,000 million years, sedimentary rocks were laid down below a shallow sea. Later, tillites – a mix of material from clay to boulders – were deposited. Tillites are products of glacial activity and can be seen in north-east Greenland; they are good

supporting evidence for the 'snowball Earth' hypothesis which suggests that the Earth was wholly, or largely, ice-covered around 650 million years ago.

Surrounding the ancient continental cores of essentially stable bedrock are areas of orogenies (massive foldings) associated with tectonic activity. One orogeny was responsible for the Urals, a mountain range that continues across Novaya Zemlya. The Caledonian Orogeny created a mountain range against the stable Baltic shield that extended from what is now western Scandinavia to the Appalachians by way of Scotland and east Greenland. In the Nearctic, further folding created the mountains of Ellesmere and Axel Heiberg islands. These orogenies all date from the Palaeozoic, though the folding that led to the great ranges of Alaska took place later, in the Mesozoic.

Mountain building was accompanied by metamorphosis and igneous intrusions. Newtontoppen, the highest peak in Svalbard (at 1,717 metres) is a granite intrusion from this period, as are some of the granite formations in eastern Greenland. By far the most spectacular intrusions form the huge granite faces of the peaks of the Auyuittuq National Park on Baffin Island. The faces on Mounts Thor and Asgard are among the tallest in the world and are a constant attraction to trek-

Winter at Brepollen, Hornsund in southern Spitsbergen. Across the sea ice are flat topped peaks indicative of the existence of a relatively hard rock band which has escaped erosion.

kers and climbers (and others – Auyuittuq was used for a famous parachute jump scene in a James Bond movie).

From the Carboniferous and on through the Mesozoic, fossil-rich sedimentary rocks were laid down in extensive basins that covered much of the Canadian Arctic, Greenland, and Svalbard. The basalt intrusions of Franz Josef Land date from this long period of continent-building. In the subsequent Palaeocene epoch at the start of the Tertiary, the Mid-Atlantic Ridge began the process of forming the Atlantic Ocean and moving the Old and New Worlds apart, laying the foundation for the present structure of the Arctic. Pangaea's break-up, of which the creation of the Atlantic was part, was also accompanied by the accretion of various terranes (well-defined pieces of land whose geology differs markedly). Terranes are a particular feature of Alaska and the far east of Russia, the geology of those areas showing that over time disparate pieces of land were 'glued on' to the existing shorelines.

In conclusion, although the general form of the land masses that surround the Arctic Ocean can be said to derive from tectonic activity, primarily the sea-floor spreading of the Mid-Atlantic Ridge and the subduction zone of the 'Ring of Fire' (see below), suggest that is a simplification. The Arctic land masses exhibit a full range of igneous, sedimentary, and metamorphic rocks; while the build-up of terranes, igneous intrusions, and local geological history has created highly complex areas at the edges of the otherwise stable continental masses, these areas unfortunately do not exhibit simple geological boundaries.

## Formation of the Arctic – volcanicity and seismicity

The interaction of the crustal plates that make up the ocean beds and continents can take one of several forms. At spreading zones new ocean floor is created. If two spreading zones meet – the meeting point of three plates – then a ridge junction can form; an example occurs close to the Azores where the North American, African and Eurasian plates are being forced apart. Where two continental plates collide then either an intervening sea can be squeezed out of existence (e.g. the Mediterranean Sea, which is being lost as Africa moves into Eurasia) or mountains can be formed (e.g. the Himalayas,

where the Indian and Eurasian plates meet). Where an oceanic plate collides with a continental plate, the lower, heavier oceanic plate dives beneath the continental plate, disappearing into the mantle where its material is recycled. Where plates slide relative to each other fault lines are created, giving rise to earthquake zones. California's San Andreas Fault and the Fairweather Fault in south-east Alaska are transform faults that occur where tectonic plates slide horizontally relative to each other. The tectonic activity of various plate interactions can be manifested as volcanoes and earthquakes; each of these phenomena has contributed to the physical form of the Arctic we see today and continues to be a feature of the area, both in the Nearctic (in Alaska and Kamchatka) and in the Palearctic (in Iceland and Jan Mayen).

Iceland's volcanic activity is well-known. The activity – indeed, Iceland's very existence – arises because of its position above a hot plume in the mantle. The cause of such plumes is not well-understood, but it is thought that they arise when hotter rock, perhaps of different chemical properties, rises from low in the mantle, leading to increased heating of the crustal layer and formation of a larger dome than at, for instance, a spreading ridge. Many such plumes have now been identified around the world. Though they are chiefly associated with spreading ridges, they are not exclusive to these regions: the Hawaiian Islands, for example, have formed above a hot spot that lies virtually at the centre of the Pacific Plate. A hot spot has pushed the Mid-Atlantic Ridge up to 2½km higher beneath Iceland than elsewhere, high enough to break the water's surface and so create an island. It is extraordinary that because of Iceland's position it is both volcanically active and home to four large ice caps, including one, Vatnajökull, which extends over 8,000 km². The juxtaposition of volcanoes and ice is such a feature of the island that it often leads to the well-worn, yet entirely reasonable, guide book description of 'land of ice and fire'.

The most famous of the Iceland's volcanoes is Hekla, east of Reykjavik. Hekla's cone, now 1,491 metres high, has built up over 6,500 years, the last significant eruption being in 2000. After a devastating eruption in 1104, when ash destroyed many local settlements, the Icelanders believed that Hekla was an entrance to Hell; this may have influenced Jules Verne, who sent the heroes of *Journey to the Centre of the Earth* into the crater of Snæfellsjökull, a volcano northwest of Reykjavik across Faxaflói. Perhaps now as famous as either is Eyjafjallajökull, the south Iceland volcano whose eruption in March 2010 caused a shutdown of air traffic across northern Europe for eight days in April, and further, less widespread, disruption until mid-May as the ash cloud from the volcano drifted southeast. Although Eyjafjallajökull's eruption was a spectacular nuisance, it could not compare with that at Laki in 1783–85, when a 32 km fissure produced a lava flow that covered more than 560 km², the largest in historical times (and the largest ever witnessed by human eyes). It is estimated that 12 km³ of rock, chiefly basaltic lava, was ejected. The eruption also produced some pyroclasts (unconsolidated material fragmented by the explosion), mainly tephra. The flow engulfed many settlements, but the direct loss of life was nothing compared to what occurred subsequently. It is estimated that 70 million tons of acid were belched out by Laki, which fell as a toxic blanket that poisoned the Icelandic soil. Millions of tons of dust blotted

out the Sun, causing crop failure in areas far from the poisoned soil and leading to the 'haze of hunger', a famine in which a significant proportion of the Icelandic population died.

Another well-known volcanic event in Iceland took place in November 1963. A plume of steam rose from the sea to the south of Vestmannaeyjar (the Westmann Islands), which lie off Iceland's southern coast. The eruption lasted three years, and gave birth to the Earth's newest land, the island of Surtsey. Surtsey is a shield volcano, one created by relatively short-term lava flows – of the order of months or years – which therefore have relatively shallow flanks (2–8°) that fall away symmetrically from a central crater. A more common volcano form is the strato or composite, which comprises of layers of lava and tephra built up over thousands or millions of years, and which are therefore of increasing thickness as the cone is approached. Mainland Iceland has many examples of both forms. Ten years after the creation of Surtsey, in January 1973 on the nearby island of Heimaey, a new

The *Strokkur* geyser at Geysir. Water hotter than 100°C, maintained liquid by extreme pressure (superheated water) rises regularly in a rock tube. As it reaches the surface, the pressure falls and steam bursts through the superheated water bubble to create the spectacular geyser.'

mountain, Eldfell was created when lava flowed from a 1½ km fissure. The island's inhabitants were evacuated, and some of their houses were saved by the pumping of seawater on to the lava's front edge to create a rock dam.

Icelandic volcanic activity may also occur beneath the island's ice caps. Ice melted by such eruptions may create a vast water-filled chamber within an ice cap which can then cause *jökulhlaups* ('glacier floods') – tidal waves of water – when the water escapes. One such *jökulhlaup* occurred following the September 1996 eruption of the Grímsvötn volcano through a 4 km fissure beneath the Vatnajökull ice cap. Fortunately, the *jökulhlaup* did not occur until November, by which time those at risk had been relocated; no lives were lost when the estimated 4 km³ of water and one million tons of ice engulfed Iceland's southern coast. At the height of the release 50,000 metres³ of water per second flowed from Grímsvötn's caldera. Damage was chiefly limited to power lines and the coastal road – the latter was re-opened within two weeks.

In addition to these intermittent, but highly explosive, indications of the hot plume that gave birth to Iceland, there are other volcanic phenomena that make the island popular with tourists. Sulphur-rich hot water has been harnessed for home heating and swimming pools – the latter most famously at the Blue Lagoon – as well as for geothermal power plants. Bubbling mud pools are a highlight of trips to Krisuvik and Hveragerði, and the original *geyser*, the one that gave its name to the form, can be seen at Geysir, to the east of Reykjavik. To the north of Iceland, Jan Mayen lies above another hot spot of the Mid-Atlantic Ridge, though in this case the activity is much more limited, being essentially confined to a single volcano.

## Subduction

When an oceanic plate converges with a continental plate, the denser oceanic plate goes beneath the lighter continental one. The movement, known as subduction, results from the same processes that drive the separation of plates at a spreading zone (i.e. currents in the mantle), but the subducting plate is also dragged by the disappearing slab as it drops into the asthenosphere. If the slab breaks away from the parent plate the drag ceases, although the slab can continue to exert an influence by suction. Where the subducting plate disappears it pulls the overlying continental plate down, creating a deep trench. The existence of such trenches had been known before the study of plate tectonics provided an explanation for their creation.

Subduction creates a zone where both earthquakes and volcanoes can occur. A subducted plate bends as it disappears, with the distortion creating fault lines, movement of which may result in earthquakes. These earthquakes may occur at substantial distances below the Earth's surface, as deep as several hundred kilometres. Earthquakes also happen as a consequence of the dragging of the overlying plate by the subducting plate. Volcanoes stem from the partial melting of the subducted plate as it reaches a critical depth, with lighter fractions melting and rising as 'blobs' of magma which reach the surface. Water may assist the process; trapped water dragged down from the ocean becomes superheated and as it rises through the mantle it causes local melting, with the rising pools of molten rock in which the water is trapped erupting if they reach the surface.

Russia's Kamchatka Peninsula, the Aleutian Islands, and Alaska are part of a volcanic arc which, with a continuation into Japan and the western USA, formed where the Pacific plate sank beneath the Eurasian and North American plates; up to 7cm of the Pacific plate is subducted annually. The arc, of volcanoes and seismically active land masses, is given the somewhat fanciful (yet not entirely inappropriate) name of the 'Ring of Fire'.

### Alaska

In June 1912 the volcanoes of Katmai and Novarupta, on the Alaska Peninsula, southwest of Anchorage, ejected 20 km³ of pyroclasts – chiefly rhyolitic-andesitic tephra – in one of the world's most violent eruptions. Before the eruption, Katmai was a 2,285 metres glaciated, cratered peak; after the event the peak's height had been reduced by around 800 metres and a caldera 4 km across and up to 1,000 metres deep had been created. On Novarupta the eruptions were from fissures as well as from the summit crater. It is said that for several days after the eruption people on nearby Kodiak Island could not see more than a metre ahead, so thick was the airborne debris. The debris

The magmatic basalt cliffs on Hooker Island, one of the islands of Russia's Franz Josef archipelago, are superb for breeding sea birds.

## The 1964 Alaska Earthquake

On 27 March 1964 (which happened to be Good Friday) an earthquake struck Alaska. The now-accepted magnitude of the earthquake was 8.6 on the Richter Scale, but seismographs registered readings varying from 8.2 to 9.2. The quake's epicentre was at the northern end of Prince William Sound, to the east of Anchorage, roughly midway between Valdez and Portage. Earth movements and landslips were accompanied by significant tsunamis, some waves reaching 50 metres, which engulfed and destroyed parts of Kodiak, Valdez, and Seward. At Portage the ground subsided by 2 metres and was swamped by seawater from the Turnagain Arm. When it became clear that Portage would be flooded at all subsequent high tides the settlement was abandoned. Pine forests were killed by the inundation of salt water. After the quake, survey teams discovered that the local area had not only subsided, it had also expanded horizontally. However, at Cordova, some 100 km east of Portage, the local area had been raised by 2 metres; high tides no longer filled the harbour, leaving some fishing boats above the new high-tide mark. Even more dramatically, on the island of Middleton, 150 km south of Cordova, a wrecked Liberty ship that could not be reached without a boat, even at the lowest tide, was lifted 9 metres, clear of the water. Large areas of the continental shelf close to Cordova and Seward were also raised permanently above the high-tide mark. The earthquake claimed the lives of 107 people.

major earthquake results. The overriding slab causes a local uplift while the compressed, domed area rebounds, causing subsidence. Investigation on Middleton Island after the 1964 quake showed a succession of uplift terraces, proving that there had been earlier earthquakes along the same fault line. Carbon dating of driftwood from these terraces indicated that major earthquakes such as the 1964 event occur on average every 800 years.

Although tsunamis initiated by the 1964 quake were significant, they cannot compare with another earthquake-generated wave that occurred just a few years earlier. In 1958 a quake released an estimated 40 million cubic metres of rock into Lituya Bay (part of the Glacier Bay National Park). Following the slump a wave crossed the bay at around 200 km/h and left a 'tidemark' some 500 metres up the opposite cliff. The wave destroyed 10 km² of forest and killed two fishermen.

Cabin destroyed and abandoned after the 1964 Alaskan earthquake.

cloaked virtually the entire northern hemisphere, lowering mean monthly temperatures by several degrees and effectively eliminating the summer. The area around the peaks now forms the Katmai National Park. One feature is the evocatively named the Valley of Ten Thousand Smokes.

The 1964 Alaskan earthquake raised or lowered more than 250,000 km² of land. Although not completely understood at the time, it is now known that the Earth's crust is not rigid but elastic, and so capable of being both stretched and compressed. If a major earthquake happens in the Anchorage area, the North American Plate becomes locked in a new position along its slip plane with the Pacific Plate. Subduction of the Pacific Plate causes the North American Plate to compress; the compression causes local doming and, hence, an uplift. Eventually a breaking point is reached, with the North American Plate unlocking and slipping – and another

The *Strokkur* geyser, Iceland.

Avachinsky volcano in southern Kamchatka.

As well as these headline-grabbing events, Alaska also exhibits less well-known evidence of volcanic activity, most particularly the *maars* that form the Devil Mountain Lakes at the northern end of the Seward Peninsula (a short distance south of Cape Espenberg and the Arctic Circle). A maar arises where rising magma reaches an area of porous rock saturated with ground water. The steam produced causes a rise in internal pressure, and eventually the ground above fails, rather in the way that a cork 'erupts' from a champagne bottle. The maar results in an essentially circular crater surrounded by the plug material, and it often fills with water to produce a lake. Most maars are about 700 metres in diameter, but the Seward Peninsula lakes are several kilometres across; they are the largest maars to have been discovered to date. They have also been extremely important in recreating the vegetation of Beringia as the ejected material was a plug of permafrost which included the surface material.

## Kamchatka

On the Russian side of the Bering Strait, subduction of the Pacific Plate beneath the Eurasian Plate has created about 60 active or potentially active volcanoes, as well as some that are now extinct, in a 700km arc along the Kamchatka Peninsula. The Kamchatka arc is responsible for about 20% of the total material ejected each year by the Earth's volcanic activity. Of recent Kamchatka events, the most spectacular was the 1956 eruption of the Bezymianny volcano in the northern peninsula. Best-estimate calculations suggest that the volcano ejected a stream of incandescent tephra at more than 500m/s (i.e. about twice the speed of sound), which reached an altitude of 45km. Volcanic dust reached the UK – almost exactly half-way around the Earth – in just 72 hours. Before the eruption Bezymianny had been a 3,085m volcano. The eruption reduced its height by 200m; a huge caldera had also been created.

In addition to its volcanoes, many of which are textbook cones, Kamchatka has an array of huge geysers in the appropriately named Valley of Geysers, and the Uzon Caldera, which exhibits mud volcanoes, hot mud pools and crater lakes, sulphur-rich hot streams and associated thermophilic micro-organisms.

## The Arctic Basin

North of Iceland, the Mid-Atlantic Ridge continues towards the North Pole, separating Svalbard and northeastern Greenland, then continuing as the 1,500km Nansen-Gakkel Ridge, which rises 2,000m above the sea bed. The seismically active area of Siberia's Verkhoyanskiy Mountains is on a continuation of this ridge (though the mountains also lie between tectonic areas of the Siberian platform and the micro-continents of far eastern Siberia). The Nansen-Gakkel is one of several ridges making up the geologically complex Arctic Basin (Figure 2.3). The Basin's ridges define deep oceanic basins sampling of which suggests they are from 60–135 million years old and were indeed formed by sea floor spreading. These young, but deep, basins – at its deepest the Fram Basin is almost 4,500m in depth and the Makarov exceeds 4,000m, while the others are all at least 3,000m deep – are ringed by shallower seas, which lie above the continental shelves.

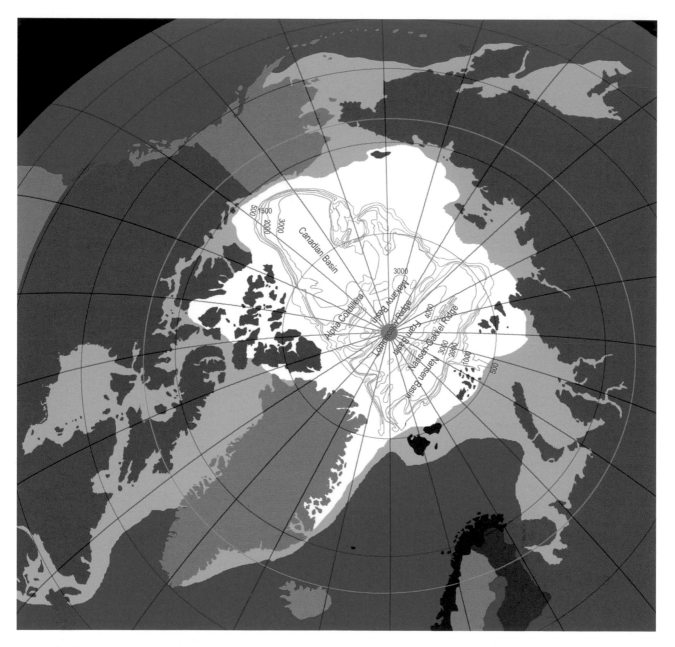

Figure 2.3 The Arctic Basin. The depth contours are labelled in metres.

Fossilised corals. Nordaustlandet, Svalbard.

Stromatolites on the eastern arm of the Great Slave Lake, North West Territories, Canada.

The continental shelves of Eurasia are extensive, reaching to and beyond Svalbard and the islands of Arctic Russia. The shelves occupy 35% of the area of the Arctic Ocean, yet account for only 2% of the water volume. The Siberian shelf, which forms part of the Eurasian shelf, is up to 900km wide, the world's widest. The seas that overlay the Eurasian shelves are shallow, only 10–20m deep in the west (Barents, Kara and Laptev seas) and 30–40m deep to the east (East Siberian and Chukchi seas). Though similar in terms of depth and the width of the underlying shelves, the seas differ markedly. The Barents Sea has the mildest climate because of the influence of the North Atlantic Drift. The islands of Novaya Zemlya act as a barrier to the eastern transfer of the Drift's warm waters, so the climate of the Kara Sea is much colder. The climate of the Laptev Sea is much more benign due to the warm-water outflow from Siberia's huge rivers: in summer the Laptev Sea can be ice-free to 77°N, its population of Walrus being the most northerly in the world. Further east, minimal river flow into the East Siberian Sea means it experiences a much harsher climate, as does the Chukchi Sea in winter, though warm water passing through the Bering Straits promotes significant summer ice melting.

To the west of the Barents Sea the deeper Greenland Sea (more than 2,000m deep) separates Greenland and Svalbard. The sea represents the widest breach in the continental landmasses surrounding the Arctic Basin. The Greenland Sea merges into the Norwegian Sea (which is just as deep), the name given to the body of water that washes the western coast of Norway. Conventionally, the Norwegian Sea extends west as far as Jan Mayen and south as far as the Faroes.

The North American continental shelf is generally considered to be less extensive than that of Asia, though as the shelf underlies the Canadian Arctic islands that contention is arguable. Baffin Bay, between Greenland and Baffin Island, is deeper (over 1,000m), as is the Beaufort Sea (over 3,000m at its northern extreme), which washes the northern coast of Alaska (east of Point Barrow) and Canada's Yukon and North-West territories: the shelf beneath the Beaufort Sea is certainly less extensive than that beneath the Chukchi Sea bordering it.

## Ancient life in the Arctic

As well as some of the Earth's oldest rocks, evidence of some of the Earth's oldest life forms has been found in the Arctic, in the banded ironstone formations of western Greenland. Until recently it was thought that about 3,500 million years ago life forms capable of harnessing the energy of the Sun by photosynthesis had evolved. These were cyanobacteria: they created stromatolites, layered calcareous structures formed from mats of cyanobacteria discovered in western Australia. However, in 2016 it was announced that stromatolites had been discovered in rocks in south-west Greenland. dating to 3,700 million years ago.

Fossils from more recent geological periods can be found all over the Arctic. Perhaps the best destination for Arctic fossil hunters is Spitsbergen, which is a geologist's dream, the lack of overlying vegetation allowing the rock to be readily examined: strata from Precambrian times to the Tertiary are exposed across the island, and they are often rich in fossils. Cambrian and Ordovician rocks yield marine invertebrates – trilobites, brachiopods and graptolites, while from the Silurian and Devonian come armoured fish – pteraspids and cephalaspids. Early 'true' (bony) fish have been found in rocks of late Devonian age. At that time the land that was to become Spitsbergen lay much closer to the Equator.

In the Carboniferous and Permian, Spitsbergen was an area of shallow seas and swamps, with luxuriant vegetation that became the island's coal measures. From Spitsbergen's Triassic and Jurassic rocks there are ammonites and fossils of marine reptiles, such as pliosaurs, while Cretaceous rocks sometimes contain the fossilised bones and footprints of dinosaurs. During the Tertiary, Spitsbergen was cloaked in forests, these producing further coal mea sures, the coal occasionally yielding beautiful leaf fossils.

But although Spitsbergen is rich in fossils, without doubt the most important Arctic finds have come from eastern Greenland and Ellesmere Island where Late Devonian deposits have yielded a stunning array of primitive tetrapods, intermediate between fish and the first terrestrial vertebrates, the amphibians. Perhaps the most remarkable of all fossils from this early period in tetrapod evolution was found in 2006 on Ellesmere. Named Tiktaalik, this animal represents a true transitional form: similar in many respects to a lobe-finned fish, with fish-like fins instead of toes, and a functional shoulder and wrist that may have been able to support the animal on land. It is believed that Tiktaalik was specialised for life in shallow stream systems, perhaps in swamps, and it may have been able to move overland using its fins.

Geological layers visible at Landnordingsvika on Bear Island, which lies between the north Norway coast and Svalbard.

# 3 Snow and ice

Although plate tectonics organised the landmasses that surround the Arctic Ocean, it is snow and ice that have shaped the landscape. Here we explore these two substances, each of which is more complex and fascinating than might at first be thought.

## Frost and snow

Add heat to most solids and they melt into liquid. With further heating, the liquid boils to form a gas. Water is different, for although it usually follows this standard 'three phase' model – ice to liquid water to steam – it can also transfer from the gaseous phase to the solid phase, and *vice versa*, directly. The transformation from ice to water vapour is called sublimation: the product of vapour transforming directly to solid is known as hoar-frost. Although hoar-frost differs from true frost, the two forms are occasionally difficult to distinguish. It is not necessary to travel to the Arctic to see hoar-frost formation: it can occur in temperate areas, though confusion can arise with frozen dew. Dew is formed close to the ground when water droplets condense from vapour-saturated air. (The highest temperature of a surface at which this can occur is known as the 'dew point'.) If the dew freezes – which usually happens only if the temperature falls below about -3°C – it forms 'silver frost', so called to differentiate it from hoar-frost. The frosting of window panes is a more definite sign of hoar-frost, spidery webs of frost forming on a dry pane.

One of the most familiar demonstrations of hoar-frost for the Arctic traveller is 'diamond dust', the gentle, glittering fall of thousands of minute ice crystals (typically less than 0.2mm across), which occurs from

Hoar-frost rose on a mountain cabin window, northern Norway

Glacial ice, Billefjorden, Svalbard.

cloud-free skies in very low temperatures (usually below -30°C). Rarer, but even more awe-inspiring and breathtakingly beautiful, is the diamond dust that sparkles in the light of the moon. Diamond dust is often the basis of parhelia.

## Snowflakes

Within clouds, small water droplets are swept upwards by rising air. As air temperature falls with height the water droplets may become supercooled: that is, they reach a temperature below the freezing point of water. These supercooled droplets may be triggered to freeze into ice crystals by the presence of minute dust particles, which act as nucleation centres. People often think that snow is frozen rain, but this is not the case. Frozen rain drops, usually mixed with melting snow, falls as sleet (a British term: in the USA the normal term for frozen rain is 'ice pellets'). However, sleet is not an entirely useful term

as there is another form of precipitation that perhaps fits the description of frozen rain better – hail. Hail is produced by the collision and coalescing of supercooled water droplets, or by snowflakes accumulating water. Hail is generally considered to consist of ice droplets with diameters greater than 5mm, with sleet or ice pellets being smaller.

The accumulation of many ice crystals as they collide and coalesce forms a snowflake. Snowflakes are usually about 1–2mm across. Their amazing beauty was first described following naked-eye observations in the early 17th century by French philosopher René Descartes and German astronomer Johannes Kepler, both of whom described the hexagonal (six-fold) shape that is such a feature of the flakes. However, the complex nature of the symmetry of individual flakes was not unravelled until the invention of the microscope: in 1665 the English scientist Robert Hooke published drawings of snowflakes from microscope observations. Over 150 years later William Scoresby made a series of superb drawings of snowflakes he observed during whaling voyages to north-east Greenland. Scoresby noted that Arctic snowflakes were much more symmetric than those he had observed in his native Britain.

The symmetry of snowflakes results from the crystalline structure of ice, which derives from the way that the hydrogen and oxygen atoms that make up a water molecule interact. Ice crystals can actually take several forms but the commonest – and the only form that can exist outside conditions of extreme pressure and very low temperatures – is a hexagonal lattice, given the prosaic name 1h by snow scientists, that gives the snowflake its six-fold symmetry. 1h also forms the basis of the rarer 12-sided and triangular snowflakes, with the former created by the coalescing of two six-sided flakes. The latter is more difficult to explain: larger triangles are probably produced by the splitting of a six-sided flake, but the smaller ones are puzzling.

The drawings of Scoresby show the variety of shapes that hexagonal symmetry offers: though many are stars, there are also columnar forms – six-sided prisms and strange dumbbells, with hexagonal 'weights' at the ends (and sometimes also in the middle) of six-sided 'bars'. Formation of snowflakes under laboratory conditions

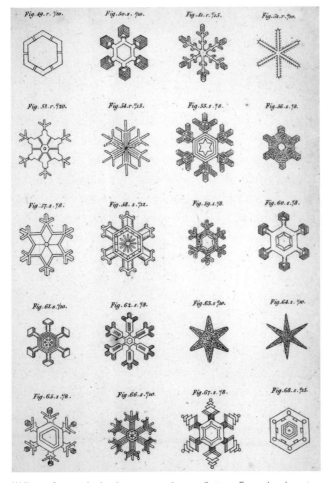

William Scoresby's drawings of snowflakes. From his book *An Account of the Arctic Regions* published in 1820.

has shown that the different shapes relate to differences in the temperature and saturation level of the cloud. In general, columnar flakes form in relatively low saturation atmospheres, with plate-like flakes being produced at high or very low saturations.

Although it is often assumed that all snowflakes are symmetrical, this is actually not the case: indeed most snowflakes are asymmetric by the time they reach the ground, since many interactions and changes in atmospheric conditions can affect the flake on its journey earthwards (almost all photographs of snowflakes show perfect symmetry, but that is only because they are the most beautiful ones and so are chosen for immortality). However, most, if not all, snowflakes do start out symmetrical. Because the flakes are very small the conditions of humidity and temperature are effectively identical across the entire surface. Therefore when growth of the flake occurs as a result of further condensation, the rate and conditions are the same at every point across the arms or faces.

## Are all snowflakes different?

An oft-heard tale is that no two snowflakes are identical. This is, of course, not provable, but is very likely to be true. The 'standard' form of the hydrogen atom comprises a single electron orbiting a nucleus consisting of a single proton. But in a small percentage of hydrogen atoms the nucleus comprises a proton and a neutron. This form (or isotope) is known as deuterium and is the basis of heavy water – $D_2O$ as opposed to $H_2O$. As there are also several isotopes of oxygen, there are different forms of water molecule. Each snowflake is composed of millions upon millions of molecules, the different forms of which vary the crystal structure in a miniscule, but cumulatively significant, way. The isotopic make-up of the molecules forming the crystal is random, so even though humidity and temperature can be considered constant across every flake, the minute differences are sufficient to create different shapes. The chances of two snowflakes being the same are, therefore, infinitesimally small.

## Falling snow

Snowfall accounts for about 5% of the total precipitation that falls on the Earth annually, this seemingly small fraction amounting to more than 30 million million tonnes per annum. The bulk (about 90%) of this snow is seasonal (or rather more temporary in temperate areas) with the rest contributing to the Earth's ice sheets.

There are different forms of snow depending on its density. Several things affect the density of snow when it is first deposited, but in general it has a specific gravity of about 0.1 (which means that 10cm of snow is the equivalent in terms of water volume of 1cm of rain). Once deposited, snow is compacted as individual snowflakes break up, the pieces pack closer together. Compaction can therefore occur entirely due to the snow's own weight if the snowfall is heavy, but it will also be aided by the wind. Melt water from the surface also percolates down into the spaces between the snow crystals. As a result of these processes the snow's specific gravity increases and its characteristics change. When the specific gravity reaches about 0.5, the snow is well-compacted and granular, a form well-known enough to skiers and mountaineers to have been given a specific name: *firn* in German or *névé* in French, though in each case the name can apply to snow with a range of specific gravities. The wind may affect the surface of such snow: it can create a crust fashioned into wave-like ridges, comparable in appearance to the waves the wind induces on the surface of a lake. This ridged surface is known as *sastrugi* and can produce an uncomfortable ride for a traveller on a snow scooter if the surface is too hard to break through, making the ride a tooth-shaking experience. The wind can also create a more level surface crust called wind slab, which can be very dangerous on slopes (and so is carefully watched for by mountaineers) as it can slide on the relatively unconsolidated base below it, causing an avalanche.

As snow compaction increases, either due to additional snowfall or through an increase in local temperature that causes further surface-melting, the specific gravity rises again. Another process is also at work deep within the snow. Known as sintering, it involves the transfer of water molecules by sublimation: a complex, lattice-like structure is formed, with the air that formed a major constituent of firn being squeezed

Sastrugi on sea ice, Barents Sea.

upwards, out of the lattice. When the specific gravity reaches about 0.8 the remaining air within the lattice becomes trapped in bubbles as escape routes to the surface are closed off. The snow has now turned to ice.

With further compaction, the specific gravity increases further, the air within the trapped bubbles becoming compressed. Glacial ice usually has a specific gravity of 0.9: the specific gravity can increase still further, but not by much, the value for pure ice is 0.917, but glacial ice always has some trapped air.

The trapped air in glacial ice has two interesting side-effects. One is that glacial ice fizzes when added to a drink, giving a moment's pleasure to passengers on a polar cruise who collect ice from a passing iceberg to add to an evening aperitif. The other is the phenomenon of blue ice. Air and other impurities in the ice scatter light at all frequencies, and as 'normal' ice is impure this means that it appears white. But water molecules preferentially absorb light with wavelengths at the red end of the spectrum. Pure ice, free of impurities, allows light to travel a relatively long distance, increasing the absorption of red wavelengths while allowing transmission of wavelengths at the blue end of the spectrum. The ice therefore appears blue, and the purer (and more compacted) the ice the bluer it appears. Blue icebergs, one of the most breathtakingly beautiful of all polar visions, are most frequently seen in the Antarctic, where they are calved from ice sheets with ice that has travelled for many hundreds of years and become highly compacted. In the Arctic, icebergs are calved from glacier fronts and, in general, these are much less ancient. Blue ice is therefore rarer in the north, though it does occur and is an equally thrilling sight. Icebergs

may also be blue-green, but this has less to do with the physics of ice, the colour coming from organic material trapped during the formation process. Travellers on pack-ice may also occasionally see patches in shades of brown: this is caused by the defecation of hauled-out seals or walruses.

Snow is extremely important for Arctic rodents. Unlike birds, most Arctic mammals cannot migrate away from the harsh northern winter (reindeer being a notable exception, though even they move to the Arctic fringe). Rodents use the snow as a blanket to protect them from the extreme cold of winter. This unlikely scenario arises from the low thermal conductivity of snow which means it acts as an insulator. The thermal conductivity of snow varies with its density, but it can be as low as that of some cavity-wall insulating materials. Indeed, so good is it as an insulator that a metre of snow can maintain a soil temperature of around 0°C in ambient air temperatures of -40°C. The warmth of the soil stimulates sublimation in the basal snow layer: water molecules rise through the snow, to be replaced by cooler, denser air which is warmed by the ground, encouraging further sublimation. The net effect is to create a basal snow layer composed of needle-like crystals, a layer with a specific gravity of 0.2–0.3. This layer is exploited by rodents, which continue searching for food while benefiting from the insulating properties of the snow blanket. Mountaineers are less enamoured of this snow form, as on slopes basal layer weakening by sublimation can result in avalanches.

## Freshwater ice

There are two main forms of ice that the Arctic traveller will meet. Sea ice is frozen sea water, and so is salty. Glaciers are formed of freshwater ice, i.e. they originate from water in the atmosphere that falls as precipitation on land. Two other forms of freshwater ice may also be encountered, on lakes and rivers.

As fresh water cools its density increases, as might be expected. But the density reaches a maximum at about 4°C (3.98°C for pure water) and then decreases. There is further reduction in density of about 8.5% when water freezes. As a consequence, ice is less dense than water, and so floats. As the surface layer of a lake cools the water therefore becomes denser and sinks, drawing warmer water to the surface. This process continues until the temperature of the entire water column reaches 4°C and the maximum density is achieved. Once this point is reached, further cooling of the surface water causes it to becomes less dense, and this cooler, less dense water can form a stable layer on top of the denser water column. If the air temperature above the lake is lower than 0°C ice can now form in a layer on the surface. Freezing starts at some point: needle-like ribbons of ice may occasionally advance rapidly from this point, an interesting feature of these being that they are eventually curtailed by their own success – as the water freezes it gives up its latent heat, warming the water ahead of the needle and so bringing further freezing to a halt. As will be familiar to anyone who has stepped on the sheet of ice formed on a puddle after a frosty night, this first skin of ice on still water can be extremely thin, perhaps only 0.2mm thick: it is essentially two-dimensional. Because the surface skin acts as an insulator, the process of increasing the thickness of lake ice is slow, with further layers forming beneath the original. There is little downward (i.e. three-dimensional) growth, the junction between the water and the ice being very smooth. Again, anyone who has ever extracted a sheet of ice from a frozen lake will know that it is glass-like, a smooth-sided pane of ice. Usually

the ice will also be as clear as glass: freshwater ice forms relatively slowly, and impurities are expelled as it grows. If the ice forms very quickly these impurities remain trapped in the lattice and the ice is opaque, but this is rare. The purity of most freshwater ice means that the water below it can usually be seen. On road surfaces this results in the occurrence of 'black ice', where the dark road surface beneath the ice is visible: as black ice is essentially invisible it can be a major hazard to drivers.

Fish survive in lakes as they can swim in the water below the ice cover. Oxygen in the water is replenished by inflowing streams, though in many ponds this is not the case: if the freezing is deep enough, or if the ice coverage prevents oxygen absorption for long enough, the oxygen dissolved in the water may be exhausted causing fish and other aquatic organisms to die.

## River ice

Although the flow of rivers involves the transfer of potential energy to kinetic energy and acts against freezing, the tumbling action of the water takes cooler water to depth and so promotes a chilling that extends across the entire water column. Ice can therefore form throughout the column. This ice is carried by

Latent heat is the heat absorbed or released by any substance as it changes phase (from solid to liquid, for example, or from liquid to gas). There are two forms usually termed the latent heat or fusion (for water this is ice to liquid water or vice versa) and the latent heat of vapourisation (water to vapour or vice versa; this is sometimes called the latent heat of evaporation). When latent heat is absorbed or released, the temperature of the substance does not change – only the phase

changes. Changes of temperature without a change of phase require the absorption or release of what is termed 'sensible' heat.

In lakes close to freezing, and leads (channels of water) in sea ice on the verge of freezing, convection currents release latent heat into the cooler air above, with water vapour condensing to form frost smoke, an ethereal, strangely beautiful phenomenon.

the flow, but is extremely 'sticky' and will plate out on any cold surface. It therefore accumulates at the river edge where rocks are exposed to the cold air. In slow-moving sections of a river, ice also migrates to the surface where it can form sheets that extend across the river. Ice sheets can also grow from each of the river banks to meet in the middle. In both cases, river ice is unusual in that, not being continuous, water may flow both above and below it. Because, as a general rule, ice forms in slower-moving water, the freezing 'front' advances upstream from the river mouth. During melting, ice-jams may then occur as large chunks of ice are driven downstream by the flow.

Frozen lakes and rivers are a hazard for the Arctic traveller, and extreme caution must always be exercised. A good general rule is to always cross a river at its widest point. This seems counter-intuitive, but it minimises the risk that the ice thickness has been eroded by sub-ice river flow. On lakes, do not walk one behind the other: if the first walker causes the ice to crack it may not break until the second man arrives. In general, 5cm of freshwater ice will support a man, but as with all general figures there are parameters that can affect the strength of the ice and this figure should not be taken as anything other than a guide.

## Sea ice

Although glacial ice has carved the land masses of the Arctic, it is the sea ice which, in both the popular

Even waterfalls are not immune to freezing, though complete freezing depends on water volume and may not always be complete before the spring thaw commences. Here at Gullfoss, Iceland, river water is freezing across the top of the falls.

imagination and in terms of extent, defines the region. The central Arctic Ocean is covered in perennial sea ice, at the fringe of which is further seasonally varying ice cover. Due to its salinity, sea ice forms in a very different way to freshwater ice.

The salinity of seawater is usually quoted as a single number that gives the weight, in grams, of salt in one kilogram of water. The average salinity of the Arctic Ocean is about 33. This is less than the average for the planet's oceans for two reasons: first, the huge continental rivers that run into the Arctic Ocean dilute it (although the Arctic Ocean has only about 1% of the Earth's volume of seawater, it receives around 11% of the total freshwater input); second, the rate of evaporation is much lower than in temperate and tropical seas – where there is sea ice, evaporation is eliminated, and at the sea ice edge the rate of evaporation from cold water to a cold atmosphere is lower than in warmer oceans. These effects also mean that the salinity of Arctic waters is highly depth-dependent, varying from 28–30 close to the surface to a 'standard' oceanic value of 35 at depths of 200–300m.

The presence of salt (chiefly sodium chloride, which constitutes 78% of the salt burden, with magnesium chloride (c.11%) and other salts at lower concentrations) lowers the temperature at which water freezes to between -1.8°C and -2.0°C. It also changes the *way* in which water freezes. By contrast to freshwater, sea water does not exhibit a maximum density at 4°C, with the density continuing to rise as temperature falls. So as the surface layer cools it sinks, a situation never being reached where cool water floats on a layer of slightly warmer water. Therefore, the whole water column has to cool to its freezing point before ice can form. However, this does not mean that the entire Arctic Ocean must freeze solid to the bottom at once: the depth of a water column is dependent on salinity, and there are sharp discontinuities in salinity in the ocean that effectively demarcate the water columns. Discontinuities occur at depths of 10–40m (though this does mean that in shallow seas the water column must be at freezing point all the way to the sea bed before freezing can occur). There is also a further effect of salts being lost from the surface layer of water as the temperature falls: this leaching leads to a surface layer of lower salinity (and therefore a higher freezing point) that aids the freezing process.

Pancake ice.

Sheets of nilas ice form on a calm sea.

In the first stage of sea-ice formation, crystals (usually plates or needles 3–4mm across) grow: this is frazil ice. The crystals multiply to form a greyish surface layer that behaves something like a thick, syrupy liquid: this is grease ice. The crystals in this thick soup now coalesce, forming plates that initially remain flexible enough to bend and move with the action of waves and winds. If there is little wind and a calm sea, the plates may form extensive sheets known as nilas ice, but in more turbulent seas the plates thicken and collide, forming pancake ice. Pancake ice usually consists of roughly circular plates, each with a raised edge resulting from rubbing against other plates. Eventually the plates coalesce to form a continuous sheet that thickens as it ages. The thickening process is again dependent on sea conditions: in turbulent seas ice usually thickens through the accumulation of frazil ice crystals on the lower surface, but in calmer seas long, columnar crystals may form. Because the Southern Ocean is, in general, much more turbulent than the seas of the Arctic, columnar ice crystals are much less common in the south, forming only 20–40% of sea ice, while they form 60–80% of Arctic sea ice.

Whether the thickening of the ice is by frazil or columnar crystals, it is, of course, temperature dependent. If the air temperature is very cold the ice may thicken by up to 20cm daily, though this rate inevitably declines because the thicker the ice becomes the more it acts as an insulator. As a consequence, thickening rates are rarely more than 40cm in a week or more than 2m in a year. Seasonal ice is therefore usually about this thick, though 'old ice' (the name given to ice more than one year old – it is also often called multiyear ice) can be up to 8m thick (though 4–5m is more usual). This thickness derives not only from freezing of seawater on the lower

ice surface, but the accumulation of snow on the upper surface, which compacts to form new (freshwater) ice. In general, sea ice of about 20cm thickness will take the weight of a person (compared with c.5cm for freshwater ice) – but again this should not be taken as a golden rule, as there are many factors that can potentially weaken the ice sheet. Since much of the salt is leached from the surface layer of sea ice, the obvious question is why there is such a substantial difference in ice strength. The answer is entrained brine pockets (which, due to their high salinity, do not freeze). A sheet of freshwater ice only 5–6mm thick can be handled as if it were a pane of glass. To handle sea ice in the same way requires a thickness of about 6cm (around ten times the thickness). The nature of the different ices is clear if the two 'panes' are dropped. The freshwater ice will shatter into myriad shards, but the sea ice will splash, rather as a ball of treacle or ice-cream would.

Another rule of sea ice that is helpful for a traveller (but again is not an absolute) is that grey sea ice is thin and should be treated with caution (the colour being due to the dark sea visible through the ice), while white ice is thick and therefore likely to be stronger. However, snow falling on grey ice can turn it white…

Sea ice has a much lower salinity than the water from which it formed, as salts leach out of the ice lattice into pockets of brine during the freezing process. The brine pockets then migrate downwards, either under the influence of gravity or because melting snow on the surface percolates down and flushes out the brine. If the surface layer of sea ice freezes very quickly the salinity can remain high, but in general freezing is slower and the salinity drops to about 5 (i.e. about 85% of the salts have gone). Further salt leaching can mean that old ice has a salinity as low as 2, meaning it can be used

## Comparing ice

Much of the central Arctic Ocean is covered by perennial ice, i.e. ice that does not melt from season to season and is, in general, more than two years old. The existence of this perennial ice is an important difference between the Arctic and the Antarctic: in the latter the majority of sea ice (about 85% – there is some perennial sea ice in both the Weddell and Ross seas) forms annually. The perennial ice coverage of the Arctic is about half the winter maximum cover. Because the winters at the two poles are out of phase, the Earth's sea ice cover is approximately the same throughout the year (although about 20% less in January–March) and averages about 25,000,000km² (though both the statement and the extent are under threat as global warming causes significant decreases in Arctic sea ice coverage). This is about 7% of the surface area of the Earth's oceans, an area roughly equal to that of North America. Although this area represents almost two-thirds of the Earth's total ice coverage, sea ice is relatively thin, so it represents less than 0.1% of the Earth's ice volume. In terms of land area covered, the Antarctic ice sheet provides more than 80% of the Earth's ice coverage (by area), with the Greenland ice sheet accounting for about 12%, and glaciers the rest.

as a source of freshwater. The leached salts increase the salinity of the water beneath the ice, helping, in part, to drive the Atlantic conveyor (see Chapter 4).

## Fast ice and pack ice

Sea ice can be anchored to the shore, forming what is called fast ice. Away from the shore sea ice is usually known as pack ice (to complete the picture, ice scientists define three forms of Arctic sea ice – pack, fast and polar cap ice, the latter being the permanent ice around the North Pole: polar cap ice, though a permanent feature, varies in thickness between summer and winter). Unbroken pack is the term for complete sea ice cover. However, heavy swells, the wind and currents can break up a continuous ice sheet, particularly during the Arctic summer when the ice thins due to melting. When the ice covers about 75–80% of the water surface it is called close pack ice. Open pack ice refers to a coverage of 50–75%. There is no accepted term for coverage below 50%, the term ice floes being used to describe large sections of broken sea ice littering the water. As the floes are broken up by wave and wind erosion, or by collision with other chunks of ice, a mass of small ice pieces is created: this

One way of crossing narrow leads on a snow scooter is to open the throttle and skip across like a stone skimmed across a lake. The technique requires strong nerves; any last minute deceleration can result in loss of the machine and a cold bath. Pond Inlet, with Bylot Island in the background.

Sea ice in Pond Inlet. In the foreground is fast ice attached to the shore of northern Baffin Island. A lead of open water separates the fast ice from the main sea ice mass. The mass is split by a lead stretching back towards distant Bylot Island.

is termed brash ice. Occasionally a vast expanse of sea ice will be eroded at its edge, so that rather than forming an area of closed or open pack, sections break free and are quickly smashed into brash. This abrupt edge, which can look as clean cut as the division between land and sea at a temperate shoreline, is often called (a little confusingly) the floe edge.

The attachment of sea ice to the shore – the fast ice – is often so tenacious that movement of the pack causes a fracture between the pack mass and a section of fast ice. The fracture line often runs more or less parallel to the shore, the pack retreating to leave a lead of open water. Leads may also form where the pack fractures due to currents or wind erosion. Leads are a nuisance to travellers on the sea ice as they often require long detours to a point where the open water narrows

sufficiently to allow a crossing, or a wait until the lead freezes over. Leads can be hazardous to boats, even to ships, and to Arctic cetaceans. They can form relatively quickly and close just as swiftly, trapping and crushing boats or marooning whales far from open water.

The fracturing of pack ice can also result in the formation of pressure ridges, as currents and waves force one section of pack to ride over another as a lead closes. Pressure ridges can also form in open or close pack ice if floes are driven against each other, a general freezing then creating unbroken pack with ridges *in situ*. Such ridges can reach 15m in height and, as with leads, are a considerable challenge for sea-ice travellers. As well as a ridge on the pack surface there is also a downward-pointing ridge below the ice, where the over-ridden flow has been pushed down. In shallow water this ridge can be driven into the seabed. Such downward ridges (occasionally given the tongue-in-cheek name of 'bummocks' – the opposite of 'hummocks') can be a hazard to drilling rigs or underwater pipes and cables.

Much more dangerous is the *ivu*, an Inuit word that describes a potentially lethal event in which a jumble of floes is pushed – by what process is still not absolutely clear, though it must involve strong onshore winds or currents – at speed onto land, rather like a frozen tsunami. Ivus can kill, but are thankfully very rare. Scientists were initially sceptical of the existence of ivus, but they have now been verified, and several have been observed and studied. The best-known of these studies followed the excavation in 1982 of an ancient site at Utqiagvik, near Barrow, Alaska, where the bodies of a family of five were discovered. It is thought the family was overwhelmed by an ivu some 400 years ago. An ivu was reported in January 2006, again at Barrow, when a wall of ice up to 12m high was pushed almost 100m on to the shore, over-riding a large protective berm and partially destroying a road.

## Ice drift

As well as the local movements of sea ice that may break up the pack and create leads, there are also macro-movements (Figure 3.1). Of these, the most famous is the Transpolar Drift, which flows from Russia's New Siberian Islands across the pole to Svalbard and the Denmark Strait. It was this drift that carried relics of the

*Jeanette* from the eastern side of the Laptev Sea to the coast of Greenland, where their discovery prompted Fridtjof Nansen to undertake the *Fram* expedition (see Chapter 6). The Drift has two well-defined branches, the Siberian ice current taking ice towards Franz Josef Land and Severnaya Zemlya, creating a small gyre (circular current) in the Laptev Sea, while the Polar ice current hauls ice towards Ellesmere Island and western Greenland. The rate at which ice drifts is dependent on wind as well as ocean currents. The *Fram* took three years to move from where it became ice-locked close to the New Siberian Islands to its point of release west of Spitsbergen. That time is less than the average, which is closer to five years (an average drift rate of about 0.1km/hour). The rate is not linear, however, since the ice accelerates from the North Pole to the Fram Strait, that part of the journey taking about one year (at an average drift rate of about 0.5km/hour). For comparison, drift rates of up to 2km/hour have been measured in storms.

Nansen's journey has made the Transpolar Drift the most famous of Arctic ice currents: the most infamous is the Beaufort Gyre, a mass of circulating ice some 1,200km across, centred on the frozen ocean to the north of Alaska. The gyre often caused problems for the American whaling fleet based at Herschel Island.

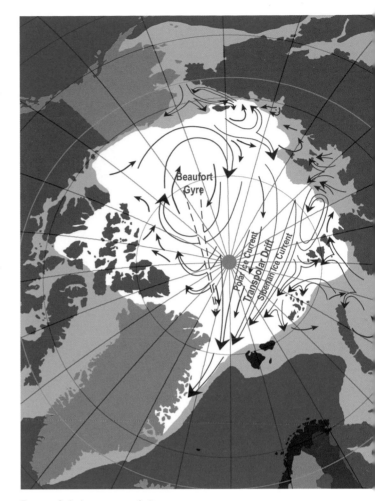

Figure 3.1 Arctic ice drift.

## Open waters

The summer thaw of Arctic sea ice is chiefly due to rain and sun, assisted by wave effects at the floe edge. Melting can start away from the floe edge, particularly if seaweed has become trapped in the ice. Dark matter absorbs sunlight more rapidly than the ice surface (where the albedo – see below – is high) and local melting begins. This forms a meltwater pool on the ice surface, the darker water absorbing more heat so the pool spreads and melting accelerates. However, in some places open water surrounded by sea ice can occur throughout the year, with some of these recurring annually in summer sea ice. These are called polynyas, the name deriving from a Russian word for a forest clearing. North Water in Smith Sound at the northern end of Baffin Bay and the Great Siberian Polynya in the Laptev Sea are classic examples. Annual polynyas result from reliable currents such as those between islands, or from reliable weather patterns, such as

offshore winds. However, some polynyas exist away from such sources and are thought to be caused by upwellings of warm water, though the precise reasons for this are not fully understood. Polynyas are important ecologically, with the annual ones being critical for some species. The open water allows oxygenation, light penetration and the surface water to warm. The salinity of open water differs from adjacent waters beneath the ice, with gradients in salinity and temperature within the water creating currents that take oxygen to depth and bring nutrients to the surface, allowing organisms to thrive. The importance of North Water is reflected by the fact that the seabird population nesting within flying distance of it is measured in tens of millions. The abundance of food also attracted Bowhead Whales to North Water where they overwintered, a fact that was rapidly understood by 19th century whalers.

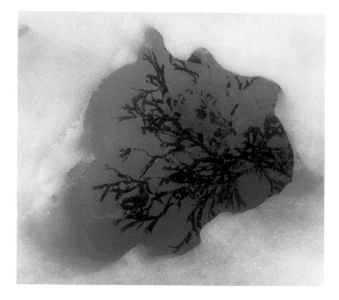

Seaweed trapped in sea ice. Because the seaweed is dark it absorbs the heat, melting the local ice.

## Glaciers

As we have seen, once snow has been sufficiently compacted it forms ice. Though ice usually appears as a very hard, but brittle, substance, glacial ice is plastic and will flow downhill under the influence of gravity (though the actual method by which glaciers move is complex, as we shall see below). Because of the effect of gravity, an ice sheet formed on flat land will be higher at the centre than at the edges, forming an ice dome. Ice radiates outward from the dome, forming glaciers (known as outlet glaciers) that move the ice to the sea or to a point where ablation (as the reduction in ice volume is termed) causes the glacier to disappear.

In general glaciers gain mass in their upper region as the average annual temperature in that area tends to be below freezing, so that the mass accumulated from snowfall exceeds that lost through surface melting. In the glacier's lower region, mass loss exceeds accumulation: this is known as the ablation zone. Mass accumulation occurs solely due to precipitation, but ablation can result from melting for one of several reasons, or by direct mass loss. Melting can be from solar or geothermal energy input, or from friction due to sliding (though that is by far the least effective in terms of energy input). Direct mass loss occurs in tidewater glaciers where the collapse of the glacier front (termed 'calving') creates icebergs. If the mass balance of a glacier is positive, i.e. accumulation

exceeds ablation, the glacier grows. If the mass balance is negative the glacier retreats. The theoretical line across a glacier where accumulation and ablation are equal is known as the equilibrium or firn line (Figure 3.2).

At first glance glacial flow would seem to be a straightforward process, but this is far from the case. The natural assumption would be that ice, which after all is well-known for being slippery, slides over the rock at the glacier's base. For this to take place, the basal ice of the glacier must not be frozen to the substrate. This may be the case if geothermal heat raises the temperature of the substrate, or if the basal ice layer's melting point is lowered due to the pressure of the overlying ice. Even if the substrate temperature is sub-zero, it may still be

Glacial ice mixed with abraded rock (moraine), Billefjorden, Svalbard.

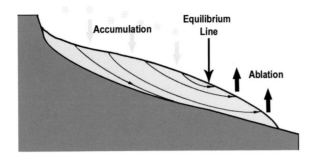

Figure 3.2 Equilibrium line on a glacier.

higher than this lowered melt temperature. In such 'warm-based' glaciers the basal layer of ice melts and the glacier slides, with the meltwater acting as a lubricant.

However, such glaciers are rare in the Arctic, most of the region's glaciers being 'cold-based', with the substrate being colder than the basal ice layer's melting point: in other words, the glacier is frozen to the substrate and basal sliding cannot occur. The advance of glaciers like this is by 'ice creep'. This differs from the flow of a liquid, as it is caused by the elongation and displacement of individual ice crystals. If a bar of ice is suspended in a room at a temperature below freezing point it will slowly elongate – that is ice creep. Under certain circumstances the bedrock may itself deform, contributing to glacial flow.

In many Arctic glaciers the situation is even more complex, as the glacier may be cold-based in its upper regions and warm-based at lower altitudes. Recent measurements of the speed of some Arctic glaciers has shown that they are speeding up due to climate change: increased surface melting causes a downward trickle of water to the glacier's base where, if it remains unfrozen, it acts as a lubricant, turning a previously cold-based glacier into a warm-based one. The measurements indicate that this rise in glacial speed occurs primarily in glaciers south of 60°N, but more recently the phenomenon has been observed in glaciers to 70°N.

Although glaciers flow downhill, the ice will also pass over large impediments to its movement, such as large bodies of rock that form part of the substrate. The flow of the ice in such situations is not straightforward, involving an enhancement of ice creep, but also regelation. Regelation is the melting of ice due to pressure and its subsequent refreezing when the pressure is released. The clearest example of this process can be seen in a cold room if a wire with a weight attached is looped around a suspended block of ice. In time the wire will cut through the block and fall away, but the block will remain intact, with little trace of the wire's passage. Where the wire touches the ice, the pressure from the weight lowers the melt temperature. The ice melts, the wire descends through the thin water layer, and the water refreezes above it. The same process occurs in a glacier; pressure from the ice mass upstream causes the ice at an obstacle to melt. The water flows around the obstacle and refreezes on the downstream side.

## Types of glaciers

There are different types of glacier, but most of these are confined to mountainous areas where the tortured landscapes produce the differing types. In the Arctic there are essentially only two types, valley (often called alpine) and piedmont glaciers. Valley glaciers are those confined to

### Crevasse diversity

Ice creep is a slow process, so if the glacier's bed changes slope it may not react fast enough to adjust to the change. Fractures then develop in the ice as ice masses on differing slope angles split from each other. Where these faults reach the surface they form a series of parallel crevasses. Crevasses are usually transverse, i.e. they form at right angles to the ice movement. However, chevron crevasses also occur: these are angled across the glacier and result from the effect of frictional drag on the ice by the walls of a containing valley. A third form, splaying crevasses, are a combination of the other two.

A particular form of crevasse, a single, wide crevasse called a bergschrund, can often be seen where a glacier separates from the valley headwall. However, despite this often being claimed as the sole position of bergschrunds, they can also form where fast-moving ice separates from slower-moving sections of a glacier.

Icefalls occur where the slope of the substrate underlying a glacier is steep, causing the ice to break up: the numerous crevasses result in the isolation of large, unstable blocks of ice, called séracs. The most famous example is the Khumbu Icefall on the approach to the Western Cwm, on the southern (Nepalese) side of Everest. In 2014 the collapse of a large sérac triggered an avalanche which killed 16 Nepalese guides. Thankfully, icefalls are rare in polar glaciers.

Heavily crevassed glacier, Blomstrandbreen, Svalbard.

the valleys that the glacier itself has carved. Such glaciers terminate in a convex ice snout, which either gives birth to a river or, if the glacier reaches the sea, calves icebergs. Those that reach the sea are often called tidewater glaciers. Tidewater glaciers often carve their base to below sea level, and the sea fills the subsequent trench if the glacier retreats. This process is the origin of fjords: the most famous fjords indent Norway's coast, but fjords are also a feature of some Arctic landscapes. Norwegian fjords are now essentially free of the glaciers that carved them, but some Arctic fjords still retain their glaciers.

Piedmont glaciers are those in which the ice reaches a plain at the base of the mountains from which they form: the name means 'mountain foot'. Having escaped from the confines of its valley walls, the ice spreads on the plain to form a characteristic lobe. Piedmont glaciers are rare, and global glacial retreat is making them even rarer. The oft-quoted 'classic example' is the Malaspint Glacier on the southern edge of Alaska's St Elias Mountains near Yakutar. This lies outside the Arctic as defined in this book, but within the Arctic boundary there is a good example at Skeiðarájökull on the southern side of Iceland's Vatnajökull.

## Ice on Earth

Though in the popular imagination ice is the dominant feature of the Arctic, over much of the land surrounding the frozen Arctic Ocean there is little or no permanent ice (with the notable exception of Greenland and, to a lesser extent, some islands of Arctic Canada and Russia). For the Earth as a whole, ice holds some 33 million cubic kilometres of freshwater. This is a vast amount: if all this ice were to melt, sea levels would rise by some 70m. Yet this volume represents only about 2% of the water on Earth. The oceans account for 93.5% of the total, rivers and lakes the other 4.5%. Of the ice, about 90% (by volume) forms the Antarctic ice sheet. Greenland's ice sheet, accounts for a further 8%, the smaller ice caps on the Arctic islands (the largest of which is on Ellesmere Island, that being twice the size of the next biggest, on west Spitsbergen) and the world's glaciers making up the remaining 2%.

Heavily crevassed tidewater glacier flowing down from the Beerenberg volcano on Jan Mayen.

## Icebergs and ice islands

Though icebergs are an oceanic feature, they form from terrestrially derived (i.e. freshwater) ice, the bergs being calved through the fracturing of the fronts of tidewater glaciers. Most Arctic icebergs therefore differ from most Antarctic icebergs, which form when the ice sheets flowing above the waters of the Southern Ocean fracture. Such fracturing creates tabular bergs, which are often vast. But the Arctic has fewer, and much smaller, ice shelves and so produces fewer tabular bergs. This key difference is related to temperature: Antarctica's ice is much colder than most Arctic ice masses, and cold ice has a much higher tensile strength than 'warmer' ice. Thus, while the tides acting on Arctic glaciers that reach the sea cause the ice to flex and, ultimately, fracture, the tides of the Southern Ocean do not impart sufficient energy to overcome the strength of the ice. Only in certain limited areas of the Arctic is the ice sufficiently cold to resist tidal forces: in most places the temperature does not drop low enough for long enough to allow ice sheets to form.

The Arctic's glacially-derived icebergs are smaller than the tabular bergs of Antarctica. Calving also produces much less uniform shapes, Arctic bergs often

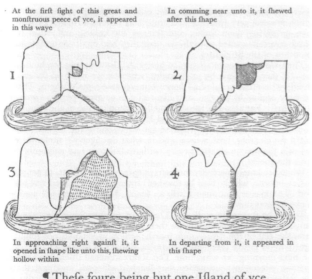

At the firft fight of this great and monftruous peece of yce, it appeared in this waye

In comming near unto it, it fhewed after this fhape

In approaching right againft it, it opened in fhape like unto this, fhewing hollow within

In departing from it, it appeared in this fhape

¶ Thefe foure being but one Ifland of yce, and as we came neere vnto it, and departed from it, in fo many fhapes it appeared.

The first depiction of an iceberg, from Thomas Ellis' book *A true report of the third and last voyage into Meta Incognita acheived by the worthie Capteine Martine Frobisher Esquire, Anno 1578*.

being misshapen in an aesthetically pleasing way. But smaller does not mean trivial, as the *Titanic* found out with terrible consequences when it collided with the underwater section of an iceberg on 15 April 1912. The *Titanic* did not, of course, have radar, but radar has its limitations as a tool for detecting icebergs: ice is only about 2% as reflective as metal (i.e. another ship) and to make matters worse a misshapen berg scatters a radar beam in many directions. Today the International Iceberg Patrol (IIP) uses aircraft to spot bergs in the North Atlantic, a white berg being readily visible by eye, even on overcast days or when there is some fog (though dense fog obviously makes spotting impossible). There is no equivalent of the IIP in the North Pacific as icebergs are only calved in south-eastern Alaska: these are few and rarely escape the Gulf of Alaska.

In 2010 an iceberg covering almost 250 km² broke from Greenland's Petermann Glacier. The object – more ice island than iceberg – rapidly broke into smaller, but still substantial pieces. For comparison, a tabular berg which broke off Antarctica's Ross Shelf in 2000 was almost 300km long and over 35km wide: covering over 11,000 km² it was larger than Jamaica. It is estimated that 30,000–40,000 icebergs calve annually from Arctic glaciers, though perhaps only 300–500 reach the open sea, as most are too small to survive for long periods. Of this total, about 3% emanate from a single glacier, the Jakobshavn Glacier on Greenland's west coast. This, one of Earth's fastest moving and most productive glaciers, moves at *c*.20m daily and calves about 40 million m³ of ice (weighing around 30 billion tonnes) annually from an 8km-wide front. The bergs are calved into Jakobshavn Isfjord (south of Ilulissat, a settlement whose name means 'iceberg'), the mouth of which (leading to Disko Bay) is restricted by a morainic bar, from a time when the glacier reached as far as the bay itself. This bar acts as a barrier to larger bergs, and such is the volume of ice produced that the fjord can be filled with stalled bergs, to the point where open water can be difficult to find. It is probable that the *Titanic*'s iceberg was calved from the Jakobshavn Glacier.

As ice is less dense than water, icebergs float, but with the larger fraction of their bulk (about 85% generally) below the sea surface. This mass distribution represents a crucial difference between icebergs and sea ice. Although both ice forms are affected by ocean currents,

Icebergs off the coast of the Liverpool Land peninsula, east Greenland.

icebergs are much less affected by the wind than is sea ice (though if the above-water section of a berg is sail-like they will catch the wind). This difference can occasionally result in the bizarre vision of a berg and sea ice moving in opposite directions, the berg ploughing a course through the ice. Historically this curiosity has been used on more than one occasion by an icebound ship as a means of escaping entrapment. Icebergs can also plough the sediments on the sea bottom, or the sea bed itself. Examples are known of trenches up to 20m deep and many tens of metres wide being ploughed, sometimes over distances of several kilometres: such ploughing would seriously endanger oil pipelines and underwater cables.

Icebergs are eroded by a combination of sunlight, wind and wave action above the surface, and water temperature and wave action below. In time, they disintegrate, often rolling over when differential erosion makes them unstable. Rolling bergs are extremely dangerous: people on them would be thrown into the sea and sucked under to almost certain death by induced currents, and boats beside them can be overwhelmed. Consequently, few scientists now ever land on large Arctic bergs and the rule for all travellers is to stay well away.

Overturned bergs can usually be spotted by the surface pattern created by underwater wave action. Some bergs disintegrate explosively, though most die more quietly. The fragments of a disintegrating berg have been given the rather banal name of bergy bits. Bergy bits are classified as being in the size range 2–5m across: smaller chunks are brash ice. The expressive name 'growler' is given to a specific class of berg debris: growlers, which can also be produced by the collapse of a glacier front, are flat-topped masses that float low in the water, rather as sea ice does. Large growlers can be a problem for ships as they can be easily missed by radar. Smaller growlers can be a problem for boats such

Collapsing front of a tidewater glacier, Tracy Arm, Alaska.

as zodiacs, because if unseen they can be steered over, causing damage to propellers.

As noted above, the ice shelves of the Arctic are smaller and far fewer in number than their counterparts in Antarctica. Arctic ice shelves also form in a different way. While some are created by the flow of glacial ice across the ocean (e.g. the Milne ice shelf off Ellesmere Island's northern coast), others form when fast ice develops over many years. The most famous of the Arctic's ice shelves, at Ward Hunt Island off the northern coast of Ellesmere Island, was formed from such an accumulation of fast ice. The Alfred Ernest ice shelf (also off Ellesmere Island's northern coast) was created by a combination of the glacial and fast ice mechanisms. Once an ice shelf has been created, snow accumulating on the upper surface and sea ice accumulating on the lower surface tend to thicken the shelf, while the accretion of sea ice on the ocean edge tends to elongate it. Arctic ice shelves are limited to Ellesmere Island's northern coast (where there are three, as mentioned above), some fjords of northern Greenland, and the Russian archipelagos of Franz Josef Land and Severnaya Zemlya. The shelves of northern Ellesmere Island are by far the biggest, though they are shrinking. Calved

ice from them is normally captured by, and contained within, the Beaufort Gyre, but has been known to escape and reach the Fram Strait.

## Glacial landforms

Although icebergs are perhaps the best-known product of glaciation, glaciers are also renowned for their influence on the landscape over which they flow. In general, this shaping of the land is more obvious in mountain areas, as cold-based glaciers (which predominate in the Arctic) are much less abrasive than those that slide – rock debris caught up in basal layers of warm-based glaciers acting as a very effective sandpaper, particularly as glacial flow is relentless. The sheer power of glaciers to transform a landscape is illustrated in the carving of U-shaped valleys, and for the alpine scenery of arêtes (thin ridges between U-shaped valleys) and cirques (circular glacier-cut basins). Cirque is French: in Britain the feature is known as a corrie or coire (Gaelic) or cwm (Welsh).

Glacial erosion also creates moraine, the often very fine debris of rock abraded from the valley sides and glacier base by the ice. This debris accumulates within, and at the edges, of the glacier. At the edges it forms narrow lines of lateral moraine. Sometimes these lines can be seen on the glacier's surface away from the edges. Termed medial moraine, this usually forms from the lines of lateral moraine where two glaciers have met, but it may be derived from entrained debris reaching the surface. Debris is also deposited at the glacial snout – terminal moraine. Terminal moraine may pile so high that it dams the valley created by a retreating glacier, with a lake forming behind the dam. Behind the terminal moraine there is often an area of hummocks where rock debris covers mounds of unmelted ice, the debris having slowed the ablation processes.

Glaciers may also leave behind other evidence of their existence when they retreat. Perhaps the most obvious is the rock over which the glacier once flowed, chamfered smooth or with striations caused by rock fragments embedded in the ice. *Roches moutonnées* are isolated, asymmetric rock masses over which glaciers have passed, with one side (the upstream side) shallow-angled and abraded smooth, the other, lee (downstream), side high-angled and roughened by ice plucking and frost erosion. The name is French – 'rock sheep' – deriving either from the sheep-like appearance of such rocks studding alpine meadows or the sheepskin wigs worn by judges and advocates of the French court in the 18th century. *Roches* are usually relatively small (a few metres across), but can be much larger, 100m or more in height and a kilometre or so long. In Sweden *flyggbergs*, vast asymmetric hills, sometimes more than 300m high and 3km long, are *roches* and may even have smaller roches studded across their surfaces. Roches that lack the clear-plucked lee side are known as rock drumlins. A further example of glacial activity is the erratic, a boulder of specific rock type carried by the glacier and then deposited in an area of dissimilar rock. Such erratics caused much head-scratching among geologists before glacial retreat and advance was understood.

Rock drumlins are named after a more common form of glacial landform. Retreating glaciers can leave behind mounds of glacial debris, generically known as till. These till mounds are called drumlins, from the Gaelic *druim*, a rounded hill. Though the exact development process of drumlins is still debated the form is, more or less, constant, an elongated mass of till, the long axis giving the direction of ice flow, with an upstream, high-angled, blunt end, the shallow, tapering downstream side ending in a point. The form is, therefore, somewhat similar to an egg: drumlins normally occur in groups known as swarms, the swarm sometimes referred to as forming a 'basket of eggs' topography.

## Glaciofluvial effects

The features discussed above are all caused by direct glacial erosion and deposition, but the landscape can also be transformed by meltwater flowing beneath or away from the glacier, by processes known collectively as glaciofluvial effects. The most conspicuous of these is the outwash fan, a lobe of till formed by numerous meltwater streams flowing over a plain. Such plains are often called *sandar* and are a feature of southern Iceland (hence the name – *sandar* is the plural of *sandur*). In Iceland *sandar* extend for many kilometres not only along the south coast, but also from the base of the glaciers of Mýrdalsjökull and Vatnajökull to the sea. Individual meltwater streams flowing beneath the ice can cut channels in the bedrock. Such channels, usually narrow (up to a few tens of metres), but surprisingly deep, and sharp-edged if the glacier retreat was recent so that there has been limited weathering, are called Nye channels (or N channels) after John Nye, the British glaciologist who first defined them. Nye channels can be very long, up to several kilometres: very much larger forms – up to 100km – are termed tunnel valleys.

Under-ice streams may also form eskers (from the Irish *eiscir*, a ridge: they are also occasionally called by their Scandinavian name, *osar*). Eskers are long, sinuous ridges of debris formed by the silting of ice-walled stream channels. A broken form of esker is termed a *kame* (from the Celtic *cam*, crooked or winding). Kames are often steep cones, though at the edges of the glacier the streams can form extended kame terraces. They differ from drumlins in being produced beyond the glacial snout rather than being exposed by a retreating glacier, but can be difficult to distinguish from eskers by non-experts. Kames are usually associated with 'kame and kettle-hole topography' a landscape in which the kames are mounds of glaciofluvial deposit, the kettle holes being intervening hollows that are often water-filled. The kettle holes form where large chunks of ice

Polychrome, Denali National Park, Alaska. In the foreground is Kettle Lake. Beyond, and to the right of it, is a large glacial erratic.

embedded in the till melt: kettle lakes are water-filled kettle holes.

## Permafrost

*Periglacial* is the term applied to cold but non-glacial landscapes (the fact that the term includes 'glacial' is unfortunate as it can cause confusion). Periglacial regions are often taken to include any area affected by freezing and thawing, as this erosional process modifies the landscape, but strictly the definition applies only to areas adjacent to ice sheets (or close to the margins of the ice sheets of the last ice age), where intense cold penetrates deep into the ground. This cold penetration often results in the development of permafrost, a major periglacial feature. Although often assumed to be ice-based, permafrost is frozen ground (though that does not mean there is no ice present), and is technically defined as rock and soil in which temperatures do not rise above 0°C during two consecutive years. Permafrost requires groundwater – in rock crevices, soil cavities or as lenses of water (which can be formed by the migration of groundwater towards a pocket

of already frozen water, or around buried, remnant glacial ice) – to be frozen. However, the leaching of salts into the pockets from the soil can increase their salinity, so they can therefore remain liquid even if the ground temperature remains permanently below 0°C.

If the annual air temperature of a locality is below about -6°C (there is a range of temperatures depending on locality, but the variation is small, perhaps -5°C to -8°C) then continuous permafrost will be found. At higher ambient air temperatures, between -6°C and -1°C, the permafrost is thinner and may be fragmented: in such an area, the permafrost is said to be discontinuous.

Despite a popular assumption that permafrost is a polar phenomenon, it occurs a considerable distance south of the present timberline, e.g. far to the south in Asian Russia and in China (see Figure 3.3). In fact, around 25% of the Earth's land mass is underlain by permafrost. Permafrost in Asia results from the intensely cold winters of the continental climate of those areas. In Europe the influence of the North Atlantic Drift largely prevented the creation of permafrost, so Europeans only became aware of its existence when Arctic

Building collapse caused by permafrost melting, Dawson City, Yukon, Canada.

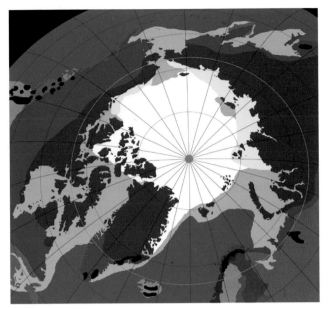

Figure 3.3 The distribution of permafrost in the Arctic and sub-Arctic. Dark red is the region of continuous permafrost. Pale red is the region of discontinuous permafrost. Yellow indicates sub-sea permafrost, while black indicates areas of alpine permafrost.

explorers attempted to bury their dead and encountered the unyielding, frozen ground beneath the shallow, seasonally thawed layer. Even after the discovery little interest was taken in the phenomenon in North America until the gold rush of the late 19th century began the economic development of the area, and the need to erect permanent structures. By contrast, the Russians knew about permafrost by the early 16th century: writings from the 18th century include references to the remains of mammoths being found in the permanently frozen ground of Siberia.

The depth of the permafrost depends largely on the geothermal heat flux into the ground below the frozen layer and the net energy balance at the surface. In parts of Siberia, the permafrost layer is almost 1,500m thick: in North America it reaches depths of about 1,000m on Baffin Island and 600m on Alaska's North Slope. Such depths are almost certainly relics of the extreme cold of the Earth's recent geological history rather than a result of the present Arctic climate. One interesting feature of the distribution of permafrost beneath northern North America is that it is much more northerly to the east of Hudson Bay than to the west. The prevailing westerly winds in northern Canada blow across Hudson Bay: as the bay remains ice-free during the early winter, these winds pick up moisture, which is deposited as snow in northern Quebec and Labrador. Consequently snowfall in those provinces is much higher than in provinces west of Hudson Bay. The layer of snow acts as an insulating blanket, preventing cold penetration into the ground.

Esker in the Bad Lands of Canada's Northwest Territories.

Strangely, ice itself, the 'progenitor' of the permafrost, also occasionally acts as an insulator, and it is thought that in some cases ice shielded the ground from extreme cold during the ice ages. As a consequence, the permafrost beneath most northerly Arctic glaciers is much thinner than in those areas of Alaska and Siberia that were not ice-covered during the last glaciation. The same is true under the sea, where unfrozen water insulated the seabed from extreme cold: permafrost occurs below the seabed only in areas where the shallow seas above the continental shelf froze solid, i.e. beneath the seas of eastern Russia and the Beaufort Sea.

In general, seasonal variations in the temperature of the permafrost do not occur at depths below about 20m. In summer the surface layer, known as the active layer, thaws. The depth of the active layer depends on the local energy balance: it may be as little as a few millimetres or more than several metres deep. Because the still-frozen permafrost beneath the active layer inhibits drainage, sections of the active layer may become saturated, forming an adhesive porridge (which rapidly accumulates in the tread of walking boots). Such areas are often called *taliks*. For the Arctic traveller in winter, taliks can be the cause of serious inconvenience as they may form hidden pockets that, if they lie beneath a thin surface crust, can overtop the boot of anyone unlucky enough to break through.

## Nivation

In periglacial areas cold also sculpts the landscape directly by the process of nivation, the frost erosion of rock, which, over time, causes it to break down. During summer, rain or melted snow seeps into cracks in the rock. As ice is less dense than water, the water expands on freezing. The pressure exerted is considerable, and over time levers chunks of rock from cliffs or bedrock (this is the same process that causes household pipes to fracture: the damage is done when the water freezes, pushing joints in the pipe apart or, sometimes, causing the pipe itself to rupture, but the problem does not become apparent until a thaw sets in).

Frost erosion occurs in both glacial and periglacial landscapes. In periglacial areas it is responsible for such distinctive features as scree slopes, created by the frost erosion of a cliff, the rock debris (or scree – the debris

Ibyuk pingo, Tuktoyaktuk. The Tuktoyaktuk Peninsula is the world capital of pingos, with Ibyuk the largest of them.

also has the more formal name 'talus', so scree slopes and talus slopes refer to the same feature) formed littering the slope below the cliff. If there is a snowfield on the slope, rocks can slide down it, piling up at the base. If the snowfield then disappears the rocks form a distinctive rampart – called a protalus rampart – beneath the cliff from which they have been prised.

Pingos are among the most impressive landforms of the periglacial zone. The name is a Mackenzie Delta Inuit word for a conical hill: use of the name is reasonable as it is estimated that about 25% of all the world's pingos are to be found on the Tuktoyaktuk Peninsula to the north-east of the delta. Pingos are mounds of ice covered with a layer of sediment, usually circular in form and occasionally of extraordinary size, up to 75m high and over 500m in diameter.

Pingos form in one of two ways. 'Open system' pingos are produced by artesian (underground) water feeding an expanding ice dome, and they are usually found in discontinuous permafrost where groundwater movement is feasible. 'Closed system' pingos form beneath a surface lake. The lake insulates the ground beneath it so it does not freeze, creating a volume of talik. If the lake drains, the talik freezes and the ice expands into the characteristic mound. Because closed pingos form from talik they have a sediment cap, the active layer of which can support considerable growth as the domed nature of the pingo allows good drainage and, if the pingo is sizeable, there will also be a sheltered side. In areas beyond, though not far beyond, the treeline, pingos occasionally have trees growing on their southern slope. Other plants also benefit, pingos having a more diverse and luxuriant growth than the neighbouring tundra. The plant life brings animal life, and the pingo becomes a small oasis. People, too, occupied pingos, though only on a temporary basis, the local Inuit using them as look-outs for spotting caribou herds. If a pingo's ice core is exposed it may thaw, the tops of some pingos having collapsed craters reminiscent of volcanoes. If, rather than draining away, the meltwater forms a lake this insulates the remaining ice of the core, extending the pingo's life.

One aspect of pingo creation that may be critical to human endeavours if warming of the Arctic opens the North-East and North-West Passages to regular

Pingo, Kongsfjorden, Svalbard.

commercial traffic, is that as permafrost underlies the seabed in some areas of the passages, underwater pingos may form. If conditions allow the passages to be used by deep draught vessels, large oil tankers, for example, these could collide with submarine pingos. Submarine permafrost will also make any attempt at drilling for oil or gas in these areas difficult.

## Palsas and patterns

Similar in appearance to pingos are palsas, mounds or ridges of frozen peat. Palsas are found in areas of discontinuous permafrost, usually in the damper areas of marshland. They are thought to be created by frost-heaving processes, or the expansion of a perennially frozen ice lens. Palsas can reach 6–8m in height in southern regions of the Arctic, though they rarely attain heights above 1m in more northerly regions. Palsa mounds are usually 10–30m across, while ridges can be up to 150m long, though lengths of 15–50m are more common. In the far north the peat forming the palsa may be 5,000 or 10,000 years old.

If pingos are the most physically impressive periglacial landform, patterned ground is the most exotic. In some soils winter temperatures can cause shrinkage of the soil and cracks to appear. Linking of these crack lines then creates a pattern of polygons or circles. In summer, water seeps into the cracks from the active layer. The water freezes in the following winter, forming wedges of ice. The annual cycle of freeze and thaw causes the wedges to expand as layers of ice are added. Wedge expansion pushes the ridge of soil at the rim (i.e. at the surface) upwards so that raised polygons are formed (see Figure 3.4). The polygons range in size from a metre or so across to around 100m, the ice wedge itself being as much as 2m wide. The widest wedges can take decades to form.

The polygons are one of several forms of patterned ground, a feature that puzzled early Arctic travellers. Most curious of them is the sorting of the ground material by size, larger stones forming the sides of some polygons, with finer material in the centre. The sorting appears man-made, but it is a natural process – frost-heaving. Frost-heaving is a consequence of the different thermal inertia of stones and finer material. Stones have a lower specific heat and so cool quicker. As the ground

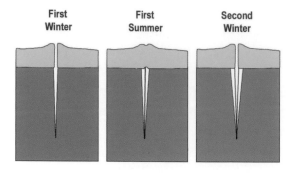

First Winter    First Summer    Second Winter

Figure 3.4 Ice wedge formation, a prelude to the formation of patterned ground.

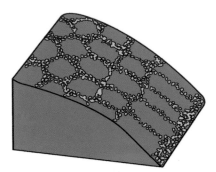

Figure 3.5 Polygons are formed on level ground, but where the ground slopes, the polygons extend to form slopes.

cools the upper surface of the stone sticks to overlying frozen material and is pulled upwards as this expands (this is termed frost-pull). As the freezing front moves downward it moves faster through the stone, reaching the material beneath which expands and pushes the stone upwards (frost-push). The net effect is to force the stone to the surface. When the thaw occurs, the finer material sinks below the stone leaving it on the surface. Doming of the ground due to frost-heave cycles now causes the stones to move outwards from the dome centre (frost-thrust). Over many freeze-thaw cycles circles of stones form, these eventually interacting to form polygons.

Stone stripes may form on sloping ground, where the polygons extend downhill (Figure 3.5) and gravity completes the sorting process. In all cases, the lines of stones are narrower than the intervening areas of finer material. The more homogenous the ground material, and the more uniform the freezing and heaving process, the more regular the polygons and patterns produced.

One form of patterned ground that can be useful to nesting birds, but may be a nuisance to the traveller, is the hummock field. Earth hummocks are essentially

Patterned ground polygons, Kvadehuken, Svalbard.

Rock glacier, Snæfell, Iceland.

non-sorted circles, though their exact creation method is poorly understood (currently popular is the idea that they form in areas where frost-heaving is irregular and concentrated in discrete areas, though why this should be so is not clear). The raised hummocks, which are often hemispherical and vary in height from a few centimetres to several metres (though smaller mounds are more common), usually support various plant colonies. The hummocks offer a convenient nesting place for birds, but make traversing an area difficult, particularly if the hummocks are around knee-high.

## Other periglacial landforms

Three final periglacial landforms are worth noting. Thermokarst is the periglacial equivalent of a karst landscape in temperate zones where water dissolving bedrock, with subsequent sub-surface flow, creates depressions. In thermokarst, localised ice thawing (due to patchy vegetation cover or asymmetric erosion) creates mounds and, occasionally water-filed, hollows, these sometimes coalescing to form thaw lakes. If sections of the area slope gently, beaded drainage may occur, with a linear series of lakes linked by small streams.

A second landform is the gradual downhill drift of soil which can occur due to frost-heaving on sloping ground, stones brought to the surface being moved downhill by each freeze-thaw cycle in a process often called frost creep. Soil drift also results from gelifluction, the periglacial equivalent of solifluction, the downhill slumping of saturated material. When the active layer of the permafrost freezes, it expands perpendicular to the frozen layer beneath, even if the ground is sloping. However, when the active layer thaws it moves with gravity (i.e. downhill), rather than back towards the layer below. Successive freeze-thaw episodes then cause distinctive gelifluction lobes to form.

The final landform is one of the most enigmatic, the rock glacier. In cold, relatively dry, high-relief landscapes in which there is a good supply of scree, this debris may flow downhill. The exact structure and creep method of rock glaciers is not understood. Some are believed to have an ice core, perhaps having begun as a debris-covered glacier (in which case the rock glacier

Drifting sea ice, south-west Spitsbergen, Svalbard.

would be a glacial, rather than a periglacial, landform). However, a rock glacier may consist of an upper layer of larger scree (which may be aggregated using ice as a 'cement') covering a layer of frozen rock sitting on a layer of smaller scree, which acts in a similar way to ball-bearings. Most rock glaciers are mid-latitude alpine landforms, there being around 1000 examples in Switzerland. But they also occur in the Arctic, one in Greenland being 5.5km long, the longest rock glacier known. In general rock glaciers are much smaller, less than 800m long and 100m wide, and travel slowly, usually at speeds of less than 1m annually.

# 4 The climate of the Arctic

The Arctic is cold. This factually accurate statement of the obvious, temperature being one of the defining features of the region, particularly in the popular imagination, belies an interesting question – why? The Arctic is cold is one of those statements which, like 'the sky is blue', is both self-evident, yet, when considered in detail, perplexing.

When the search for the North-West Passage began, one idea was to sail north directly over the North Pole, the reasoning being that since the Sun was visible throughout the summer close to the pole, sunlight should melt the sea ice. Only further south where there were significant periods of darkness could the sea ice persist. All a captain needed to do was to find a gap in the ring of sea ice and sail through; beyond would be the open polar sea and (apart from the need to find another gap on the other side) the journey to the east would be straightforward. But it was eventually discovered that the long Arctic summer did not melt the sea ice: the extra sunlight did not compensate for the long Arctic winter.

There are several reasons why not. As already noted in Chapter 1 the low-angle Arctic Sun illuminates a greater area of the Earth's surface than it does at more southerly latitudes so that there is a significantly smaller energetic input per unit area. Despite the polar summers, the North and South Poles receive only 60% of the insolation (incident solar radiation) of a point on the equator. The low angle also means that radiation from the Sun (i.e. light) must pass through more of the Earth's atmosphere before reaching the surface, some sunlight being absorbed or scattered by the molecules that make up the atmosphere – carbon dioxide and water vapour absorb and re-radiate some of the incident radiation, and dust particles scatter it.

Some of the radiation that does reach the Earth is absorbed, warming the surface, but some is reflected. This effect, termed albedo, varies with the make-up

The Arctic is cold. Winter night, Kapp Dufferin, Svalbard.

of the surface. Darker surfaces absorb more of the incident radiation than lighter ones. Dark soils absorb 90% of incident radiation and reflect only 10%, but for Greenland's Inland Ice these figures are reversed. Clouds may also have a high albedo, reflecting as much as 80% of the incident radiation, so that much of the insolation does not even reach the Earth. As summer cloud cover in the Arctic tends to be high this can have a significant effect: the low-level stratus cloud that dominates the Arctic sky during the summer has an albedo of 60–70% (though direct absorption by the cloud cover is minimal). High surface albedo then reflects much of what insolation reaches the Arctic surface, though some of this can be reflected back by the cloud base. Multiple reflections by the Earth's surface and the cloud base creates a 'flat' light in which people can find travel difficult: on occasions, this flat light can even produce white-out conditions. However, as noted below, the extensive cloud cover which once characterised the Arctic is now reducing. While for the traveller this is good news in terms of the view, the reduction in cloud cover reduces cloud albedo,

Moonlight over Aghard Bay, Svalbard.

and has, inevitably, increased the sunlight reaching the ice, enhancing melting.

The radiation absorbed by the Earth is partially re-radiated, but at longer wavelengths than the incident radiation. Incident radiation is a spectrum, with all but about 0.1% in the wavelength band 0.15–4μm (the peak intensity is at 0.5μm (500nm): visible light, in the wavelength band 400–700nm, accounts for about 50% of the total). The re-radiated energy is at wavelengths of 4–300μm with a peak intensity at about 10μm. While the Earth's atmosphere is essentially transparent to the

incident radiation (though there is, as noted above, some absorption and scattering), it is only semi-transparent to the longer wavelength, re-radiation, which is absorbed by water vapour and some atmospheric gases and warms the atmosphere. Cold air holds up to ten times less water vapour than warm air, one reason that the atmosphere of the Arctic is often amazingly clear. That means that less of the re-radiated energy is absorbed by the atmosphere above the Arctic: the energy escapes into space.

By contrast to the Arctic, at the Equator the Earth receives more energy from the sun than is radiated back.

There is, therefore, an energy surplus at the Equator, just as there is a deficit at the North Pole. In fact, there is a surplus of received over radiated energy at all latitudes below about 38°N. The laws of thermodynamics require the redistribution of that energy, and air and ocean currents transfer heat from lower latitudes to the Arctic. Latent heat is also gained each time water vapour is converted to snow. However, although the air and ocean currents go some way in redistributing tropical solar energy to the polar regions, their effect is not large enough to entirely compensate for reduced energy input from insolation. As a result, the Arctic is cold.

## General weather patterns

The rising of warm, buoyant air at the Equator and the sinking of cold, dense air at the poles, together with the deflection of resulting air streams due to the rotation of the Earth from west to east (the Coriolis force) would, in an ideal world, create a stable, easily understood wind pattern. However, such an idealised view quickly breaks down when both macroscopic and local effects are imposed on the simple pattern. But there are some permanent and semi-permanent features of the Arctic atmosphere that consistently influence the region's weather patterns, so some general comments on the weather the traveller might experience can be made. One such feature is the polar vortex, a stable, low-pressure system in the middle–upper atmosphere (at altitudes between 15km and 80km) around which westerly circumpolar air flows. The vortex, which usually sits above the pole (though it can move hundreds of kilometres from it) steers both the cyclonic and anticyclonic systems that affect the Arctic traveller. More local features are the winter low-pressure systems that form over the Aleutians and between Iceland and south Greenland, the winter high-pressure system above eastern Siberia, and the persistent highs above the Beaufort Sea and the Greenlandic ice sheet.

The polar vortex and the Icelandic and Aleutian lows are associated with climatic oscillations: these are well-documented but poorly understood. The wind speed of the polar vortex influences the Arctic Oscillation (AO), which affects the entire Arctic area. When the winds are high there is lower pressure near the pole and cold temperatures in Arctic Canada, though warmer temperatures in Eurasia (and in Alaska) and more winter ice in the Bering Sea. The AO is very closely linked to the North Atlantic Oscillation (NAO) – some experts claim the NAO is one of the components of the AO – which depends on the pressure difference between the winter low over Iceland and winter high over the Azores. This influences the flow of westerlies over the Gulf Stream and, therefore, how much heat is picked up and carried to northern Europe and Siberia. When the pressure difference is high, European winters are warmer and wetter and there is less ice in the Barents Sea. However, eastern Canada is colder and there is more ice in Baffin and Hudson bays, and in the Labrador Sea. When the pressure difference is low, the reverse is true. The oscillation between the two states takes about 10 years. A third oscillation is the Pacific Decadal Oscillation (PDO), which is associated with the Aleutian low (though it is less well-correlated with air

## A general overview of the Arctic climate

The effects of the circumpolar distribution of land masses in the Arctic on prevailing winds, together with the influence of ocean currents, mean that the climate of specific places is influenced by local conditions much more than in, say, Antarctica. However, some general comments can be made. The Greenland ice sheet is cold and dry, both in summer and winter, though the coastal regions of the island are, of course, influenced by the adjacent seas. The Atlantic Arctic is dominated by the North Atlantic Drift, which gives rise to cool winters and warm summers (even in Svalbard and Franz Josef Land), with wind speeds higher in summer than in winter. In Siberia, the wind speeds are reversed, increasing in winter. The continental climate of Siberia means cold winters and cool summers. The northern Pacific is warmer than the adjacent land masses (but colder than the Atlantic). The area is cloudier and has higher wind speeds and precipitation. Arctic North America is generally cold in winter and warm in summer (often surprisingly so). Summers tend to be cloudy with higher precipitation. Eastern Canada is influenced by the extent of Baffin Bay (as is western Greenland): the Bay is climatically similar to the Atlantic but cooler. The Arctic interior is cold, and cloudier in summer than in winter.

pressure than the NAO). When the PDO is in a positive phase the surface temperature of the North Pacific rises (and, interestingly, the Alaskan salmon catch is higher). Both the NAO and the PDO are associated with the AO, though the nature of the correlation, as with the oscillations themselves, is poorly understood. Like the AO and NAO, the PDO exhibits a ten-year periodicity.

## Ocean currents

Since the North Atlantic Oscillation is associated with heat pick-up from the Gulf Stream, it is worth considering what drives ocean currents around the Arctic, and the effect they have on climate. The most important of the currents is undoubtedly the North Atlantic Drift, the warm water current that drives the Arctic boundary northward in Europe relative to North America.

The rotation of the Earth induces a prevailing west–east wind in the northern hemisphere, which induces surface currents in the Atlantic and Pacific Oceans of similar direction. In each case, this results in sea levels being higher on the eastern side of the oceans: as an example, the Pacific side of the Panama Canal is several metres higher than the Atlantic side. This difference is less marked in the Arctic, but sea level on the Pacific side of the Arctic basin is still about 45cm higher than on the Atlantic side. This height difference generates the trans-Arctic drift, with water flowing into the Arctic basin through the Bering Strait. The northern Pacific Ocean is less salty than the North Atlantic. The Bering Sea is less salty again due to the inflow of freshwater, chiefly from the Yukon River, to a relatively small, shallow sea. With the massive freshwater run-off of the rivers of Russia, and of the Mackenzie River in Canada, the salinity of the Arctic Ocean on the Bering Sea side falls further, creating a surface layer of cold, low-salinity water on the Pacific side, the leaching of salt during sea ice formation not compensating for the inflow of freshwater. This surface layer is 10–60m thick: below it lies a layer of water in which temperature remains relatively constant but salinity increases with depth.

In the North Atlantic the Gulf Stream – 100km wide and 1,000m deep, flowing at 100km/day – carries warm water north-eastwards, the salinity of this water being high due to evaporation. Close to the Azores, at about

Borgefell, Norway (above) and Canada's Barren Lands near the Mackaye Lake (below) at 63°N. The difference in vegetation indicates the effect of the Gulf Stream on northern Europe.

47°W, the Gulf Stream bifurcates, one arm (the Azores Current) turning south, while the North Atlantic Drift heads towards northern Europe. It is the heat flux of the drift (which is about 5°C above the average temperature for its latitude, and adds an extra 35% to the heat input from the Sun), and its associated winds, which push the European Arctic north. As the drift heads north it cools. At the temperatures of seawater in the Arctic, water density is dependent more on salinity than temperature: the salty waters of the drift therefore dive below the layer of increasingly salty water (known as the cold halocline), which itself lies beneath the less salty surface waters. The cold halocline prevents the warm Atlantic waters from melting the sea ice. The Atlantic waters circulate in the Arctic Basin, cooling further and sinking as they do.

This deep, cold, salty water forms the Arctic Ocean deep water current, which flows south at depth to balance the northward flow of the North Atlantic Drift. This exchange is known as a thermohaline circulation, so called because it depends on both temperature and salinity.

The exchange of warm surface water from the western mid-Atlantic to the Arctic Ocean and the reverse flow of deep, cold water is occasionally called the Atlantic conveyor (though the correct scientific term is the Atlantic meridional overturning circulation). It is known that historically the conveyor has been halted, and it is thought that a weakening of it caused Europe's 'Little Ice Age' in the late 17th/early 18th centuries (though what would have caused such a weakening is unclear). One effect of global warming might be the switching off of the conveyor again, as freshwater from the melting Greenland ice sheet dilutes the salty waters of the halocline. Such a stoppage would be of limited duration as there is a finite volume of water in the ice sheet, and the Earth's rotation would always push warm waters west–east, but it would have a very significant effect on the climate of northern Europe, producing winter temperatures more comparable with those of similar latitudes in North America. It is worth bearing in mind that 'limited duration' in this context is relative to geological timescales, not to those of the human lifespan.

There is no similar thermohaline circulation in the Pacific, though why this should be is not completely understood. Clearly the net flow of water into the Arctic basin is an influence, though another effect seems to be that the waters of the North Pacific have a more stable stratification pattern than those of the Atlantic. Consequently, the North Atlantic Drift is the major ocean current affecting the Arctic climate. Away from the drift, in Asian Russia and North America, ocean currents have less effect on local temperature than do air masses moving from close to the pole. However, the Labrador Current brings cold water and air through the Davis Strait, chilling eastern Canada and western Greenland.

## Temperature

Temperatures in the Arctic are not as low as those of Antarctica, the Arctic climate being moderated by the Atlantic and Pacific Oceans, while Antarctica is insulated from the effect of warmer water masses by the Southern Ocean circulating around the continent. At Russia's Vostok research station in Antarctica a temperature of -89.2°C has been recorded, making it the coldest place on Earth. In the Arctic, the lowest temperatures are recorded at points away from the seas, at the North Ice station in Greenland (-66.1°C) and at Oymyakon in the Verkhoyanskiy region of north-east Siberia (-67.8°C). Surprisingly, Oymyakon lies to the south of the boundary used to define 'the Arctic' in this book, its extreme temperature arising, in part, from its continental (as opposed to maritime) climate. Climatologists occasionally divide land masses into zones of approximately equivalent temperature. On that basis, the islands of Arctic Russia are classified as 'cold'. The northern coast of Russia from the White Sea to the Chukotka border is classified as 'moderately cold', though Chukotka itself is 'cold'. However, inland Siberia east of the Yenisey river is 'very cold'. To this classification must be added the effect of temperature inversion. Normally, temperature decreases with increasing height above the Earth's surface, but within continental land masses 'cold air lakes' can arise, layers of cold air being trapped below a layer of warmer air created when sinking air heats adiabatically (due to compression) as it descends. Such inversions are both more intense and longer-lived where the cold air is trapped within a valley, particularly one remote from the sea. Cold air lakes occur in both North America, particularly in the area to the west and south-west of Canada's Great Bear Lake, and in Siberia. The combination of a 'very cold' zone and a cold air lake gave Oymyakon its record-breaking temperatures.

One other effect of temperature inversion with which Arctic travellers soon become familiar is ice-fog. This is produced when water vapour is released into air at very low temperatures (in general around -30°C), the water vapour freezing to form ice crystals (usually about 30μm in diameter, but even smaller in some very dense fogs) which create the fog. Because of the requirement for water vapour such fogs are usually associated with areas of settled population, the vapour derived from vehicle exhausts and industrial sites. Famously, Fairbanks, Alaska suffers from such fogs, with visibility reduced to a few metres within the fog bank, though the bank itself may only be 10–15m thick. Fairbanks' ice fogs have been known to cover an area of around 200km² and to persist for up to 15 days.

Over such a large area and with macroscopic effects such as the North Atlantic Drift to be taken into consideration, together with more local effects, it is difficult to generalise about the variation of temperature through the Arctic year. The effect of the drift on northern temperatures can be seen in the temperature difference between Oymyakon and the Norwegian coast at the same latitude. In January the difference in mean temperature is about 50°C, about the same as the difference between the North Pole and the Equator. The warming effect of the open sea alone can be seen when considering the difference between inland and coastal Greenland. Though less dramatic than the difference between Oymyakon and Norway it is, at about 30°C, still very pronounced.

Figure 4.1 shows the mean surface air temperature of the Arctic in January and July. It is interesting to note the change in the difference between mean summer and mean winter temperatures across the region. As well as the descriptive definitions given above for dividing the Arctic into regions of similar winter temperature, three climatic regions are also defined and these can be inferred from Figure 4.1. In maritime areas, a classic example of which is Jan Mayen, the summer–winter variation is of the order of 10°C. In general, the maximum summer temperature in such areas occurs in August, with the minimum in March. In continental areas, such as inland Siberia, this variation is 40°C, perhaps even more. In these areas the maximum and minimum temperatures are seen in July and January respectively. Between the two lie continental coastal areas in which the variation is transitional, usually about 20°C. In these areas maximum temperatures are observed in July or early August, and minimum temperatures in January or early February.

Gullfoss, Iceland: left, it is high summer and rainbows form in the spray of water plunging into the river canyon; right, mid-winter and ice crusts the rocks.

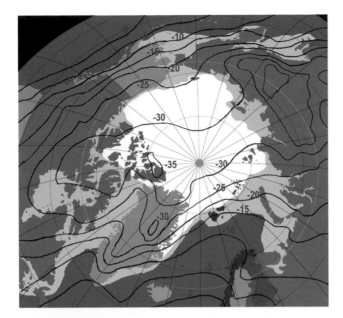

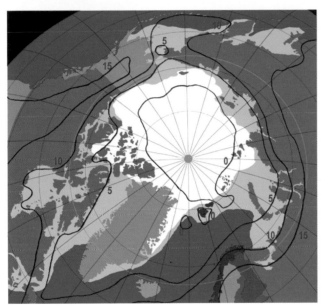

Figure 4.1 Temperature contours (°C) for January (upper) and July.

by many degrees, turning otherwise cool Arctic days into idyllic, warm ones. However, clear skies in winter permit increased heat loss from the Earth and, if the Sun is below the horizon so there is no corresponding heat input, can lead to intense cold.

## Cloud cover

The Arctic is a cloudy region, particularly during the summer when grey, low-level stratus dominates (Figure 4.2). The cloudiest area is the Atlantic Arctic, which

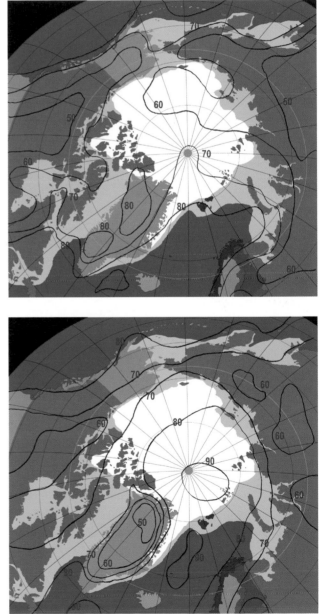

Figure 4.2 Cloud cover contours (% of month) for January (upper) and July

Though the data of Figure 4.1 are of use for a traveller who wonders what temperatures might be encountered at a particular time of the year, local conditions play a significant role in how cold it will actually be. Air temperature is only one factor in this equation. Wind speed is a dominating factor, leading to wind chill that can be considered as an effective reduction in ambient temperature (see Table below). Wind chill can, of course, be mitigated by reducing or eliminating skin exposure. Cloud cover is also important. On clear days in summer the Sun can enhance effective temperature

Winter, Kongsfjorden, Svalbard.

averages 80% cover almost constantly throughout the year. By contrast, cloud cover in the Eurasian, central and Canadian Arctic is highly seasonal. For these regions, the summer average is 80%, reducing to 60% in winter. The coverage is relatively stable for each season, with the change from summer to winter coverage and *vice versa* being remarkably sudden, and occurring over the space of a month (May and October). Further analysis of the data on cloud types indicates that the variation in cover is largely the result of changes in low cloud cover, low cloud constituting the 'normal' Arctic cloud cover of stratus plus cumulus, cumulonimbus and stratocumulus. Low cloud cover varies from about 25% in winter to 70% in summer. Medium (altocumulus, altostratus and nimbostratus) cloud is relatively constant throughout the year (at around 35%), but does show an autumnal peak. High cloud (cirrus, cirrocumulus and cirrostratus) also shows an autumnal peak, but otherwise reverses the low-level variation by reducing during the summer months. Observations during the latter years of the 20th century initially indicated that the seasonal variation in cloud cover seen over most of the Arctic was accentuating, with an increase in summer and a reduction in winter. However, the most recent study has shown that in fact overall cloud cover has declined, leading to both a reduction in cloud albedo and an increase in the sunlight reaching the sea ice and ice-covered land. Calculations suggest that a 1% reduction in cloud cover over Greenland results in the melting of an additional 27 billion tons of ice all of which is converted into a rise in sea level.

As cloud cover obscures the Sun, on average the Arctic traveller sees the Sun for only about 20–25% of the time when it might be visible. Accumulated data suggests that the sunniest place in the Arctic is the inland ice of Greenland, followed by central Canada and east central Siberia (with about 80% of the inland ice sunlight total). Next come Alaska and Canada's Arctic islands (about 67%). Coastal Greenland, Iceland, western Siberia and eastern Canada have about 50% of the inland ice sunlight, while Svalbard, the Bering Sea coast of Siberia and the Atlantic coast of Canada reach only about 40%. The cloud cover above Svalbard associated with this minimal sunshine means that southern Spitsbergen receives the lowest insolation of any place in the Arctic. That said, one of the most idyllic days I ever spent in the Arctic was in southern Spitsbergen on a calm, sunny day with a temperature of about 20°C: as elsewhere, Arctic weather can confound both a traveller's plans and expectations.

## Precipitation

The central Arctic basin is an arid area, with precipitation comparable to Antarctica, which is frequently referred to as a polar desert. In Chapter 1, the polar desert was defined in terms of temperature. It can also be defined by precipitation and plant coverage – an area where the annual precipitation does not exceed 130mm, with plant cover of less than 25% and active soil of depth 20–70mm. As we have already seen (Figure 1.5), the polar desert area of the Arctic covers virtually the entire Canadian Arctic archipelago, together with coastal areas of northern Greenland, Svalbard, the Russian Arctic islands and a section of the Russian mainland. Some authorities define the desert against annual precipitation of <250mm if plant coverage remains below 25%. The central Arctic receives 100–250mm of precipitation annually, as do Svalbard, the Russian Arctic archipelagos and islands, much of the Russian northern mainland and the northern mainland of Canada, yet most of these areas would be considered desert-like in terms of plant cover. Annual precipitation is higher on the eastern side of Canada (up to 700mm in Labrador) and on the Pacific coast of eastern Siberia (up to 600mm): Kamchatka receives as much as 900–1,000mm annually. This is exceeded by southern Greenland, where the south-eastern tip of the island receives 1,500–2,500mm (though the south-western tip sees 900–1,000mm). The higher figures relate, as would be expected, to the proximity of the sea, and to wind direction, the aridity of the central Arctic deriving from the reduction in insolation of the area relative to the Equator: at the Equator the solar energy causes hot air to rise, while cold air sinks in the Arctic. The sinking air inhibits precipitation.

Though the closeness of an area to the sea creates local distortions in precipitation rates, some general comments can be made. Precipitation tends to decrease as one moves north. It is lowest in spring, since the frozen sea limits moisture take-up. That said, the seasonal variation of precipitation is rather muted. As an example, on average there are the same number of days with precipitation greater than 0.1mm in Svalbard in both January and July. Something similar is observed in virtually all parts of the Arctic (though the number of days varies). Those looking for dry days in January should head for inland Greenland or, since that is not an entirely practical suggestion, for Nunavut or northern Asian Russia. Those looking for dry days in July should head for the same areas, but to their surprise will find that the monthly average of days with more than 0.1mm of precipitation will have risen.

One obvious Arctic precipitate is snowfall, and it is instructive to look at the persistence of snow cover across the region. However, it must be remembered that there is no correlation between days of snow cover and days of precipitation, as the extent of cover also depends on temperature: given a low temperature and limited sublimation, snowfall can persist for a long time. Persistence of snow cover peaks close to the North Pole, where there is snow cover for about 350 days annually. Heading south this persistence inevitably falls. In Severnaya Zemlya it averages around 300 days, with about 240 days on the other islands of Russia's western Arctic and 250 days in Svalbard. The northern coast of

Russia sees 260 days of snow cover. In the New World the northern areas of Canada's Arctic islands see 300 days, while Alaska's northern coast and the southern Canadian Arctic archipelago see 260 days. However, these figures relate to data averaged over the last half of the 20th century. With temperatures increasing in the Arctic the day rates are likely to change, perhaps significantly, in the decades ahead.

## Wind

The cold, dense air of the Arctic draws warmer air northward, the Earth's rotation combining with this air stream to create an anti-clockwise vortex over the pole. Superimposed on this global pattern are the effects of land mass topography and local climatic conditions (e.g. proximity of the sea). Though these effects mean that it is very difficult to be specific about winds in particular areas, some generalisations are possible. As with precipitation, the Arctic wind pattern does not exhibit a pronounced seasonal variation. Winds tend to blow from the central Arctic towards Arctic Canada, sweeping east across Hudson Bay (the effect of which is to reduce the extent of permafrost on the eastern side of the Bay). Winds are also funnelled through the Denmark Strait, though in summer the wind heads towards the

Windswept ground free of snow, Van Mijenfjorden, Svalbard.

pole from northern Greenland. The prevailing wind is also northward from the Russian Arctic. The Barents and Norwegian Seas are infamous for the winter 'polar easterlies' that sweep across them. In general, the Nearctic is less windy than the Palearctic: during the last half of the 20th century, Canadian Arctic weather stations reported calms on about 120 days annually.

In winter, the average wind speed over much of the Arctic is 4–6m/s, with higher average speeds experienced between Svalbard and Fennoscandia (8–10m/s) and in the North Atlantic, particularly on the western side. In summer, there is little difference in average wind speed over much of the Arctic, though it tends to be lower between Svalbard and Fennoscandia (4–6m/s) and in the North Atlantic. Summer winds are only 50% as strong as winter winds in the western North Atlantic. In the central Arctic wind speeds above 25m/s (90kph) are rare: for comparison, speeds of up to 50m/s (180kph) have been measured in the North Atlantic Arctic, which is generally the windiest part of the region.

## Katabatic and anabatic winds

One type of wind that is a feature of the polar regions is the katabatic. Katabatics, from the Greek for 'going downhill', develop where cold, dense air descends from an upland area into a valley under the influence of gravity. Katabatic wind aided Nansen's team on the first crossing of Greenland's Inland Ice.

Anabatic winds – the opposite of a katabatic, with the airstream moving uphill when the air in a valley is heated by the Sun – also occur. As might be expected, they are much weaker than katabatics.

The illustration below is from Nansen's book on his Greenland crossing.

As well as katabatic winds, the upland areas of the Arctic can also experience *föhn* winds.

Descending air in the lee of a mountain is compressed and so warms adiabatically, causing a sudden rise in temperature at the mountain base. Such winds are relatively common in Greenland where they are called *neqqajaaqs*. They can be disastrous for ungulate herds in winter as the local increase in temperature can melt snow, which then subsequently refreezes as an ice-coating on vegetation that the animals cannot break through.

Although wind may be the bane of the Arctic traveller's life it does have advantages. For one, it makes life difficult for mosquitoes – always a positive – and it is also responsible for one of the Arctic's most beautiful phenomenon, orographic clouds. Water vapour in Arctic air is usually close to saturation point. Consequently, when the air is forced to rise over a mountain the resultant cooling creates a cloud of condensed water droplets. The cloud produced is a shape-replica of the underlying land. As wind moves the air across the mountain, the droplets at the leading edge re-vaporise as the air mass falls on the lee side. Since the vapour of the new air in the trailing edge is condensing, the cloud shape remains stationary, irrespective of the wind speed.

Wind speed is directly related to wind chill, the name given to the enhanced cooling of exposed flesh caused by the wind. In still air, hot objects lose heat primarily by conduction (although radiation and convection do contribute). Convection is a more effective method of heat transfer than conduction, and convective heat loss dominates in moving air: warm-blooded animals therefore lose heat more rapidly if skin is exposed to the wind. One way of considering this extra cooling is to evaluate the still air temperature which produces equal cooling by, chiefly conductive, losses. This, much lower, temperature is the one which is now often quoted in winter weather forecasts.

One further effect of wind is the creation of blizzard conditions. Generally, blizzards occur as a result of wind and falling snow, but this is not always the case. In wind speeds above about 10m/s, lying snow that has not been completely compacted will be picked up and sent scudding across the landscape. As wind speed increases, the height to which the snow is lifted also increases:

Orographic cloud, Helgeland, Norway.

## Wind chill

The table below has been compiled from an equation relating wind speed at 1.5m above the ground and ambient temperature to effective temperature. The original equivalent temperatures were derived in Antarctica by observing the freezing rate of water. These rates were then converted into an empirical formula. A revised equation has now been adopted by the US and Canadian weather services, and it is that equation that has been used to derive the data below. Note, however, that other tables also exist and there may not be complete correlation between those and the data presented here. The red 'danger area' is also open to various interpretations. The US and Canadian weather services note only that frostbite may occur within 30 minutes if conditions are within the red zone. Other tables are more specific, noting, for instance, that exposed flesh can freeze in less than 30 seconds at temperatures of -40°C and below in wind speeds greater than 25km/hr. The table should, therefore, be taken as indicative only and is useful only for pointing out the dangers of exposed flesh in cold, windy conditions.

| Wind Speed (kph) | Temperature (°c) | | | | | | | | | | |
|---|---|---|---|---|---|---|---|---|---|---|---|
| *Still* | 10 | 5 | 0 | -5 | -10 | -15 | -20 | -25 | -30 | -35 | -40 |
| 5 | 9 | 3 | -3 | -8 | -14 | -20 | -26 | -32 | -38 | -43 | -49 |
| 10 | 8 | 2 | -4 | -11 | -17 | -23 | -29 | -35 | -41 | -48 | -54 |
| 15 | 7 | 1 | -6 | -12 | -18 | -25 | -31 | -37 | -44 | -50 | -57 |
| 20 | 7 | 0 | -6 | -13 | -20 | -26 | -33 | -39 | -46 | -52 | -59 |
| 25 | 6 | -1 | -7 | -14 | -20 | -27 | -34 | -40 | -47 | -54 | -60 |
| 30 | 6 | -1 | -8 | -15 | -21 | -28 | -35 | -42 | -48 | -55 | -62 |
| 35 | 5 | -1 | -8 | -15 | -22 | -29 | -36 | -43 | -49 | -56 | -63 |
| 40 | 5 | -2 | -9 | -16 | -23 | -30 | -36 | -43 | -50 | -57 | -64 |
| 45 | 5 | -2 | -9 | -16 | -23 | -30 | -37 | -44 | -51 | -58 | -65 |
| 50 | 5 | -3 | -10 | -17 | -24 | -31 | -38 | -45 | -52 | -59 | -66 |
| 55 | 4 | -3 | -10 | -17 | -24 | -31 | -38 | -46 | -53 | -60 | -67 |
| 60 | 4 | -3 | -10 | -17 | -25 | -32 | -39 | -46 | -53 | -61 | -68 |

above about 15m/s the snow layer created is deep enough to overtop a man – a blizzard has been created in the absence of falling snow. Even in the absence of falling snow, the whirling snow lifted from the surface can make travel impossible.

Late evening light over the Strongbreen Glacier, Kvalvaagen, Svalbard.

Aurora, northern Scandinavia.

# 5 Solar and atmospheric phenomena

For people living below the Arctic Circle – and that means almost the entire population of the Earth – the Arctic is a place where for six months of the year the Sun shines all day, and then for the next six months the darkness of the Arctic winter is alleviated only by the flickering light of the aurora. Though broadly speaking these statements are true, the play of light on the Arctic landscape is actually far more subtle.

## The Arctic day

Arctic travellers arriving from temperate regions bring with them the assumption that the Sun rises in the east and sets in the west, and can be disorientated when they discover that this 'law of nature' no longer applies. A traveller at the North Pole on Midsummer's Day sees the Sun move across the sky along a flat circle at a constant angle of 23.5°. On that same day, a traveller standing at the Arctic Circle will see the Sun higher in the southern sky at noon and touching the northern horizon at midnight. A short distance south of the Circle the Sun rises and sets in almost the same place, rising a little east of north and setting a little west of north. Only much further south does the familiar eastern rise, western set become established. One further difference for the temperate traveller is that the effect of the 'Arctic' Sun rising and setting in the north but being southerly at noon means that it appears to go across the sky rather than around it; although the Sun is following a curved path, to the human eye it seems to follow a straight line from sunrise to noon, then reverses along the same line to set. One other interesting aspect of the sun's motion at the North Pole is that because of its constant elevation, the shadow cast by a stick is always the same length, the shadow tip tracing a circle over the course of 24 hours.

At the Arctic Circle on Midwinter's Day the reverse occurs, with the Sun touching the southern horizon at noon. Now a short distance to the south the rising and setting Sun appears in the southern sky.

An issue related to the Sun's movement across the Arctic sky is whether, with 24 hours of daylight, the day can be divided into periods of 'day' and 'night'. In practice the answer is yes, as it is only at the pole (and there only for a relatively short time) that the Sun is at a constant elevation in the sky. Over much of the Arctic – and all the area in which wildlife is normally found – the Sun's passage across the sky includes a dip towards, and a rise away from, the Earth. Because of this, and a small but usually discernible temperature variation as a consequence, most Arctic animals maintain a recognisable day-night activity pattern.

## The Arctic night

While it is true that all places above the Arctic Circle do not see the Sun for a period varying from one day at the Circle to half the year at the pole, the idea that this means that there is continuous total darkness is incorrect as there are extended periods of twilight. Twilight is divided into three types, the difference depending on the position of the Sun below the horizon. If the Sun is less than 6° below the horizon there is 'civil twilight'. At this time, many activities that generally require daylight can still be carried out. This is not usually the case during the next type of twilight – 'nautical twilight', which occurs when the Sun lies between 6° and 12° below the horizon. The name derives from the fact that during this period the brighter stars are visible, allowing celestial navigation: this type of twilight was therefore of benefit to sailors. When the Sun lies between 12° and 18° below the horizon there is 'astronomical twilight', when the fainter stars are visible and astronomers can begin their observations of the night sky.

At the North Pole at the winter solstice (a time of total darkness) the stars do not rise and set. A time exposure photograph of the night sky would reveal a series of parallel rings, the radius of each diminishing until Polaris, the North Star, was reached. Polaris does not actually sit above the North Pole: it too would create a ring, but one only 2° across. The blackness within that circle is the region of the sky that truly sits above the North Pole. The behaviour of the moon during the long Arctic summer and winter is rarely mentioned, but that too can surprise the traveller. The moon's orbit around the Earth lies about 5° from the plane of the ecliptic. Consequently, at places to the north of 72°N there are periods each month when the moon does not set, and other periods when it does not rise. During the winter the moon does not set during the full moon (and periods close to it), with moonlight enhancing the otherwise bleak winter darkness. In summer the moon does not set when it is new (and in periods close to the new moon). As these are periods of continuous Sun

the moon is rarely seen, so the fact that these are also periods of minimal moonlight is of little consequence.

## Magnetic fields

Planet Earth has a magnetic field, and as with all such fields there are north and south magnetic poles, these being aligned, more or less, with the geographical poles. The alignment is not perfect, with Earth's North Magnetic Pole, to which compasses point, wandering about the Canadian Arctic. The pole was first reached in 1831 by a team under the leadership of James Clark Ross, the discovery being heralded as a triumph for British exploration: it featured as a delightful, but entirely incorrect illustration in a book popular at the time.

At the time it was first reached, the pole was on the western coast of the Boothia Peninsula. By a curious

James Clark Ross at the North Magnetic Pole, an illustration from Robert Huish's book *The Last Voyage of Capt. John Ross* published in 1836. Despite the delightful idea behind the engraving almost every relevant feature is incorrect: the aurora borealis is shown, but Ross travelled during the continuous daylight of the Arctic summer when it would not have been visible; a telescope is used to spot the pole, but

the pole is not visible, being detected by the use of a dip circle, essentially a magnetised pointer suspended by a fine thread which points vertically downwards when it is positioned above the pole; and a party climbs to the top of the 'pole peak' implying a mountain to be conquered, whereas the pole can sit beneath any feature, including the sea – Ross actually found the pole more or less at sea level.

## The Earth's poles

Looking at an atlas it seems that the Earth has two poles, a north and a south. But there are several more, some fixed, some moving, and all have their uses.

Geographical poles: *These are fixed. They are the North and South Poles in the atlas, with the equator lying equidistant from them. In 1793, a few months after the execution of Louis XVI, the French Revolutionary Council appointed the scientist Joseph Lagrange – Italian born but of French ancestry – to head a commission whose task was the creation of a new system of weights and measures. The Council hoped that this new system would be adopted worldwide: their hope was eventually realised as the metric system is now used almost everywhere. The new unit of length was to be defined as one ten-millionth of the distance between the North Pole and the equator, measured along the line of longitude that passed through France's most northerly town, Dunkerque. But measuring that distance was not easy and an approximation had to be made. Because of this it was decided to create a length standard: it would be the distance between two marks scratched on a bar of platinum-iridium alloy in Paris. The name chosen for the new unit of length was the metre, and not surprisingly the distance between the North Pole and the equator is about 10,000km.*

Rotation poles: *These are movable. They are the poles around which the Earth rotates. They lie within 20m of the geographical poles, around which they precess with a period of 435 days.*

Magnetic poles: *These are movable. These are the ones at which a dip circle measuring the Earth's magnetic field points straight down, i.e. the magnetic field at that point is at right angles to the Earth's surface. Compasses point to the North Magnetic Pole.*

Geomagnetic poles: *These are movable. On what might be termed an astronomical scale, that is, considering the Earth's magnetic field from space (ignoring the 'microscopic' effect of rock asymmetries in favour of a 'macroscopic' view), the Earth behaves as a simple bar magnet, with north and south poles that align themselves through the planet's centre, but do not, of course, align with the magnetic poles. As with the magnetic poles these drift, but on much longer timescales. At present the North Geomagnetic Pole is close to north-eastern coast of Ellesmere Island.*

Poles of inaccessibility: *These are fixed. These are the points that are furthest from land (in the case of the Arctic) or the sea (in the case of Antarctica). The term is contrived as the points are not really any more inaccessible than others, and in the case of Antarctica they depend on the extent of the sea ice. The Arctic pole of inaccessibility is at 84°03'N, 174°51'W. The southern pole is, on average, at 85°50'S, 65°47'E.*

coincidence, Ross's discovery was at the southernmost point of the pole's wander: since that time it has migrated north across Bathurst Island into the Arctic Ocean. In 2017 it was at 86.5°N, 172.6°W: during the 20th century it had travelled over 1000km, and recently the pole's wander has become more rapid, moving at about 40km/year. The wander is thought to be caused by changes in electric currents in the Earth's metallic core (the same differential movements responsible for the planet's magnetic field). Short-term changes in these electric currents also cause the magnetic pole to oscillate with a period of hours, the oscillation being along an essentially oval path which, over a 24-hour period, may have a long axis of c.100km.

## The aurora borealis

The Earth's magnetic field is responsible, in part, for the *aurora borealis* or northern lights, the most spectacular

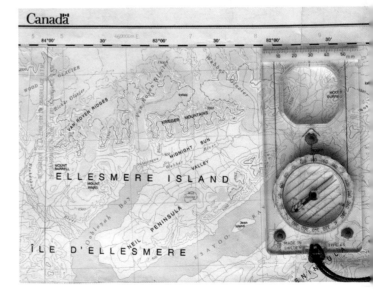

For walkers on Canada's Ellesmere Island the magnetic deviation which must be catered for to ensure correct map bearings is very large as the island lies almost due north of the North Magnetic Pole.

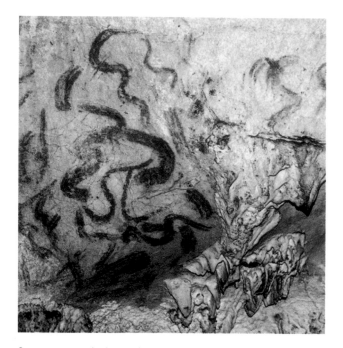

Some experts believe that serpentine rock drawings – such as those on the walls of the La Pileta Macaronis cave in Spain, which date from about 20,000BP – may represent the aurora, though that is a somewhat controversial view.

of all Arctic phenomena. The first documented observation of the aurora is found in a Chinese text from about 2600BC, which tells the story of Fu-Pao who witnessed strange lightning at night high in the northern sky and soon after became pregnant. The tale is an early indication of the mystical properties with which early observers imbued the aurora. Chinese texts include a list of auroral observations dating from 687BC, but by then – indeed, long before Fu-Pao's witnessing of the lights – the aurora would have been well-known to northern peoples (and those of the south, too, though they were much less likely to see it as the southern auroral oval is entirely encompassed by Antarctica). The lack of extraneous light at night – which tends to obscure the aurora in northern cities today, as auroral light is very faint and easily overwhelmed by, for instance, street lighting – allowed the lights to dominate the night sky.

It is no surprise that these curious, moving lights became entangled with myth and legend. In New Zealand, the southern aurora was believed to be the reflection in the sky of fires lit by ancestors of the Maori who had paddled their canoes south and become trapped in a land of snow and ice. For North American

Inuit, the lights lit the path of the recently dead to the heavens, emanating from torches held aloft by ravens. The Inuit (among others) claim to occasionally hear the lights, the faint crackle they describe being identified as the feet of the dead walking across the crisp snows of heaven. Greenland Inuit saw the lights as the souls of babies who had died soon after childbirth, peering down at a world they never knew and offering comfort to their parents. The Sámi peoples spoke of a 'fire fox' that raced across the sky each night, his coat sparking each time he clipped a mountain. They also wondered if, far away and out of sight, huge whales were spouting, with the jets of exhaled breath scattering the light of the stars. For the Chukchi of north-eastern Siberia, the lights were the spirits of those who had died a violent death, perhaps from murder, suicide or childbirth. In the sky these wounded spirits played a suitably violent game, kicking and hurling a walrus skull around the sky. Later, people saw the lights as heralding great events or foretelling disasters, while the medieval Christian church saw the occasional red light mixed with the more usual green as being an indication of the blood of martyrs.

Early scientific theories on the aurora concentrated on the burning of gases: not until the work of the Norwegian scientist Kristian Birkeland did it become clear that the phenomenon was linked with electricity and magnetism and, therefore, the Earth's magnetic field. Even today, although the basis of the aurora is well understood, there are aspects of it that remain perplexing.

Most of the Earth's atmosphere (about 75% of it) is confined in the troposphere, which extends to about 12km above the surface. Above this is the stratosphere, which extends to about 50km and includes, towards its top edge, the ozone layer that protects life on Earth from harmful ultraviolet radiation (see Chapter 18). Further up are the mesosphere and the thermosphere, extending to about 1,000km. At that height, the Earth's magnetic field creates an envelope called the magnetosphere. This is the lower edge of the Van Allen belts, shells of solar particles trapped by the magnetosphere. During the second half of the 20th century in the early days of space travel, rockets carried Geiger counters into the upper atmosphere: they stopped transmitting at a height of about 1,000km. Most scientists assumed that the

A depiction of the aurora seen over Bamberg, Germany in 1560. Rarely seen so far south, the aurora terrified the locals who believed they were witnessing a battle between celestial armies, the light flashing from their armour and swords.

counters had failed, but US physicist James Van Allen suggested that the counters had been so overloaded with incident particles that the were unable to function. He was correct and the belts were named after him.

Each year the Sun discharges about 50 million million tonnes of material into space as a 'wind' of charged atomic and subatomic particles. When the solar wind, which is travelling at many thousands of kilometres per second, encounters the magnetosphere it distorts it, compressing the sunward side and drawing the opposite side into an extended tail (Figure 5.1). The charged particles of the solar wind spiral along the field lines of the Earth's magnetic field, reaching their lowest altitude in the polar regions where the magnetic field lines are almost perpendicular to the Earth's surface. As the particles penetrate towards the Earth they are accelerated in the magnetic field and collide with atoms of gases of the high atmosphere (chiefly in the lower thermosphere, but also in the mesosphere). The collisions excite (energise) the atoms (strictly the electrons of the atoms). As they return to their unexcited ground state, they emit energy in the form of light. This light is the aurora.

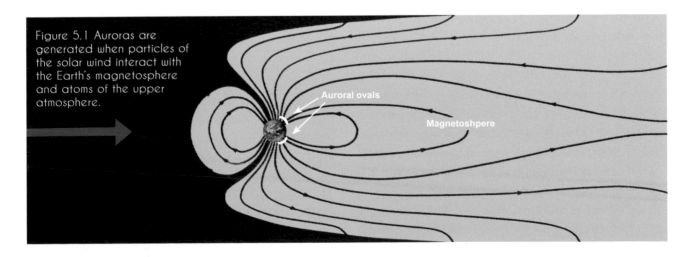

Figure 5.1 Auroras are generated when particles of the solar wind interact with the Earth's magnetosphere and atoms of the upper atmosphere.

Auroral ovals

Magnetoshpere

The electrons have fixed 'excitation states' between which they move, and emit light at different wavelengths accordingly. Light emitted by electrons in oxygen atoms moving from a second to a first excitation state is at a wavelength of 557.5nm – yellow-green, the general colour of the aurora. Movement of electrons in oxygen atoms from the first excitation state back to the unexcited state leads to the emission of red light at 630.0nm or 636.4nm. However, this transition is delayed if the atom has already emitted green light and it is likely that the oxygen atom interacts with other atoms in the atmosphere before the red light is emitted. Red light is therefore rarer than green. The upper sections of some auroras can be red, but auroras that are red overall are very rare.

Excited electrons in nitrogen atoms emit a pale blue or violet light (at wavelengths of 391.4nm and 427.8nm), this light usually being overwhelmed by oxygen's green. However, nitrogen also emits red light (at wavelengths between 661.1nm and 686.1nm), this occurring mostly at low altitudes, so green auroras sometimes have a red basal fringe.

## Types of auroras

Auroras are classified into six colour forms (of which type C – overall green – is the most common) and four light intensities, with around 33 shapes also having been identified, though these are essentially combinations of a small number of basic shapes. An arc is an even crescent of light stretching across the sky. If the arc is folded, usually towards its base, it is known as a band. Veils are arcs without defined edges that appear to fill the sky, fading away at the top and bottom. Rays are beams of brighter light that may be seen within arcs and veils, or may occur independently. Coronas are the most spectacular; they are arcs or bands seen from directly below so the light seems to erupt from a central, bright, linear area. Auroras also move, the light appearing to shimmer in the way that rustled curtains might. Both the shapes and the movement are caused by variation in the flux of solar wind particles.

Although auroras can, like rainbows, sometimes appear to touch the Earth, the lower edge is rarely less than 60–80km from the surface. At lower heights, the density of the atmosphere is such that oxygen atoms

Coloured auroras are rare. This one, which includes the full spectrum, was photographed in northern Scandinavia.

have no time to emit light before they collide with other atoms. However, nitrogen's red light is emitted very quickly, so at the lowest heights, if the incident particles get that far, it is this red light that may be seen. Auroras extend to about 400km above the Earth's surface, the brightest light usually occurring at a height of about 110km.

Because the solar wind 'blows' constantly, the conditions for the development of auroras are, in principle, present at all times. (As an aside here, it is worth noting that auroras are not restricted to Earth, but would be visible on any planetary body with a magnetic field: auroras have recently been seen above the poles of Jupiter.) But while they may occur at any time, certain factors affect aurora visibility. The intensity of the solar wind is not constant, so displays are sometimes too faint to observe. Auroral light is faint and so is not visible during the day, and is affected by background light during the night: a bright moon can make the aurora invisible. Visibility is also, of course, affected by the weather.

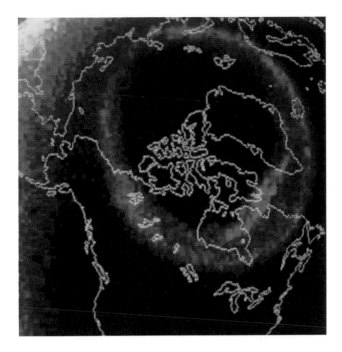

The northern and southern auroras occur in oval belts. The oval remains constant with respect to the position of the Sun, the Earth rotating beneath it. This satellite photograph shows the northern aurora oval.

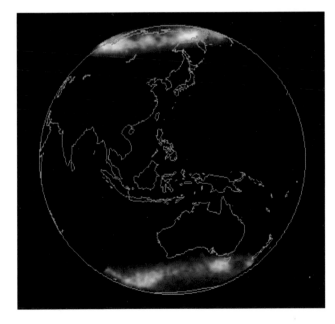

Because the solar wind bathes the Earth symmetrically, auroras occur at the north and south polar regions at the same time. This photograph shows such conjugate auroras.

Auroras primarily occur in a flattened oval belt. The shape is akin to that of a guillemot's egg and is caused by the distortion of the Earth's magnetic field: the flattened section of the oval faces the Sun, the elongated part is away from it. The Earth rotates beneath the oval.

The aurora oval has a 'diameter' of about 5,000km, lies at about 60°–65°N, and is centred on the North Geomagnetic Pole. At latitudes above and below the oval the aurora is less likely to be observed. At the Geographical North Pole the probability is about 20% (assuming 90–100% visibility in the oval). In London, the probability is less than 5%, about the same as it is in New York. An equivalent auroral oval exists in the southern hemisphere, centred on the South Geomagnetic Pole. As would be expected, auroras occur symmetrically in the Arctic and Antarctic as the solar wind is equivalent at all points of the Earth's magnetosphere. In September 1909, an aurora was visible at Singapore and Jakarta, which lie close to, but on opposite sides of, the equator.

Though symmetrical at the magnetosphere, the particle flux of the solar wind is not constant, varying with the 11-year sunspot cycle. At the peak of solar activity auroras are more intense and may then be visible a long way south. There are also shorter cycles: if there was a good aurora last night, there will probably be a reasonable one tonight, as the solar wind rises and falls in intensity gradually rather than being switched on and off. As the solar wind is also dependent on solar activity, if there was a good aurora last night, there will probably be another good one in 27 days, the rotation time of the Sun.

Although auroras are now reasonably well understood (though the exact production mechanisms for the various identified forms of aurora is unclear), there is still much to intrigue scientists, one example being whether the lights can be 'heard'. As noted above, native Arctic peoples claim to be able to occasionally hear the lights, a claim that was echoed by some European explorers, and is still contended by the occasional modern day Arctic traveller. One of the earliest descriptions of the noise associated with the aurora was that of Samuel Hearne, (see Chapter 8), who heard 'a whistling and cracking noise, like the waving of a flag in a fresh gale', a description with which most other claimants concur. However, there is no known reason for the light production mechanism to produce sound. Indeed, there are good reasons to believe that auroras cannot generate sound waves: the light is too high for any sound from it to reach the Earth, as the density of the air at the altitude of auroras is too low for the efficient transmission of sound waves; the light is also too high for the suggested synchronicity of changes in light intensity or pattern with sound; even from the lowest level of the auroral base, sound would take several minutes to reach the observer while the light arrives virtually instantaneously. But many explanations have been suggested for the noise. One of the most popular is coronal discharge – the ionisation of the air close to the observer. As this would be strongest at sharp points it has been suggested that discharge from the observer's hair might be the cause. Piezoelectricity in nearby rocks has also been proposed, as has a mechanism that again depends on the observer. In this, leakage of electrical impulses from nerves within the eye would be 'heard' by the brain. Obviously, such leakages would occur all the time in the observer, the reason they would only be picked up by the brain in aurora watchers being that the aurora is usually viewed in a quiet, wilderness area. In support of this, many observers claiming to have heard the aurora notice that the noise stops if they cover their eyes. As the sound has yet to be definitely recorded, most of these mechanisms – indeed, for many scientists

the whole idea of auroral sound – have been dismissed as fanciful. However, there are other aspects of aurora that were also once decried but are now supported by scientific observations. Auroral light should be contained within the oval, the centre of the oval – known as the polar cap – being light-free. But in the early 20th century Australian scientists in Antarctica gathered evidence of what appeared to be auroras crossing the oval. This work was either dismissed or ignored, but in the 1980s photographs taken from satellites clearly showed a thin, linear aurora crossing the polar cap, linking with the oval on both sides. This so-called cap aurora was aligned with the Sun, apparently sweeping across the sky as the Earth rotated below it.

Equally perplexing are the gaps that occasionally occur in the aurora's oval. Some gaps mean that the oval is not complete, others that, though complete, the oval has a curious notch in its otherwise smooth outline. The gaps are assumed to occur because of discontinuities

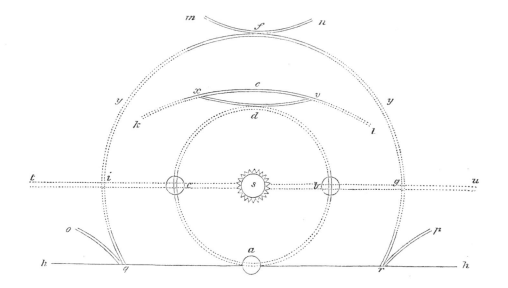

Using the notation on Parry's sketch:

a, b, c, d is the 22° halo. Outside it is a 46° halo. This is formed by light passing through the side and end faces of ice crystals.

t–u is a parhelic circle, produced when light is reflected from near-vertical faces of the ice crystals. Usually the light is reflected from an outer crystal face, but it may be due to one or more reflections from internal faces. Parhelic circles may encircle the whole sky.

k, e, l is an upper tangential arc, formed by light entering a crystal through a side face and leaving through another inclined at 60° to the first. When the Sun is close to the horizon, the upper arc forms a tight V, but as the Sun climbs, the arc opens to form 'gull wings'.

x, e, v was a previously unknown arc lying above the V notch of the upper tangential arc. It is now known to be caused by light interacting with crystals of a specific orientation. Known as the Parry arc, this feature is strongly prismatic (i.e. features all the colours of the spectrum).

a which Parry drew as a circle and probably thought was a sun dog (see b and c) is actually a lower tangential arc. This also forms a tight but inverted V, of which Parry only observed the apex.

m, f, n is a circumzenithal arc, created by light refracting in horizontal, plate crystals. The light enters the crystal through the horizontal upper face and leaves through a side face. Circumzenithal arcs are always prismatic, the inverted arc appearing as a 'smile' in the upper sky.

o–q and r–p were long assumed to be a mis-drawing by Parry, but it is now known that he was absolutely correct and had depicted 'subhelical arcs'. Two forms, supralateral and infralateral arcs, appear infrequently and are caused by light passing through a side and the base of hexagonal ice columns. The arcs are prismatic.

b and c are sun dogs or mock suns. Together with the 22° halo these are the most frequently seen parhelia. They are formed by horizontal plate crystals (those that form the circumzenithal arc) and are strongly prismatic.

in the magnetosphere, but what these are, and what mechanism drives them, is not understood.

## Parhelia

*Parhelia* is the general term given to the range of solar and lunar haloes, arcs and sun dogs that occasionally surround the Sun, the effects being created by light refraction in ice crystals suspended in the troposphere. Most tropospheric ice crystals are either hexagonal flat plates or hexagonal columns. Light traversing these crystals is refracted at angles varying from about 22° to about 50°. As smaller angles are the most probable, the most frequently seen parhelion is a 22° halo surrounding the Sun. The minimum angle is 21.7° for red light, so the inner edge of the halo often appears red, with the remaining spectral colours becoming fainter (or being absent) as the halo fades away.

One of the most famous examples of parhelia is that sketched by William Parry at the Melville Island winter quarters (1819–20) of his ships during his first voyage in search of the North-West Passage (Figure 5.2). Some of Parry's parhelia were prismatic, the rainbow colours adding to the beauty and wonder of the sighting.

Parry's example includes most of the frequently seen parhelia phenomena (though 'frequently' here is a relative term). As well as the phenomena that Parry drew so assiduously, further, very rare, parhelia have also been observed and given names – Hastings, Tricker and Wegener arcs among others. All these phenomena are produced by subtle variations of the path of light through ice crystals.

One other related phenomenon may also be seen, the sun pillar, a form of sun dog, produced when the plate-like crystals that produce sun dogs are aligned vertically. Sun pillars usually form directly below or above the Sun.

As parhelia are produced by the refraction of light by ice crystals they can occur at night as well as during the day. Full moons are best as they provide more light. A full moon behind thin cirrus cloud will often produce a 22° halo and moondogs may also be seen. However, as moonlight is so faint relative to sunlight, the more exotic halos are much rarer at night, although many have been observed.

Parhelion, Isfjorden, Svalbard.

Mirage at Mackenzie Bay, Hudson Land, north-east Greenland.

In the Arctic, good viewing days (i.e. with a low Sun and plenty of ice crystals) are much more common than at southern latitudes, and a 22° halo and associated sun dogs will be seen as often as one day in four. Tangential arcs occur on about one quarter of the days when the halo/sun dogs are seen, 46° halos on about one occasion in 25, Parry arcs about 1 in 100, and the rarer arcs on perhaps one occasion in 300.

## Mirages

Mirages are caused by the refraction of light from distant objects to the observer due to the light travelling through layers of air at different densities, the differences resulting from changes in humidity and temperature. The refraction results in the light being bent into a curved path, but the observer assumes it is following a straight path, as is usual, and is fooled into seeing distorted images of objects. Refraction, which allows the Sun to be seen when it is below the horizon, gives rise to two forms of mirage. If the temperature falls from a warm surface then 'inferior' mirages occur, the image appearing below its 'true' position. Desert mirages, and the 'water pools' frequently observed by drivers above hot metalled roads in summer are examples of 'inferior mirages': such mirages are rarely encountered in the Arctic. If the temperature change is an increase from a cold surface, then the mirage is said to be a 'superior mirage', as the image appears above its 'true' position. The more usual name for this form is *fata morgana* (from Morgana le Fay, sister of King Arthur, who was said to be able to conjure such images). Fata morgana are relatively common in the Arctic, where warm air above ice sheets and the cold sea set up the right conditions. In superior mirages, the light follows a convex path so the observer sees a much taller object than is present. The range of mountains seen by John Ross during his first attempt to discover the North-West Passage, and the non-existent lands, such as Crocker Land, famously 'discovered' by Robert Peary are examples of such 'superior' mirages. The shimmering quality of most mirages results from turbulence in the atmosphere.

The purity of Arctic air and the clarity of the light, together with the usually flat terrain, also allow objects that are far away to be observed, a phenomenon that can occasionally lead to misidentification due to the difficulties of depth perception. In one of his books the experienced Canadian Arctic traveller Vilhjalmur Stefansson wrote that he once spent an hour trying to get close to a Brown Bear only to finally discover that it was a marmot.

## Ice blink, water sky and white-out

When sunlight is reflected from sea ice or an ice sheet it brightens the base of overlying clouds. For the observer, too far away to see the ice in question (which may be below the horizon) this brightening is an indication of the existence of the ice and is termed ice blink. Ice blink can add a delightful element to sunsets viewed from eastern Greenland when out of sight, sunlit sections of the inland ice act as a mirror for the setting Sun, which may also be out of sight.

The opposite of ice blink is the darkening of sections of cloud when it overlies the dark ocean between areas of pack ice. These dark streaks are called water sky. Early polar explorers, who spent relatively long periods exploring the pack ice because their sailing ships were slow and vulnerable, became experts at reading the base of the clouds, looking for the dark lines in the silver glare that indicated worthwhile leads that could be utilised to continue their journey.

A final atmospheric condition that can assail the traveller, one that is more climatic than atmospheric but represents a very real hazard, is the white-out. Though often assumed to be associated with blizzards, white-outs can occur in more benign conditions. If a traveller is journeying across snow under complete cloud cover that diffuses the light, then the conditions for a white-out are almost in place: adding snowfall and a light wind enhances the effect. In a white-out, there is a loss of orientation because of the lack of a discernible difference between the colour of the ground and the sky. The falling snow reduces visibility so that estimation of distance, already difficult, becomes even more so. A light wind enhances the problem – a strong wind would aid orientation (while adding different hazards of its own). Loss of orientation can be so pronounced that nausea can result as it becomes difficult to register up from down. In such events, it is best to stop and wait, as the chances of blundering over a cliff increase dramatically.

Water sky above Forlandsundet, Svalbard.

Whale Alley, Yttagran Island, off Chukotka's eastern coast. In the 1960s Russian archaeologists discovered this remarkable site. The island is sited close to routes followed by Bowhead Whales during their spring and summer migrations through the Bering Straits. Though nomadic people, the local Yuppiat Eskimos, or their ancestors, gathered twice each year to hunt the whales and, it is believed raised, a processional pathway of dozens of upright whale ribs and skulls for ceremonies to ask for the whales to return. It is believed the site is over 1000 years old.

# 6 The human history of the Arctic

To do justice to the human history of the Arctic would require a book of its own. Below only brief outlines of the early human history of the area, and the subsequent exploration by, initially, Europeans and, later, Americans, will be given. The interested reader is directed to:

Hoffecker, John F., *A prehistory of the north: human settlement of the higher latitudes*, Rutgers University Press, 2005.

McGhee, Robert, *The last imaginary place: a human history of the Arctic world*, Oxford University Press, 2005.

Vaughan, Richard, *The Arctic: A History*, Sutton Publishing, 2007.

Those interested in the search for the North-West Passage will enjoy:

Savours, Ann, *The Search for the North West Passage*, St Martin's Press, 1999.

While the classic, if contentious, book on the race for the pole is:

Herbert, Wally, *The Noose of Laurels*, Hodder and Stoughton, 1989.

Also worth considering is the superb atlas of Arctic exploration:

Hayes, Derek, *Historical Atlas of the Arctic*, Douglas and McIntyre, 2003

In this short historical survey we look first at the original Arctic dwellers, then consider the development

Koryak shaman, northern Kamchatka.

Polar Eskimo Massutsiak Eipe in search of narwhal near Thule north-west Greenland.

of the Arctic nations, firstly those of Greenland and the European Arctic, then tracing the explorations which led to the development of Canada.

## The native peoples of the Arctic

In about 330BCE, when Aristotle was teaching at his school in Athens and his former pupil Alexander the Great was campaigning in India, a Greek named Pytheas set sail from Massalia, a trading port on Mediterranean coast, now the site of Marseille. He sailed through the Pillars of Hercules (the Gibraltar Strait) and turned north. As far as Brittany he was probably following a known route; it is thought that the Carthaginian explorer Himilco reached the area around Quiberon in around 500BCE, while there is evidence that Phoenician vessels had made regular voyages to the Cornish coast to trade for locally mined tin. Pytheas was away for six years and exactly where he went has long been the subject of debate. He seems to have passed the Orkney Isles, then continued north again sailing for six days to reach 'Thule' where the summer day was 21 or 22 hours long. There he heard that further north the sea stiffened or congealed. Had he reach reached Norway? But wherever he was he had so outpaced the understanding of the day

that his tale was dismissed, its wonders lost for almost 1,000 years.

But whether Pytheas' journey was real or not, and wherever he reached, the lands he saw were already inhabited and had been for millenia. Archaeological finds have been slowly pushing back the time when man first arrived in the Arctic, recent findings in northern Russia (near Yenesey Bay at 72°N), published in 2016, have suggested mammoths were being hunted by people 45,000 years ago, during an interglacial period, and that occupation of the Arctic has been continuous, or nearly so, in the Asian Arctic since the end of the last ice age. In the Nearctic recent genome studies of native Americans (published in 2015) have also pushed back man's occupation of the continent to about 23,000 years ago. Early man, it seems, was a resourceful and restless traveller. This finding was confirmed (2017) when bone fragments in caves in the Yukon were carbon dated to be 24,000 years old.

However, while these finds indicate human habitation, the first true Arctic culture is named for finds at Independence Fjord in north-east Greenland. There, from 2,500BCE lived a folk who had spread right across the Arctic. They hunted Musk Ox and Caribou with bows and arrows, and caught fish. They presumably cached food taken at times of plenty during the long hours of summer daylight for use during the winter: Independence Fjord lies beyond 80°N, where the Arctic winter brings months of sunless days. They lived in tents of skins draped over driftwood year round, with a central fire for warmth and cooking. But their history does not appear to have been continuous. At some point the Arctic cooled which would have put an end to an existence based on mammals, these moving south in search of forage, and fishing once the sea was more permanently frozen. A hunting culture based on small pockets of humans was also vulnerable to accidents, the dangerous business of hunting Musk Ox perhaps leading to a fatal or disabling injury to a hunter and the death of his family. A second culture, known as Independence II, arose in the last millennia BCE, but while this flourished a new wave of settlers were moving into the western Arctic and heading east.

Known as the Dorset culture from finds initially made near Cape Dorset on Baffin Island, the folk were culturally very different. They hunted sea mammals, the change to a cooler climate being to their advantage as they were used to hunting on the sea ice. Perhaps they originally lived in areas such as the Foxe Basin – there are finds on Igloolik – where there are few Caribou, but an abundance of Walrus and seals. The Dorset people used harpoons, devices with a barbed head that detached from the throwing handle but remained attached by a rope to a float made from an inflated seal bladder or skin. Once harpooned, the animal grew tired attempting to drag the float under, and each time it surfaced it could be struck again until finally it could be speared to death.

Dorset people had more substantial winter quarters than the Independence folk. These were partially sub-surface, dug into the ground to provide extra insulation, with walls of stones and turf over which skins were draped on driftwood frames, though the interior was similar to that of the earlier peoples. The houses were heated with blubber oil burning in soapstone lamps. Yet for all their clear advantages for survival in a colder Arctic that they possessed, the Dorset folk had lost some things that

## What's in a name?

Collectively, the pre-Dorset and Dorset peoples are known as Paleo-Eskimo. Eskimo was the name Europeans gave the Arctic peoples of North America when they first encountered them, having heard the name from the 'Indians' of southern Canada. The derivation is usually said to be from the Athabaskan or Algonquin for 'eater of raw meat' and so was later viewed as derogatory. But there is no consensus on this derivation, some considering an Ungava origin meaning 'snow-shore hunter'. Whatever the origin, it is thought that the word 'eskimo' was used contemptuously by native Americans, who had a history of conflict with the people of the north. Today the Arctic peoples of eastern Canada refer to themselves as Inuit, which means simply 'the people'. Inuit is plural, a single person being an Inuk. In Greenland, where the people share a common ancestry with the Inuit, the preferred name is Kalaallit. In western Canada the native people are the Inuvialuit.

However, the peoples of the Bering Sea are less bothered by the name, and often refer to themselves as Eskimos, though they usually attach a prefix to acknowledge their specific tribe: the Yuppiat (sometimes rendered as Inupiat; the singular of the name is Yupik or, occasionally Yu'pik,

the Siberian form) who live in north-western and northern Alaska and across the sea in Chukotka; the Yuit live in south-west Alaska; the Alutiiq on the Alaska Peninsula and east to Kodiak Island; and the Aleuts, who live to the west. In most cases, these names mean 'the people'.

In Eurasia, too, there has been a trend away from received names to ones based on a common heritage. Lapps now prefer the term Sámi, which means 'ourselves' or 'the people', and is probably the basis of the Finnish word for their own country, Suomi. It may also be the basis of Samoyed, the collective word given by early explorers to the Nenet, Enet and Nganasan, peoples living in European Russia from the eastern shore of the White Sea to the edge of the Taimyr Peninsula. It was widely assumed that these peoples practised cannibalism, and the term Samoyed, given to the northern tribes by their southern neighbours, is thought to have derived from that practice. Consequently, the Nenets and other peoples have rejected it in favour of their own words. Both Nenet and Enet mean 'man' in the local language, while Nganasan means 'people', the older tribal terms again having a basis in isolation. The Nenets also occasionally refer to themselves as the Khasava, 'the people'.

Nenet woman preparing reindeer leg skin for shoe making.

Nenet annual migration on the Yamal Peninsula, Russia.

made life for their predecessors more tolerable. They did not have dogs and, therefore, dogsleds, nor did they have boats. They did not have the bow and arrow, nor did they have drills, making the manufacture of needles a much more difficult task. Yet despite these limitations the Dorset people spread across the central and eastern Arctic and were the dominant culture for about 2,000 years. Then the Arctic warmed again, and their sea-ice hunting techniques became less well-suited to the new climate: by about 1500 the Dorset people had died out.

The Dorset people were themselves replaced by a new wave of settlers from the Bering Sea. The newcomers were efficient hunters of Bowheads and other large whales, animals much bigger than those taken by the Dorset people (who occasionally took Narwhal and Beluga, but do not seem to have ever taken the larger whales). The Bering Sea hunters used kayaks (from the Inuit *qajaq*), formed by stretching animal skins over a simple wooden framework, a vessel light enough to carry and that could be rolled if it capsized. On land sledges, drawn by dogs, were used to hunt Polar Bears and other animals with bows and arrows. In summer tents were still of animal skins stretched over driftwood or bone frames, with stone and turf winter houses. The people wore skin clothing and slept under skin bedding. Most importantly, these people had iron tools, the metal presumably originating from the southern civilisations of the Pacific rim.

The newcomers are now called the Thule people and are the ancestors of the modern Inuit. Inuit folklore tells of their forefathers driving the 'Tunit' from the lands they occupied as they moved east, and many have suggested that the Tunit were Dorset peoples. However, it is not clear whether the takeover of the Arctic was entirely hostile. It could have been a combination of conflict, new diseases brought by the Thule, and the fact that Dorset peoples were outcompeted when it came to exploiting the Arctic's scarce resources. There is also the intriguing possibility that Dorset peoples might have continued to thrive in small areas. Perhaps when the Norse reached Greenland they met both Thule and Dorset peoples. Even more remarkable is the suggestion that Dorset people survived until the early 20th century on Southampton Island. A people called the Sagdlermiut lived there; they spoke a different language to the Inuit and lacked kayaks, going to sea on inflatable sealskins.

A 'tent ring', a circle of stones used to hold down skins draped over a frame of driftwood. This ring, on Canada's Ellesmere Island, could be 1000 years old.

Partially subterranean winter house at Nutat, north-west Greenland. The house was occupied annually until the 1950s.

They were wiped out by an unknown disease, apparently brought ashore by a sickly sailor from a whaling ship. An examination of DNA from skeletal remains suggests a mixed ancestry, both Dorset and Thule, which makes the Sagdlermiut even more intriguing.

## People of the Eurasian Arctic

The Inuit are the native folk most commonly associated with the Arctic. But there were other peoples to the south, First Native tribes of both Alaska and Canada, and the Aleut of the Aleutian and Commander islands. What they had in common was isolation from others for centuries. By contrast, the northern native folk of Europe were in contact with southern peoples from earliest times. Northern Scandinavia had been inhabited from late Stone Age times, a continuous occupation, though with influxes of different peoples. Of these the most important were the Norse, a Germanic people who arrived in about 500AD and the Sámi who they met when they arrived. The Sámi were reindeer herders from east of the Urals, and who now occupy Fennoscandia. Beyond the Urals there were, and remain, a collection of peoples who occupied the north of what is now Siberia. Sharing a common, central Asian ancestry these were, moving east, the Nenets (who also occupy land around the White Sea, west of the Urals), the Nganasan (the most northerly of all Russia's indigenous peoples, whose range extended onto the Taimyr Peninsula), the Enet, Selkup, Yakut, and Evenki, as well as smaller groups such as the Dolgan and Yukaghir. Finally, in north-east Siberia, were the Chukchis and Yuppiat Eskimo. To the south of them in northern Kamchatka lived the Koryak and the Eveni. All these peoples were essentially reindeer herders, though the Bering Sea communities were sea mammal hunters. Some of the groups have survived the advance of Russia and then the years of USSR rule better than others, but in all cases their traditions have lived on. As with the other northern dwellers they had beliefs which are now collectively termed shamanism, the word derived from *saman* – wise. The shaman was the tribe's spiritual helper, a priest. To the Sámi he was the *noaidi*; in Inuktitut, the language of the Inuit, he was the *angakok*.

Shamanism was viewed with suspicion by Christian missionaries as contact between 'civilisations' of southern Europe and the northern European 'pagans and savages' grew, a suspicion that was later passed to North America as Europeans explored the American Arctic. Native languages and traditions were supressed and it is only in the last 50 or so years that the cultural sophistication and relevance of the northern peoples has been recognised. A degree of self-governance, sometimes more, many times less, has been granted across the north, and while the main religion is now Christianity, often there are echoes of the old beliefs, especially in the arts. Shamanism is also being viewed more sympathetically, particularly as the west has seen an increase in what might be generalized as 'Mother Earth' beliefs.

## Greenland

The first Europeans to encounter Nearctic northern peoples were Norsemen from Iceland. The Norse reached, and settled, Iceland in about 860, though the Iceland they settled was already inhabited, the southern coast being home to Irish monks who had arrived in *curachs* (boats of cowhide stretched over a wooden frame) via the Faeroes, perhaps as early as the middle of the 8th century.

From Iceland, the Norse made journeys both north and west. To the north they discovered Svalbard – 'cold edge'. For political reasons the Norwegians have claimed that this was the Svalbard archipelago, but most experts believe it is much more likely to have been north-east Greenland. To the west of Iceland Norse sailors certainly saw the east coast of Greenland. The first landing on Greenland was by Eirik the Red in 982. Eirik spent three years there (exiled for the murder of a neighbour),

Siberian Yupik woman in traditional costume.

returning with tales of a lush, green land. This was true: even today, the coastal plain of Greenland is vegetated, and it was likely to have been even more so in Eirik's time, as his trip coincided with a period of Arctic warming.

On his return to Iceland, Eirik persuaded many Norse to return with him and two settlements were established, Østerbygd (East Settlement), at Qaqortoq, and Vesterbygd (West Settlement), at Nuuk (formerly Godthåb). The ruins of Eirik's own settlement of Brattahlíð, close to the shore of Tunulliarfik (Eiriksfjord) can still be seen. At the height of Norse occupation, the population was perhaps 4,000. From Greenland one Norse ship, blown off course on a journey to Iceland, saw land. Later this was reached by Liefur Eiriksson, son of Eirik the Red. While it is known that the Norse occupied a site (at L'Anse aux Meadows) on Newfoundland, it is likely they also explored southern Baffin Island and the Labrador coast perhaps even travelling as far north as Ellesmere Island, and south to the Cape Cod area of New England. In the winter of 1002–03, Snorri Thorfinnsson was born in a Vinland winter camp, the first non-native American born in the New World.

On west Greenland, the Norse certainly reached 73°N where three cairns, one hiding a stone inscribed with runes, were discovered on an island north of Upernavik. The Norse also definitely had contact with native Greenlanders, calling them *skrællinger*, a word that might derive from *skral*, small or weak, or from Karelia, a district of northern Finland/Russia whose native inhabitants were short, stocky and dark. It seems there was both trade and conflict between the Norse and the Greenlanders. Some speculate that the latter ultimately killed off the former, as the fate of Norse settlers is a mystery. The last Bishop of Greenland died in 1378, while the last recorded ship to have sailed from there left in 1410: nothing is heard after that. The reason for the abandonment of the settlements and what happened to the settlers is still debated. Until recently it was believed that a change to a colder climate made farming more difficult and animal husbandry marginal because of the lack of hay for winter fodder. The cold would also have increased local sea ice, making travel to Iceland and Scandinavia more difficult. As agriculture failed, life became untenable. Some settlers may have left: those that stayed probably died of starvation. However, recent geological studies do not entirely support the climate change theory, suggesting that the climate was similar when settlement ended to how it had been when the first settlers arrived, so it may be that a decline in contact with Iceland/Scandinavia, perhaps because walrus ivory became more difficult to obtain so trade diminished, meant ships arrived less often and without required supplies settlement life became less agreeable. Isolated communities might then have struggled against inbreeding and disease, and perhaps even in conflict with native Greenlanders.

A Chukchi yarang or choom. Unlike the tents of the Sámi and Nenets, the choom is constructed using long poles which extend from a series of tripods.

The destinctive onion domes of Russian churches are seen throughout Alaska. This one is on Kodiak Island.

# 7 The exploration of the Russian Arctic

Following Columbus's discovery of the New World, Pope Alexander VI granted the Spanish the western hemisphere and Portugal the eastern, under the terms of the Treaty of Tordesillas in 1494, making things difficult for the English, French and Dutch. Therefore when, shortly after the signing, Giovanni Cabato arrived in Bristol offering to lead an expedition to Cathay the excited English made him an honorary Englishman, and, with a name change to John Cabot, he set sail from Bristol on 20 May 1497. On 24 June Cabot landed in Newfoundland. Finding evidence of inhabitants and fearful of confrontation because of his limited numbers, Cabot took on water and left, exploring the local coast before returning to Bristol.

There are references in the Bible and early writings in India that may refer to unicorns, but at some stage a real narwhal tusk must have given credence to the existence of an animal no one had ever seen. Consequently, when a tusk was found on the shore of the Kara Sea sometime in the 16th century its appearance reinforced the legend. The tusk caused a great deal of excitement among Europeans 'knowing that Unycorns are bredde in the landes of Cathaye, Chynayne and other Oriental Regions.' The discovery added extra impetus to the search for a northern passage to the Orient: the merchants of northern Europe were anxious to find an alternative to the dangerous and heavily taxed land routes that brought silk and spices from the East, and the long and hazardous sea journey around the Cape of Good Hope. A second voyage by Cabot had also failed to find a route to Cathay, so perhaps it is not surprising that when the English decided to look again in 1553 they went east. Three ships set sail, but, separated by a storm, only one, the *Bonaventure* under Richard Chancellor reached the Barents Sea and, eventually, what would become Arkhangelsk (Archangel). There Chancellor discovered to his amazement that he was not in Cathay but Muscovy (Russia). He was enthusiastically welcomed by officials of Tsar Ivan IV and taken 2,400km to Moscow by sled. At the capital he was equally warmly received and negotiated an Anglo-Russian trade treaty that made his London backers (who subsequently formed the Muscovy Company) rich.

## Barents heads east

The relative success of Chancellor's expedition halted English exploration of an easier route to Cathay, but others maintained an interest. In 1594 the Dutch sent three ships that way, one being commanded by Willem Barents. The idea was for Barents to head north, rounding the northern end of Novaya Zemlya, whose southern end was known to the seafaring Pomores of the White Sea coast, while the other two would sail between the island's southern tip and the mainland. Barents' attempt failed, sea ice blocking his path at about 77°N – probably a record northing at the time. He returned to find that one of the other two ships had made it through the Kara Sea and, finding open water, had turned back assuming that a passage existed. The delighted Dutch royal house sent an expedition of seven ships the following year. But the sea ice in 1595 was very different, little progress was made and no passage was found. Disappointed the Dutch royalty lost interest, but an Amsterdam merchant financed a two ship expedition in 1596. Barents went again, but not as commander of either ship. Sailing north the Dutch found Svalbard – or refound it as perhaps the Norse and possibly Russian Pomores had already seen it – reaching 79°49'N, the north-western tip of Spitsbergen, before turning south. One ship now headed home, but the other, with Barents on board, turned east and rounded Novaya Zemlya's northern tip. In a sheltered bay the Dutch were forced to overwinter. The cold was intense and scurvy began to take its toll: of the 17 man crew four died. When spring came, the remainder headed south, converting two rowing boats to skiffs as their ship was no longer seaworthy. Barents died of scurvy, but the remaining crew made it to Novaya Zemlya's southern tip

There is clear evidence of Pomore journeys as far as Svalbard, but the earliest date of such voyages is disputed. This photograph shows a Russian cross at Murchisonfjorden. This site can only be accurately dated to the late 17th/early 18th centuries, but Russian scholars claim earlier dates on the basis of disputed dendrochronology. Such claims are considered to be politically motivated by a nervous Norwegian government.

where a group of Russian fishermen rescued them. The survivors returned home as heroes, but there were no further Dutch expeditions.

## Siberia

The country ruled by Ivan IV, the Tsar who had signed the Anglo-Russian trade treaty, extended as far as the Urals. Ivan is known as 'the Terrible' as the cruelties of his reign were gross even for a period of history not noted for benign treatment of those considered enemies of the state. His reign of terror, aided by disease, ill-considered military campaigns and, in particular, a decline in the supply of furs, had brought Russia to the edge of disaster. In the absence of central heating fur was needed to ward off Europe's winter chills, and its export represented Russia's major source of income. The Sable, a member of the weasel family with a much-prized thick coat, had been all but exterminated in northern Muscovy: without new sources Ivan faced economic ruin. He therefore, encouraged (or, at least, failed to discourage – Ivan's support was vague and ambiguous) one of Muscovy's most powerful mercantile families, to probe eastwards beyond the Urals.

## Siberia and Dezhnev's journey

The returning Russians brought thousands of Sable, and the taiga was rapidly populated by fur trappers, with towns set up on the area's navigable rivers to collect and ship the booty. Within a century the Russians had reached Kamchatka, leaving only the extremities of Taimyr and Chukotka beyond the Tsar's grasp. Taimyr was reached in 1620, the Russians rounding Cape Chelyuskin, the northernmost point on the Eurasian mainland. By 1630 the Lena had been reached, followed by the Yana, the Indigirka and the Kolyma. By 1639 the Russians had seen the Pacific and the Sea of Okhotsk.

Then, in 1648, one of the most significant of all expeditions to the Arctic took place. Despite the vast wealth of Siberia, new sources of Sable and other fur-bearing animals were always being sought. Spurred by rumours of that in southern Chukotka there were yet more furs and that its shore offered walrus ivory, an expedition set out to explore. Although led by an agent for a Moscow merchant, the real leader was Semen Ivanovich Dezhnev. Dezhnev was probably a Pomore

## Introducing Siberia

Across the Urals, traditionally the boundary between Europe and Asia, lay Siberia, named after the Mongolian word *siber*, meaning beautiful or pure, or, perhaps, from the Tartar *sibir*, which translates as 'sleeping land'. The sheer scale of Siberia is breathtaking. Trains on the Trans-Siberian Railway take eight days to chug their way from Moscow to Vladivostok, six of those spent east of the Urals. East of the obelisk which traditionally marks the 'boundary' between Europe and Asia the train crosses five time zones, while east of Vladivostok, a traveller would cross three more while edging around the Sea of Okhotsk to the Bering Sea. Siberia stretches from the Arctic Ocean to the Mongolian steppe; it covers almost 8% of the world's land area. The whole of the United States, including Alaska, together with all the countries of Europe (excluding European Russia) could fit comfortably into Siberia. Each of Siberia's three great rivers, the Ob, the Yenisey and the Lena, drains a basin bigger than western Europe. Of the vast Siberian forest – the taiga – Chekhov wrote that that only migrating birds knew where it ended.

and so familiar with the *koch*, the ships in which the expedition set out. *Kochs* were superbly adapted for exploring Arctic waters, being small with a flat bottom. They had a single mast and sail, but were light enough to be rowed. This, and their shallow draft, made them highly manoeuvrable in icy waters, the flat-bottom allowing then to be readily freed if pack ice threatened entrapment. There was a double hull, an inner layer of boards sewed together with juniper roots, the outer layer nailed. Interestingly, the *koch* looked somewhat like *Gjøa*, Amundsen's North-West Passage ship built three centuries later.

In favourable ice conditions, but poor weather, five of seven expedition ships were lost. But the remaining two rounded what is now Cape Dezhnev, Eurasia's north-eastern tip, and sighted the Diomede Islands. Another ship was lost, but the last, that of Dezhnev sailed south of the Anadyr River. It had travelled over 3,000km in 100 days and had passed through the Bering Strait. Dezhnev returned home, but over subsequent years he returned to the Anadyr and collected more than two tons of walrus ivory.

The memorial to Semen Dezhnev which stands at Cape Dezhnez, the most easterly point of Eurasia.

## Bering and Russian America

Strangely, despite both its significance and its value as an epic tale of adventure and survival, Dezhnev's journey was forgotten for almost a century. In 1724 as Tsar Peter the Great lay dying he sanctioned an expedition to explore Siberia's eastern shore. Its leader was to be Vitus Bering, a 44-year-old Dane recently retired from Russia's Imperial Navy. His orders were to look for lands where fur could be exploited and to see if Asia and America were joined – for although a vague memory suggested that to the north they were not, no one knew if they were to the south. Bering's expedition is famous, but actually achieved little.

Sailing north from Kamchatka he discovered St Lawrence Island, but failed to travel far enough to reach the Strait which now bears his name (and which Dezhnev had negotiated): his voyage lasted just 51 days.

Back in St Petersburg Bering's cautious voyage was not well received, but he was given command of a second expedition and sailed again 1741. Sailing east Bering discovered some of the Aleutian Islands, and also Kayak Island (off Alaska's southern shore) where the expedition's naturalist, the German Georg Steller leapt ashore to become the first European to land on Alaska. Returning towards Kamchatka Bering and his crew became sick with scurvy. On one of the Commander Islands, the

A memorial to Bering's men has been erected at their grave site on Bering Island. In 1991 a Russo-Danish expedition exhumed the bodies on the island. The skeleton of Bering himself was identified from the description of the way he had died, in a pit of sand which had collapsed over him. The sand had been warm and Bering refused to be moved. When his head was reconstructed from the exhumed skull it bore no resemblance to the assumed portraits of him.

leader died and was buried in a shallow grave. The island is now called Bering Island, and thanks to James Cook's exploration of the area 50 years later, the leader's name is now also attached to the sea he partially explored and the Strait he never reached. The other legacy of Bering's second expedition was the discovery of the Sea Otter, the exploitation of which is dealt with in Chapter 18.

## The Great Northern Expedition and the North-East Passage

Bering's Second Kamchatka Expedition was one part of what became known as the Great Northern Expedition, an enterprise that surveyed the entire north coast of Russia from the White Sea to Chukotka and the east coast as far as Japan. This was a monumental exercise, completed between 1734 and 1741 by five separate teams despite the capricious nature of the Arctic ice, the vile weather, scurvy and the hazards of overwintering in mind-numbing cold. Completion of the work inevitably led to the revival of interest in a North-East Passage, and a Russian expedition set out in 1764. But rather than heading east, the ships went north, heading between Svalbard and Greenland, reaching 80°26'N, a new record, before being stopped by ice. As often before, failure dulled the enthusiasm for further attempts and when, a century later, the next attempt was made it was by Adolf Erik Nordenskiöld, Finnish-born to Swedish parents. He moved to Sweden when he was 25, and took part in several Arctic expeditions, mostly to Spitsbergen, but also to the Russian Arctic, particularly the Kara Sea and as far east as the Yenisey River, journeys which taught him a great deal about the ice conditions in the area. In June 1878 Nordenskiöld set out from Karlskrona, southern Sweden, in the *Vega*, a 300-ton, three-masted whaler with a steam engine and a crew of 30. By early August the expedition had reached the Kara Sea, finding it ice-free. Cape Chelyuskin was passed on 19 August, the Lena delta on 27 August. Ice conditions then worsened, and the *Vega* was stopped on 28 September, just two days sailing from Cape Dezhnev.

Because of excellent preparations the winter was spent comfortably. On 18 July 1879 the *Vega* was released from the ice. Two days later she passed Cape Dezhnev: completion of the North-East Passage had been a masterpiece of good organisation and seamanship. The *Vega*

sailed on to Japan, then around China to the Indian Ocean and across it to reach the Suez Canal. She sailed across the Mediterranean to the Straits of Gibraltar, then around Portugal, Spain and France to the English Channel and the North Sea, finally reaching Sweden in April 1880. On 24 April (still Vega Day in Sweden) the ship reached Stockholm and a hero's welcome. Nordenskiöld was even made a baron.

### Amundsen and the Maud

Arctic travellers who reach Cambridge Bay, on Victoria Island's south-eastern shore, may be surprised to see the wreck of an old ship, just breaking the surface. This is the *Maud*, built by Roald Amundsen initially to repeat Fridtjof Nansen's attempt to reach the North Pole in the *Fram*. Amundsen's plan was to repeat Nordenskiöld's trip and then seal the ship in the ice and drift to the North Pole. Setting out in June 1918, Amundsen had bad luck and bad ice, overwintering twice before making the second west-to-east NE Passage transit, and the third overall. By July 1920 when the *Maud* finally rounded Cape Dezhnev, Amundsen had given up the idea of immediately repeating the *Fram* drift and headed for Alaska. But after resupplying the ship he went north again intent on the Pole. But another overwinter and damage to the ship's propeller meant Amundsen finally lost interest in the Pole. He was also low on funds and the *Maud* was seized by his creditors. She was bought by the Hudson's Bay Company, and renamed *Baymaud*. But drawing too much water to be a useful supply vessel she was abandoned at Cambridge Bay, where she sank at her mooring in August 1930.

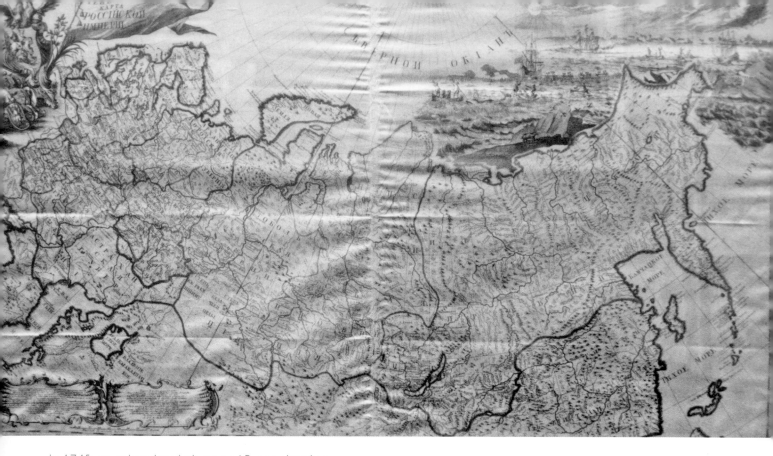

In 1745 an embroidered silk map of Russia detailing the findings of the Great Northern Expedition was finally completed and presented to the Tsaritsa Elisabeth. The achievements of the expedition cannot be overstated – it was another 100 years before the surveying of the northern Canadian coast was completed. Today the silk map, somewhat forlorn and neglected, is stored in the attic of a St Petersburg museum.

This extraordinary rock formation, looking strangely similar to the Easter Island statues, is on Bolshoy Lyackovskiy, one of the New Siberian islands.

## Russia's Arctic islands

The achievements of the Great Northern Expedition in surveying Russia's northern coast were commemorated in a silk map of Russia that was presented in 1745 to the Empress Elisabeth, the last surviving child of Peter the Great. The map shows Novaya Zemlya, which had been known from very early times, but none of the other islands and archipelagos of Arctic Russia.

Of the others, Novosibirskiye Ostrova (the New Siberian Islands) were first recorded in 1770 when a fur trapper noticed a herd of reindeer heading south towards him across the sea ice and followed their tracks northwards to discover two islands, and reindeer tracks coming from even further north.

In 1848 Henry Kellett, captain of the *Herald*, sailed through the Bering Strait as part of a Franklin search expedition and discovered Herald Island, naming it after his ship. He climbed to the top and saw land to the west: this was called Kellett's Land on early British maps, but was later renamed Wrangel by an American whaling ship captain in memory of Ferdinand Petrovich von Wrangell (the American misspelled the name, but his version has stuck) who had learned of its existence from Chukchi natives during an expedition he commanded in 1820.

Franz Josef Land was the next island group to be discovered, though its formal discovery in 1873, by an Austro-Hungarian expedition who named it for their Emperor, was almost certainly preceded by a sighting in 1865 by a Norwegian sealing expedition. The final Russian archipelago, Severnaya Zemlya, was not discovered until the Arctic Ocean Hydrographic Expedition of 1910–15, though its existence had been predicted by studying sea currents. The archipelago was thoroughly explored by George Ushakov, a proud communist, which explains the names of October Revolution and Bolshevik islands and why a significant peak is Mount Hammer and Sickle. The most northerly point on the archipelago is Cape Arktichevsky (Cape Arctic), the starting point for many expeditions to the North Pole.

## Voyages in the eastern Arctic

Russian exploration which revealed the existence of its Arctic islands, and further attempts/transits of the North-East Passage continued until the present day, but several expeditions, not always by Russians, need special mention,

### *The* Jeanette

In 1879 George Washington De Long led a US Navy expedition aboard the *Jeanette* to the north-east Russia Arctic. The nominal purpose was to discover if Wrangel Island was truly an island or part of the Asian continent. But De Long also hoped to find the open polar sea (see below) and to sail it to the North Pole. The ship became beset in sea ice, drifting east rather than north to the hoped-for open water. The drift went close to Wrangel – proving it was an island – but lasted two years, by which time the ship began to take on water. The crew moved

Cape Tegetthoff, Franz Josef Land. The Austrians named the headland, with its distinctive rock towers, after their ship.

everything they could onto the ice and set for the New Siberian Islands hauling three heavy boats. On reaching Bennett, the archipelago's most westerly island, De Long turned south towards mainland Russia. Open water was reached, but a storm separated the boats: no trace of one or its crew of eight was ever found. The other two reached the Lena delta. One found the delta's main stream and soon met locals, but the other, De Long, 12 men and Snoozer, the ship's dog, became lost in the maze of delta channels. As they struggled through the endless delta marshland they ran out of food, finally eating Snoozer. But that meal offered only temporary respite. De Long sent the two strongest men ahead to seek assistance, but those that remained with him began to die of exhaustion and frostbite. De Long made a last diary entry on 30 October, the 140th day since leaving the *Jeanette*.

The two men De Long sent ahead met locals who took them to join the other boat party, but by the time these had returned to the delta all that remained were bodies and diaries. In all 20 men died. Of the survivors one committed suicide, one went insane.

Years later driftwood discovered on Greenland's south-west coast was found to have come from the *Jeanette*, leading Prof. Henrik Mohn to suggest that a current flowed across the Arctic Ocean from Siberia. As we shall below, that idea was the inspiration for Nansen's famous *Fram* voyage.

## Albanov and the Chelyuskin

*Jeanette* was American, but Russia could provide its own tragedies, and a near disaster which was to point a way to the future of polar exploration. In 1912 the *Saint Anna*, commanded by Georgi Brusilov left Arkhangelsk on 4 September – much too late in the year for Arctic travel. Brusilov had been delayed in Murmansk and had also failed to sign on the crew he needed, finding only five experienced sailors, one, Valerian Albanov, acting as his deputy. The objective of the trip was to discover new whaling and sealing grounds, and to make a second transit of the North-East Passage.

Brusilov may have been seduced by Nordenskiöld's suggestion that the Kara Sea was ice-free in the late summer. If so he was abruptly brought back to reality when his ship became ice-bound close to the Yamal Peninsula on 15 October. The crew walked to the peninsula and saw the tracks of local reindeer herders. These locals offered salvation, but the crew decided to stay with the ship, understandable but very wrong. During the next 17 months the ship drifted slowly north, finally reaching 82°58'N off the northern tip of Franz Josef Land. There a simmering conflict between Brusilov and Albanov, who believed the leader was incompetent, finally boiled over. Albanov decided to leave the ship when, with food and fuel running low, Brusilov had no plan other than to hope the ship would break free. When he left, 13 men decided to go with him.

Sledges and kayaks were built and loaded with supplies (that Brusilov itemised and made Albanov sign for). It was only 120km to land but progress was slow and the weather appalling. When the weather improved, the ice drifted north faster than the men moved south, and after 11 days three men gave up and returned to the ship. The trek of the remaining 11 now became a nightmare, with seals for food and fuel being in short supply. One man went off on his own and failed to return. Progress slowed, the men became listless, scurvy adding to exhaustion, but in late June land was finally reached. Albanov could now work out his position and headed east for Cape Flora. One man died, then another. The team then divided into two, with Albanov and three others taking the supplies and paddling two kayaks while the other four skied cross-country with no loads: the skiers failed to make the rendezvous. Then one kayak went missing. Finally, Albanov and the sole other survivor reached Cape Flora where they were lucky to find a ship which saved them from overwintering. No trace was ever found of any of the other men who set out, or of the *Saint Anna* and those who remained with her. Later Albanov was to write a marvellous account of his journey (*In the Land of White Death*), but it is always worth remembering that only survivors write memoirs, the remainder die dreadful deaths.

In July 1932 the ice-breaking steamer *Aledsandr Sibiryakov* (a converted sealer) headed east, going around the northern tip of Severnaya Zemlya. Heavy ice off Chukotka smashed the ship's propeller, but using a makeshift sail the ship reached the Bering Strait, the first one-season transit. Encouraged, the Soviets sent other ships east. One was the 4,000t *Chelyuskin*, which had no ice-breaking capacity but was considered large

The first plane that landed on ice to rescue those on the *Chelyuskin*.

enough to nose through significant ice. In late 1933 the ship became ice-bound, eventually drifting north-west towards Wrangel Island. After wintering in the ice, the ship was crushed and sank, one man drowning as he attempted to jump to safety. The 104 survivors – including a baby girl born on 31 August 1932 in the Kara Sea and named, of course, Karina – set up camp on the ice, the expedition leader Professor Otto Schmidt citing Albanov's journey as the reason for not attempting a crossing of the ice to Chukotka or Wrangel. Ample supplies had been removed before the ship sank and

the campers had a relatively comfortable time awaiting rescue by air, their stay enlivened (or perhaps not) by a non-stop series of lectures by Schmidt, a devout communist. From 5 March to 13 April, seven pilots made repeated flights to a makeshift ice runway and safely rescued all the survivors. The pilots (Lyapidevski, Levanevski, Molokov, Kamanin, Slepnev, Vodopyanov and Doronin) were the first to receive the award of Hero of the Soviet Union. The rescue had been copybook, and pointed the way to future exploration, by air, of both polar regions.

Cairn at Winter Cove, Victoria Island, Canada. It was built by sailors of the *Enterprise* in 1851 during one of the searches for the Franklin Expedition.

# 8 Canada and the North-West Passage

Having first failed to find a passage to Cathay by going west, then failed by heading east instead, the English went west again in 1576, Sir Martin Frobisher, leaving London in June 1576 with three tiny ships, one of which sank off Greenland's southern tip. A second returned to England, but Frobisher continued, finding Baffin Island and the bay that now bears his name. He sailed along it, convinced that to his right was Asia, to his left America. Frobisher met an Inuit, the first European since the Norse to do, and gave the first account of meeting another Arctic inhabitant, the mosquito. But when five of his crew went missing, not knowing if the men had mutinied or been captured (and perhaps eaten – the English had seen the Inuit eating raw

fish and raw seal), Frobisher took a hostage, and departed. Three centuries later the American Charles Francis Hall was told by the Inuit that the missing men had stayed in the village for some time before departing in a boat they had built, never to be seen again: they had probably disobeyed orders to go ashore for some private trading, overstayed and were left behind when a rattled Frobisher took his hostage back to England, where the poor man soon died, probably of pneumonia.

Frobisher also brought back a lump of black rock, so like coal that it was thrown onto a fire. It glistened. Retrieved and tested it was claimed by several assayers to be pyrite, but another maintained it was high quality

The southern coast of Meta Incognita Peninsula, Baffin Island. It is spring, and in the foreground the sea ice of Hudson Strait is breaking up. Martin Frobisher reached the bay which bears his name, on the northern side of the Peninsula.

gold ore. Consequently in 1577 Frobisher was back in his Bay. Not surprisingly the Inuit were less keen on their visitors this time and in a fracas five Inuit were killed and one sailor badly wounded. Frobisher captured a man, woman and child, and a kayak, loaded 200 tons of ore into his ship and sailed for home. Back in London the male Inuk entertained Queen Elizabeth by killing swans from his kayak, but all three Inuit soon died.

Despite expert misgivings over the gold ore, the Queen underwrote a huge expedition of 15 ships that sailed, again under Frobisher, in May 1578. This time over 1,000 tons of ore was returned, but all attempts by the London alchemists failed to turn it into anything valuable.

## Henry Hudson

The English proved more persistent in the northward search than the Dutch had been in the east. After optimistic reports from an expedition in 1585 the Muscovy Company financed Henry Hudson to lead a trip. But Hudson chose to go north over the Pole, sailing the Open Polar Sea. Hudson, and others, believed perpetual sunlight would ensure the sea was clear of ice, ice forming a ring to the south. All a captain needed to do was find a suitable gap in the ice circle to gain open water. Hudson sailed north to Svalbard but soon discovered there was either no open water or no break in the ice circle. He placated his merchant paymasters by noting that the bays of Svalbard were home to whales, of vast size and in great numbers, and the Company sent him back again, this time to try for the North-East Passage. The journey was a failure, and the Company much less impressed. But Hudson found another sponsor and crossed the Atlantic to discover the Hudson River, Coney Island and Manhattan. The finds pleased his sponsors and he set sail again in 1610. This time he followed the northern Quebec shore and discovered the vast, calm waters of what he believed was the Pacific Ocean: it was the bay which now bears his name. Turning south, Hudson watched the shore, waiting for the cities of Japan to appear. But there was nothing but bleak Arctic tundra. He reached James Bay where a very unoriental winter froze both the sea and the ship. When the ship was finally released in June 1611 the crew mutinied, putting Hudson, his son, four sick

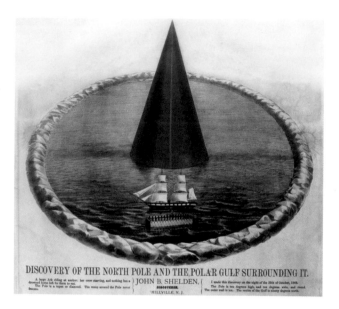

In 1869 John Shelden claimed to have reached the North Pole, a huge cone of diamond or topaz, by sailing the Open Polar Sea.

men and three non-mutineers in an open boat: they were never heard of again.

The thirteen mutineers were reduced by a fight with the Inuit and scurvy to just eight under the command of Robert Bylot. On reaching England they might have expected to be hanged, but Bylot's claim to have solved the riddle of the North-West Passage meant economics trumped the law, and he was financed to sail again.

## The search continues

The expedition that included Bylot and other mutineers, and used Hudson's ship (the *Discovery*), was the first of a number that mapped the islands of Arctic Canada. On one William Baffin calculated the longitude of his ship by taking a complete lunar observation on 21 June, a feat that earned him the admiration of all that followed him and is one reason why many consider him the greatest of the early Arctic explorers. The Dane Jens Munk crossed Hudson Bay, wintering at the mouth of the Churchill River, the 65 man expedition having a torrid time, scurvy killing all but Munk and two others. In a remarkable feat of seamanship the three managed to sail back in one of the expedition's two ships. But on reaching Denmark, King Christian IV ordered him to go back to retrieve the other ship. Not surprisingly Munk failed to raise a crew.

In 1631 one of the more interesting expeditions set out from England. In two ships named *Charles* and *Henrietta Maria* for the King and Queen the two captains, Luke Foxe and Thomas James, were sponsored by rival merchants in London and Bristol. As much a race as a quest, the two made minor discoveries. But the trips were memorable in two ways. Firstly, the fact that each man carried a letter from King Charles to the Emperor of Japan. The letters were in English as it was reasoned that the Emperor, being both a king and a cultured man, would obviously speak the language. The second reason is that each man wrote a book on his journey. Though each is interesting, in different ways, James' became a classic and is said by some to have inspired Coleridge to write the *Rime of the Ancient Mariner*. James' trip was also notable for his crew having overwintered. Scurvy again took its toll, but the crew managed better than Munk's men. As spring approached James, in the hope of attracting natives who might bring food, climbed a tree to watch for an approach while his men built a fire of brushwood. The fire got out of hand and the men were lucky to escape with their lives. After months during which they had all but frozen to death, the possibility of death by burning must have seemed an ironic twist of fate.

## The Hudson's Bay Company

In England, the Civil War stopped further exploration for many years. With the Restoration, eyes turned west again and to their horror the English found Frenchmen had occupied the northern continent of North America. The French were exploiting Canada's wealth of furs, particularly Beaver, and the English, wanting to take their share formed the Hudson's Bay Company to that end. From a Canadian exploration point of view the Company was important, but as its concern was profit rather than a North-West Passage, its efforts in search of the latter were grudging and not always significant. But in its search for new lands to exploit, the Company was much more successful. Samuel Hearne was sent west to see if there was any truth in the rumour that there was a river that flowed between banks of solid copper. In 1771 he explored the Coppermine River to its mouth, claiming everything he saw for the Company.

Next to head west was the Hebridean-born Alexander Mackenzie, who initially worked for a rival company – the North-West Fur Trading Company: it later merged with the Hudson's Bay Company. Hearing stories of a river which flowed west from the Great Slave Lake to the sea Mackenzie wondered if it might reach the Pacific, and be navigable. In 1789 he followed it to its mouth, but there all he could see was ice: it was not the Pacific Ocean. Mackenzie called the river he had explored Disappointment – today it bears his name.

## Canadian graffiti

Close to the mouth of the Churchill River, The Hudson's Bay Company built the Prince of Wales Fort, a trading post, but fortified in case the natives became restless. Close by is Sloop Cove, an inlet that empties each low tide and so could be used as a natural dry dock for repairing ships. The iron rings set into the rocky cove bank are still visible, as is the graffiti carved by bored sailors on a nearby smooth rock face. The inscriptions include a 'signature' of Samuel Hearne.

Frustrated, Mackenzie headed west again in 1793 and on 19 July he followed the Bella Coola River to the sea. He had reached the Pacific Ocean, the first European to gaze at it from the American shore, and the first to have crossed the continent. It was a very significant moment, but it must be recalled that by then the north Pacific had been explored by sea, James Cook exploring the western end of the Passage during his 1776–80 expedition.

## The Royal Navy heads north-west

By the close of the Napoleonic Wars in the early 19th century, Britain was the major world power, its pre-eminence based on naval might. Such status brought fears as well as benefits, and these were, in part, exploited to persuade the British Admiralty to use naval ships to search for an Arctic seaway. Russia, a competitor on the world stage, lay at the other end of such a passage and seemed intent on expanding into America: the Russians had already taken the Aleutian Islands and Alaska.

A search for a North-West Passage therefore had political aims. It would also keep the navy active,

maintaining a pool of experienced officers and seamen. And for John Barrow, senior civil servant at the Admiralty and a geographer and historian, it would be exploration for its own sake, a worthwhile endeavour.

Barrow's enthusiasm for exploration coincided with a sudden break-up of the Arctic ice, icebergs spilling down into the Atlantic and cooling Europe. Whalers, notably William Scoresby of Whitby, informed the authorities that if a search was to be made this time of minimal sea ice would be auspicious. In 1818 Barrow sent out two expeditions. John Ross and William Edward Parry were to search for the North-West Passage, David Buchan and John Franklin were to sail over the North Pole (the notion of an Open Polar Sea still had many advocates). The two expeditions were then to meet off the Siberian coast. Ross's trip is notable for an illustration in his expedition book portraying the meeting of the British and Greenlanders. To one side are skin-clad locals with a dog-drawn sled, on the other naval officers in dress uniform of tailed coats, buckled shoes and cocked hats. Britain was in its imperial phase and had just become (in its own eyes at least) the master of Europe, and the innate feeling of superiority that Britain's position engendered in its naval officers prevented them from really seeing

The British meet the Inuit, an illustration from John Ross's book of his 1818 expedition.

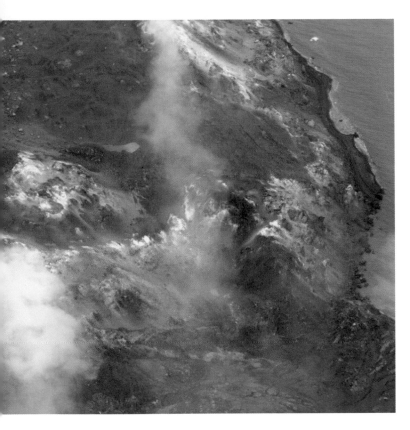

Franklin Bay, the Smoking Cliffs are bituminous shale, probably ignited by a lightning strike several thousand years ago.

The grave of one of Parry's men on Igloolik Island.

what was in front of them. No illustration could better depict the failure of the British to understand what they were getting themselves into. The Inuit were superbly adapted to the local environment in terms not only of clothing and equipment, but also in terms of the size of their group. The British missed that and consistently sent ill-clad men in parties too large to feed themselves off the land. When trips went reasonably well these things became annoyances, but when they went badly, they were disastrous, as the British were later to discover at appalling cost.

The trips were the first of many, with explorations by both land and sea, each adding a piece to the jigsaw that was the map of northern Canada. John Franklin made two land expeditions mapping part of the northern coastline. On one his team almost died of starvation, surviving by eating the leather of their boots, which made Franklin's book of the trip a best-seller and him being feted as the 'man who ate his boots'. On one trip, sponsored by private interest and led by John Ross, because the navy eventually lost interest as it had become clear that even if a passage

existed it would not be useful as either a trade or military route, James Clark Ross sledged to the North Magnetic Pole where his team raised a cairn and the Union Flag as Ross solemnly claimed the territory for the British Crown. The claim led to another surreal encounter with the Inuit, the British telling them that they now owned the land. To a nomadic people this was meaningless – the British might as well have said they owned the sky – and the baffled Inuit responded by noting the shortage of seals and fish hooks.

## Franklin's Last Expedition

Back in Britain, the success of Ross's expedition allowed Sir John Barrow, now in his eighties, and perhaps wishing to complete unfinished business before he died, to persuade the Admiralty that Britain would be a laughing stock if, having expended so much effort and having found both the eastern and western ends of the Passage, she did not explore the last remaining section in the middle. The Admiralty agreed, but finding a commander proved difficult. They eventually selected the clearly unsuitable John Franklin. Now almost 60 and overweight, he had just been dismissed (a dismissal engineered by vested interests) from his job as Governor of Tasmania.

Franklin took two ships, *Erebus* and *Terror*, which were strengthened to withstand pressure from the ice and fitted with railway locomotive steam engines that turned screw propellers, a radically new idea. The total crew was 133 – John Ross noted how hard it had been to feed one-sixth that number when his ship had been lost. But Barrow considered the food problem had been solved, the ships being stocked with tinned food, a revolutionary new idea that promised to eliminate both hunger and scurvy. The canned meat and vegetables were supplied by Stephen Goldner, the man with the lowest tender and a production line that left much to be desired in terms of the quality of the food in the cans and the cleanliness of his production methods. It was said that the only part

of the pigs that did not go into Goldner's can was the squeal. With the slaughter of pigs, sheep and cows being carried out on the premises and within sight of other animals, the filth that reached the cans was indescribable. Goldner's cans arrived only hours before the ships sailed, too late for samples to be taken to check on the quality of the contents or the adequacy of the can seals. On 19 May 1845 the ships sailed. On 26 July they were seen by a whaler, moored to an iceberg close to the entrance to Lancaster Sound. It was the last time either ship or any of the crew – reduced to 129 after four men had been sent home from Greenland – were seen by European eyes.

History had told the British not to be too concerned if nothing was heard of Arctic expeditions for several years, but by 1848 there were demands for a rescue mission. Many trips were then made, some official British, some American, some private. Collectively the Franklin search expeditions solved the problem of the

This watercolour was painted by Lt S. Gurney Cresswell, an officer on McClure's Franklin search expedition. Cresswell was in command of the sledge party which he has illustrated setting out from HMS *Investigator*.

North-West Passage, though none completed a full transit. One trip, led by Robert McClure, did complete a transit, by sailing east from the Bering Strait, but then having to sledge from an ice-bound ship to another ship which had arrived from the west. Technically McClure found not one, but two Passages: he received a knighthood and the reward for completion – which he refused to share with the captain of the ship that rescued him and his crew – but history does not record him as leading the first transit.

But while the search expeditions mapped the north, they did not find any trace of Franklin until 1850 when, on Beechey Island (not actually an island, as it is joined by a narrow causeway to Devon Island) a stack of Golder's meat cans and the graves of three men who had died in 1846 were discovered. Clearly Franklin had spent the winter here. But there was no clue as to where he had gone next.

The next news of Franklin's fate came from Dr John Rae, a Hudson Bay's Company surgeon. The Orcadian Rae had rapidly learned that the adoption of Inuit methods made survival much easier than it was for naval officers who insisted on maintaining naval methods in an environment for which they were entirely unsuited. This idea – 'going native' as the British Establishment contemptuously called it – almost certainly added to the venom with which Rae's news was greeted. In 1853 Rae met Inuit who told him of their meeting with a large group (perhaps 40) white men, who were dragging a boat along the western shore of King William Island. By signs and pidgeon-Inuit they learned that the white men had abandoned ships crushed in the ice and were looking for animals to hunt. Later, the Inuit found the remains of many men further south, telling Rae there were clear signs of cannibalism, human flesh being found in a kettle.

Rae brought back relics traded from the Inuit. These proved beyond doubt that the men were Franklin's. But Rae's suggestion of cannibalism, particularly as the account was based exclusively on Inuit testimony, brought out the worst in the British establishment. The *Times* thundered that no one could take the testimony seriously as the Inuit 'like all savages are liars'. The author Charles Dickens was equally outraged. Egged on by Franklin's wife, Lady Jane, who orchestrated a campaign against Rae and his account, Dickens published articles claiming the story was bound to be false as it was based

on the word of 'the savage' and 'we believe every savage to be in his heart covetous, treacherous and cruel'. Dickens went on to hint that it was more likely that the Inuit had murdered Franklin's men, and that if there was indeed human flesh to be found in kettles it was probably an offering by the Inuit 'to their barbarous, wide-mouthed, goggle-eyed gods.' The reaction is a lesson in Victorian values. The belief in the innate superiority of the Briton to any native is manifest (though the reality is that the upper classes in Britain put the working class in much the same category as the 'savages' they so despised). The logical extension of this xenophobia was that the British officer class had nothing to learn from natives and everything to teach them. It was an attitude that had sent men in dress uniforms to the Arctic and would, in a few years, send others equally unprepared to Antarctica.

The controversy hurt Rae immensely. He felt humiliated, a feeling heightened when the government did not honour him with the knighthood they had bestowed on much less deserving individuals, and quibbled over the payment of the reward for finding the fate of Franklin, with Lady Jane Franklin consistently lobbying against him. Lady Franklin had been a staunch supporter of her husband all along, and now, with the government more concerned with the Crimea, financed her own trip. Led by Francis McClintock it was to find real evidence, in the shape of two notes left, dated a year apart, on a single sheet of paper in a cairn on King William Island, and a ship's boat and several skeletons, which pointed to the truth of Rae's story.

Despite their brevity, the note reveals a great deal. The two ships had left Beechey in summer 1846 and headed south, presumably through Peel Sound, becoming beset off King William Island in September and abandoned two years later. Lt. Gore, who left the first entry, was dead, as were 23 others (three on Beechey, 20 later, including Franklin himself). That is a much higher death rate than on any previous expedition.

Next in 1864 the American Charles Francis Hall, who believed he had been chosen by God to lead Franklin survivors back to civilisation, gathered information from the Inuit, and found a skeleton buried on southern King William Island. The skeleton, believed to be that of Henry Le Vesconte, an officer on the *Erebus*, was eventually taken to Britain and buried at Greenwich. Another skeleton, believed to be that of John Irving, an

## A message from beyond the grave

The 'McClintock note' is actually two notes, written one year apart, on an official form that requested, in six languages, its return to the Admiralty or the local British Consul. The first note, signed by Lt. Graham Gore and Charles Frederick Des Voeux, Mate was written in May 1847. It reads:

*HMS Ships Erebus and Terror 28 May 1847 wintered in the Ice in Lat 70°05'N, Long 98°23'W. Having wintered in 1846–7 at Beechey Island in Lat 74°43'28"N Long 91·39'15"W after having ascended Wellington Channel to Lat 77° and returned by the west side of Cornwallis Island. Sir John Franklin commanding the Expedition. All well. Party consisting of 2 officers and 6 men left ship on Monday 24th May 1847.*

The date of the wintering on Beechey is wrong – it was 1845–46 – but the remarkable information is that Franklin had sailed around Cornwallis Island, finding open water in Wellington Channel but presumably being stopped by ice at about 77°N. What was the condition of the ice in the Parry Channel? Could he have sailed west from Cornwallis and gone beyond Melville and Banks islands towards Alaska?

The second note reads:

*[25th April 1]848 HM Ships Terror and Erebus were deserted on the 22nd April 5 leagues NNW of this [hav]ing been beset since 12th Sept 1846. The officers and crews consisting of 105 souls under the command [of Cap]tain FRM Crozier landed here – in Lat 69°37'42" Long 98°41' [This] paper was found by Lt Irving under the cairn supposed to have been built by Sir James Ross in 1831 – where it had been deposited (4 miles to the northward) – by the late commander Gore in June 1847. Sir James Ross's pillar has not however been found and the paper has been transferred to this position which is that in which Sir J Ross pillar was erected – Sir John Franklin died on the 11th June 1847 and the total loss by deaths to this date 9 officers and 15 men.*

This note was signed by James Fitzjames, Captain HMS *Erebus*. It was also signed FRM Crozier, Captain and Senior Officer, Crozier adding a postscript 'and start on tomorrow, 26th, for Backs Fish River'.

---

officer on the *Terror*, was also repatriated, and buried in Edinburgh.

## The fate of Franklin

The evidence discovered by McClintock, coupled with the Inuit observations garnered by Rae and Hall formed the basis of a coherent story that gives a good impression of what happened to Franklin's men. The ships having been entombed in ice at the northern end of King William Island, and with food and fuel running low, the men abandoned the ships and started south, intending to use boats to follow a river to a Hudson's Bay Company outpost. Then, overcome by hunger and (perhaps) scurvy they died one by one, the last groups dying close to the island's southern shore. But further, remarkable evidence, was gathered later. In 1984 autopsies on the three corpses (almost perfectly preserved by the permafrost) at Beechey Island were carried out. These showed high levels of lead (though death was from natural causes), as did analysis of bones from King William Island: the latter also showed distinct signs of scurvy. More recently (2011), the skeleton of 'Henry Le

The body of John Torrington, one of Franklin's seamen exhumed on Beechey Island in 1984.

Vesconte' was examined when it was exhumed during renovation work at Greenwich. Facial reconstruction showed that it was more likely to be that of Harry Godsir, the expedition's doctor. His bones also showed signs of elevated lead, but none of scurvy, the lack of which casts doubt on the suggestion of scurvy in other bones.

Lead could have derived from the filtration system used to convert seawater to drinking water or, perhaps more likely, from the solder used to seal Goldner's food cans. Canning was a new technique in the 1840s and the poisonous potential of lead was not understood. Another suggestion is that the food within the cans was itself contaminated because of poor hygiene at the canning factory; one writer has even suggested botulism as a cause of death. Certainly the death rate was very high and the preponderance of officer deaths also suggests a can-based mechanism: as a rule officers ate better than crew, and the canned food might have been considered superior fare. Roald Amundsen also related a story told to him by Inuit during his passage transit, that Inuit retrieving cans of food from the Franklin trip died after eating the contents. However, hundreds of empty cans were left on Beechey Island – botulism is staggeringly toxic, and the cans were large and would have fed many men at once. If the cans were really infected with botulism then hundreds of cans could not possibly have been opened as the expedition would have been swiftly wiped out.

An alternative scenario, one supported by Inuit stories and now gaining credibility suggests that with men dying, perhaps from a combination of lead poisoning, contaminated food (but not botulism) and (perhaps) scurvy, Crozier (in command after Franklin's death) decided to abandon the ships. He brought as much as possible ashore, then set out south. At first his crew managed to shoot a lot of game (after the Franklin expedition the Inuit abandoned King William Island because its animal life had been exterminated), but eventually the food ran out. One group of men now tried to regain the ships, dying along the way – this scenario would explain why the boat found by McClintock's teams was pointing north not south – while others continued south, driven eventually to cannibalism.

But this long-held view of the tragic end of the expedition has been called into question by the discovery of both the *Erebus*, in 2014, and the *Terror*, in 2016. *Terror* was found in a bay off King William Island's west coast, well south of where the two ships were assumed to have been abandoned, while the *Erebus* was found even further south, close to Kirkwall Island off Adelaide Peninsula's western coast. In each case the ships were lying on the sea bed. *Terror* seems to have prepared for winter, suggesting both ships had originally been reached and sailed south, but the crew then abandoned *Terror* and continued in *Erebus* which again became ice-bound causing the remaining expedition members to abandon her for the second time. The Inuit had long told of finding the ship, with one dead man on board, and that it had sunk when they cut into it to plunder its contents and let in the sea: finding *Erebus* seems to have confirmed their story. But while this account of the two ships seems much more plausible, mysteries remain: why did Crozier go south when Fury Beach, where there were still supplies, was closer? Why were so many luxury items hauled by a sick, starving crew? And why did the survivors, if they were heading for the Great Fish River, chose to cross Simpson Strait at its widest point rather than at its narrowest? The latter offers the intriguing possibility that the very last survivors may have been heading for Repulse Bay, and there are tantalising Inuit tales of white men surviving for many years, and even that some almost made it to Hudson's Bay Company forts.

When McClintock returned, Lady Jane Franklin ensured that the Westminster Abbey memorial to her husband, and another in Waterloo Place, London, bore inscriptions stating he was the discoverer of the North-West Passage. If the discovery requires only the identification of a waterway then it was John Rae, who made that discovery in 1846. However, the recent discovery of one of Franklin's ships to the south of King William Island suggests that one ship was indeed sailed to Kirkwall Island, a little way north, and then drifted south in the ice before sinking. That would mean Franklin's expedition may also, unwittingly and unknowingly, have found a passage: McClure's discovery of the northern passage post-dated both. But if the requirement is to have completed a navigable route rather than merely identifying it, then Norwegian Roald Amundsen wins the race.

## Amundsen and the completion of the passage

The Norwegian Roald Engebreth Gravning Amundsen (1872–1928) was brought up on the outskirts of Oslo (then Christiania), acquiring the skills of skiing and seamanship at an early age. He was on the *Belgica* which carried the first expedition to overwinter in Antarctica, completed the first ship transit of the North-West Passage, led the first team to the South Pole, made the second west-east transit of the North-East Passage, and made a successful North Pole flight (being, perhaps, among the first to see the pole). It is a list which gives him a strong claim to the title of the greatest of all polar explorers.

Amundsen had learned well from the early accounts of the British in the Arctic. He realised that safety lay in small, shallow-drafted, highly manoeuvrable ships, in fewer men, and in adopting Inuit methods of dress. In 1901 he bought *Gjøa* (pronounced *you-ah*) a tiny (47-tonne) fishing boat. Refitted and with a crew of six, and a number of dogs, he started from Oslo for the Canadian Arctic in June 1903. *Gjøa* went through Lancaster Sound to Beechey Island, then headed south through Peel Strait to the channel between the Boothia Peninsula and King William Island. Amundsen found the channel shallow and shoal-filled – had Franklin also found this? – and *Gjøa* ran aground damaging her keel. The ship was refloated, but grounded again, this time more seriously. Amundsen decided to abandon ship, but Anton Lund, ship's mate, suggested jettisoning cargo to reduce the draft. This worked and the ship floated free.

*Gjøa* rounded the southern tip of King William Island, but even though Simpson Strait was ice-free, winter was approaching so Amundsen stopped at what is now called Gjøahaven, a Canadian Historic Park. During the winter Amundsen attempted to sledge to the North Magnetic Pole, but failed. The sledging took a lot of time, and a second winter was passed at Gjøahaven. Then, on 13 August 1905, *Gjøa* left its harbour and sailed west. On 26 August *Gjøa* met a US whaler off the southernmost point of Banks Island. Further east, near Herschel Island the crew were forced to overwinter for

In the past Bellot Strait was very rarely traversed except by ice-breaker, though the predicted decrease in sea ice coverage and thickness will allow easier future access. This photograph was taken during a west-east transit. To the right is the northern end of the Boothia Peninsula. Ahead is Zenith Point, the actual northern extremity of the North American continent.

The *Gjøa* today. It is now out of the water, outside the Fram Museum at Bygdøy across the water from Oslo.

a third time. In October, using dog-sledges Amundsen travelled south and formally announced his completion of the passage, though technically it was not completed until 1906. The ship reached San Francisco on 19 October 1906 where it stayed until 1972: it was then returned to Norway to stand close to *Fram* on Oslo's Bygdøy museum's site.

Not until 1940 was Amundsen's traverse repeated. Various firsts followed – first west to east transit, first in one season (both by Canadian Henrik Larsen in the

*St Roch*); first submarine transit (USS *Skate* in 1962, travelling east-west; first commercial transit (the 155,000 tonne US tanker *Manhattan* in 1969); first single-handed transit (Dutchman Willy de Roos in the 13m ketch *Williwaw* east-west, 1977). The first commercial passenger transits were made in 1984 and 1985 (east-west, then west-east). Today there are regular passenger trips and, with the possibility of the Arctic ice reducing due to global warming, there are again whispers that a commercial route might become a reality.

The statue of Leifur Eiriksson outside the Hallgrimskirkja, Reykjavik, Iceland.

# 9 Greenland

After the failure of the Norse settlements there was no contact between Europeans and the Greenland Inuit for more than 50 years. An expedition searching for Norse survivors sponsored by King Christian I of Denmark in 1472 or 1473 contacted east coast Greenlanders, but it was a hostile meeting. There was sporadic contact and bartered trade during the 17th century, with several Greenlanders being abducted and taken to Denmark and Norway: none survived for long, succumbing to disease, climate and the hopelessness of their situation. The oral tradition of the Greenlanders was such that when Hans Povelsen Egede, a missionary from Bergen, arrived in 1721 the native people he met could tell him the names of the abductees. Egede was searching for Norsemen and seeking to convert the locals. The conversion of the locals was successful, but smallpox brought from Copenhagen by a young child taken for a short stay in Copenhagen killed 25% of the Greenland population.

Egede had hoped that Greenland would be settled by worthy Christians, but when the Danish king decided to reinforce his sovereignty with settlers, it was convicts and 'women of easy virtue' who settled Godthåb (now called Nuuk), their behaviour being so appalling that Egede had to be protected from the Greenlanders, who held him responsible for the newcomers. Scurvy decimated the Nuuk colony and fortunately later settlements were better controlled. The climate and limited agricultural land also meant that it could sustain far fewer people (than, for example, Australia) and consequently native Greenlanders were never overwhelmed and marginalised and have maintained a large measure of control over their country to this day, though the growth of settlements did result in poverty and misery in many places initially (and in the longer term in some communities).

## Crossing the Inland Ice

Greenland's coastlines are substantially different, the

The rune stone dicovered in north-west Greenland.

west having a more expansive coastal plain, the east coast being more rugged, with fewer native settlements which were also smaller and poorer. The west coast was explored early by Europeans, the east coast having only sporadic visits by whalers with full explorations not being completed until the 20th century. Exploration of the interior, a vast ice sheet covering almost 2,000,000km², began tentatively. In 1751 Lars Dalager, a Danish trader, and five Inuit penetrated about 15km from near Paamiut. A century later (the 1860s) Edward Whymper, conqueror of the Matterhorn, made two attempts but barely got out of sight of the ice edge near Ilulissat. Next, in 1870, Adolf Nordenskiöld, later to make the first transit of the North-East Passage, led a team of four 57km inland from Auleitsivik Fjord, south of Disko Bay. He returned in 1883 and, from the base, penetrated 116km. From a camp there, two Sámi members of his team skied on. When they returned 2½ days later they claimed to have reached 42°51'W, having travelled 230km. But they had no means of measuring distances, apart, apparently, from optimism and a misplaced view of their capabilities: modern opinion favours a turning point at about 46°W, some 100km from Nordenskiöld's camp. The two men reported seeing no exposed land during their whole journey, surprising Nordenskiöld, who believed that Greenland had an ice-free, perhaps even wooded, heart.

In 1886 Robert Peary claimed to have pushed 160km inland, but this is viewed with scepticism. However, there is no doubt about the achievements of the next great explorer to arrive in Greenland. In 1888 Fridtjof Nansen, one of the greatest polar explorers, most famous for his *Fram* expedition, arrived on Greenland's east coast. Interestingly, given that later Amundsen beat Scott to the South Pole in part because of his use of dogs, Nansen decided against their use. He had no experience with them. He considered using reindeer, as the Sámi did, but Greenland seemed no place for a vegetarian animal. He therefore decided to ski and man-haul sledges. He did take a pony, but that seems to have been as meat on the hoof rather than beast of burden. He also decided to travel east-west. Both seemed odd choices given the history of British attempts to haul ships' boats and the fact that Greenland's west coast was more populated. Of his projected route, Nansen said it cut off the possibility of retreat – it would be 'death or the west coast'.

Nansen took a team of five expert skiers. The plan had been to start from Sermilik, but difficult ice meant the expedition set out from Umivik to the south. On 15 August, after safe-guarding the boats and leaving some supplies (in case 'death or the west coast' proved a questionable slogan – sensation is one thing, stupidity quite another) they set off. Reaching the inland ice involved a hard climb of 2500m, but once on the flat ice sheet they could ski. Man-hauling was hard work, the weather was poor, the ice monotonous and the psychological burden of being first must have been considerable. However, by mid-September they began to descend off the ice, and could rig sails on their sledges to use katabatic winds to aid progress (see Box on p61). On 24 September they were off the ice, for the first time in 40 days: they had skied 560km. They then constructed a makeshift boat and sailed the fjord to Nuuk, arriving in early October.

Nansen's crossing was a bitter blow to Peary who felt cheated, his 1886 trip giving him proprietorial rights. He responded by taking an expedition to NW Greenland in 1891. After overwintering a small team set out for the inland ice. Peary used dog sledges, having learned much from the Inuit (he also built igloos for sleeping) eventually reaching Independence Fjord. Peary believed this to be a channel (Peary Channel) separating Greenland from a more northerly island (Peary Land), a view which later was to have tragic consequences.

In the following years, Peary made other journeys, including another to Independence Fjord. But these trips concentrated on other issues. Peary searched for the iron deposits that were the source of the Greenlanders' metal. What he found, on an island in Melville Bay, was several large, iron-rich meteorites. These he appropriated (stole?) transporting them back to the US and (it is said) selling them. As the only source of local iron this was a

Drawing from Nansen's book on his Greenland crossing. This gives an indication of the vastness and emptiness of the Inland Ice, and of the team's isolation during the expedition.

remarkable theft by someone who learned so much from the people. But Peary's attitude to the locals was as proprietorial as that to the Inland Ice and, later, the Pole. He once asked of the Inuit 'of what use are they?' and answered himself by suggesting that they existed only to help him discover the pole. Peary felt that taking the meteorites, which had supplied the Inuit with iron for generations, was acceptable, since he now gave them all the iron goods they required. Peary also took six Greenlanders back to the US on one trip. They were displayed as circus freaks. Four died quickly, but one, Minik, was an eight-year old boy. His father had been one of those that died early, and the boy attended what turned out to be a fake funeral, his father's skeleton being displayed in a glass case. Minik campaigned for the return of the bones to Greenland: they were never returned. Minik himself did go back, but unable to hunt or speak the language he returned to the US where he died aged 28. Peary also sold Greenlandic corpses to the American Museum of Natural History for study, obtaining them by digging up fresh graves, and took images of live Greenlanders, male and female. His interest in doing so seems not to have been entirely anthropological as he fathered children by an Inuit woman, as did his black 'valet' Matthew Henson: their part-American, part-Inuit descendants still live in north-west Greenland.

Today all this seems very distasteful. And it is of course, but should perhaps be seen in the context of the times with Peary's attitude towards 'natives', both Arctic and those of the American West, mirroring that of society as a whole. It is best, therefore, to end with Peary's 1900 expedition which followed the northern coast reaching and naming Cape Morris Jesup (mainland Greenland's most northerly point, at 83°39'N) which effectively disposed of the idea that Greenland might extend all the way to the Pole.

## The Danes complete the coastal map

Peary's exploration left only the north-eastern coast of Greenland unexplored. A Danish expedition to fill the gap set sail in 1906. From a base at Danmarkshavn (at about 75°N), after overwintering, a series of supply depots was established and then two teams set out to survey the area. One went west to find the Peary

Tukemeq Peary, Robert Peary's grand-daughter, photographed at her village of Querqertat on Inglefield Bredning, north-west Greenland. The man beside her is wearing traditional Polar Bear fur trousers.

Channel, the other headed north, passing Independence Fjord. When they met later the first team reported that the Channel did not appear to exist, but not knowing the geography of Independence Fjord, the first team of three felt obliged to check.

When the team had not arrived back by September search parties went out: nothing was found. After another winter the search was resumed in spring 1908. Eventually they found one body, and a note saying the others had died when winter had caught them unprepared. Determined to find the other two bodies the Danes searched again in 1910. After overwintering a team of two – Ejnar Mikkelsen and Iver Iversen went north in 1911. They found the two bodies, and a note saying that Peary's Channel did not exist: there was no Peary Land. The two men's return turned into an epic struggle against appalling weather. At one point Iversen was so desperate for food that he asked Mikkelsen to carry their rifle, as he feared he might shoot and eat him. Eventually, 30km from Danmarkshavn, the two men had to leave everything behind as they no longer had the strength to carry anything except their own weight – and barely that. As they approached the hut they had

At Brattahlíð the ruins of Eirik the Red's farm can be seen. It is usually claimed to be the first settlement in Greenland, though in recent years that claim has been challenged. There is also a reconstruction of what is believed to have been the first Christian church built in the Nearctic. The church is named after Eirik's wife Þjóðhildur. Legend has it that Þjóðhildur was a devout Christian and that she withdrew Eirik's conjugal rights, stating that they would only be restored when he was baptised. Eirik was a commited pagan and resolutely declined. In time the position became unsatisfactory for both parties and a compromise was reached, Eirik building his wife a church in exchange for a return to the marriage bed. His only stipulation was that the church should be built out of sight of the farm.

to have a rest less than 50m from it as they were too exhausted to walk those last few steps. But the ship that had brought them was gone, and a note in the hut they found there said their colleagues had gone too.

The two men survived the winter by hunting, and by eating the supplies left at the hut. In the spring, having retrieved the records they had left 30km away the men were too weak to attempt the journey south to the Inuit settlement at Scoresbysund to the south. They waited all summer for rescue. It did not come. After another winter things were looking grim; they were short of food. But this time a rescue ship did arrive.

Mikkelsen and Iversen's survival against almost overwhelming odds is arguably the greatest in the history of polar exploration. Others have endured similar, perhaps worse, conditions, but never for as long; the pair also managed to hang on to the records that completed the mapping of Greenland.

The discovery of Peary's errors was confirmed in 1912 by the first expedition in a series that would become legendary – the Thule expeditions led by Knud Rasmussen, born in Greenland of Danish parents, a remarkable series of trips which combined adventure, exploration and anthropology. Also in 1912 a scientific expedition (the team included Alfred Wegener, who had also been on the 1906 expedition) crossed the inland ice westwards from Danmarkshavn to Laxefjord, a distance of 1,100km, about twice as far as Nansen's transit. The team of four also overwintered on the Inland Ice, the first men to do so: Wegener was later to die in another overwintering (1930–31) on the ice, exhaustion overtaking him as he struggled to get off the plateau.

## Eirik Raudes Land

At about the same time as these first winterings on the inland ice, Greenland was at the centre of an international court case. Norwegian fur-trappers had been overwintering on Greenland's east coast since the early years of the 20th century, their presence fuelling a dispute that had rumbled on since Norway had transferred from the Danish to the Swedish crown in 1814. The dispute was over the sovereignty of Greenland as Eirik Raudes (Eirik the Red) had been born in Norway. On 27 June 1931 four men raised the Norwegian flag at the Myggbukta trapping station on east Greenland and Norway claimed Eirik Raudes Land, encompassing east Greenland north of the Inuit settlement of Illoqqortoormiut (Scoresbysund). Ownership of Eirik Raudes Land was contested at the International Court in The Hague in 1933. The Danish claim was supported by Greenland and by the United States, whose foreign policy was still dominated by the Monroe doctrine.

The court found in favour of Denmark. Today, though Greenland has been granted Home Rule, the Danes still maintain a military presence at Daneborg on the north-east coast (within 'Eirik Raudes Land') and annually patrols the east and north coasts by dog sledge (the Sirius patrol) to reinforce its sovereignty.

## Recent journeys on Greenland

Greenland's vast ice sheet has been a magnet for adventurers ever since the first traverse by Nansen. Today, crossings of the inland ice are organised by commercial companies, but were once virtually a rite of passage for polar explorers. The Inland Ice has been crossed south-north and north-south along its long axis, while the coast has been circumnavigated by two men who used kayaks and dog-sledges in an epic multi-year (though not continuous) journey. But these are only some of the journeys made, and doubtless there will be countless more in the years ahead.

An old Norwegian fur trapper's hut at Moskoksfjorden, north-east Greenland.

The *Norge* airship mast at Ny Ålesund, Spitsbergen, Svalbard.

# 10 To the Pole

It is debatable which expedition has the right to be termed the first to attempt to reach the North Pole, as many of the early travellers, Henry Hudson, for example, assumed that to reach Cathay it was necessary merely to sail north over the pole, the pole itself being incidental. Arguably, the first to sail north with the specific intention of reaching the pole was the Englishman Constantine Phipps in 1773, a trip memorable for the near loss 14-year-old midshipman Horatio Nelson to a Polar Bear.

## Early attempts

William Parry, hero of several attempts at the North-West Passage, was next to try in 1827, though his main aim was to win a reward for reaching 83°N – his crew dragging ship's boats north from Spitsbergen – but failed at 82°45'N.

The next attempts were American, Elisha Kent Kane leading an expedition in 1853, and Isaac Israel Hayes another in 1860. Each did good exploratory work but did not push the record northing further. The Civil War prevented further American efforts until 1871 when Charles Francis Hall, already noted for believing he had been chosen to rescue Franklin's men, now believing he 'was born to discover the North Pole. That is my purpose. Once I have set my right foot on the pole, I shall be perfectly willing to die'. It was a partially prophetic comment, Hall dying, perhaps murdered, at wintering quarters on Greenland: his expedition broke up in confusion and near disaster.

The British returned in 1875, George Nares taking an expedition to the American Arctic. But he had learned nothing from Parry's attempt, and again man-hauled heavy ship's boats across the ice. This was slow and time-consuming, and frustrating as the ice often moved south faster than the men moved north. Nevertheless, a new northing record (83°20'26"N) was established. It lasted two years, being broken by the American Adolphus

An illustration from Parry's book on his North Pole expedition. It is entitled *The boats drawn up for the night* which, together with the composition, was perhaps an attempt to impose a pleasant domesticity on to what was in reality an unpleasant experience for the team.

Hall's body, exhumed in 1968 for autopsy. The US flag in which Hall had been wrapped had stained his body. Though humour has little place in a possible murder, it is difficult to avoid a smile when hearing that Hall was laid to rest in a structure created by the ship's carpenter, Nathaniel Coffin.

The *Proteus*, the ship sent with supplies for Greely's expedition, sinking before anything could be unloaded.

*Fram* in the ice. The ships had both a steam engine and sails for motive power, and a wind turbine to provide electrical power.

Greely, whose expedition managed to go 6.5km further (to 83°24'N) heading north from NW Greenland. But the aftermath of the success was a disaster. Moving his men to Cape Sabine on Pim Island off Ellesmere Island's east coast, in the hope of meeting a relief ship, Greely discovered a note saying that the ship had sunk. Winter was a nightmare. There were two Inuit hunters, but one died of scurvy and the other drowned. The remaining team were forced to eat their buffalo-hide sleeping bags, heated by burning old rope. One man, who stole food, was executed after a makeshift court martial, the expedition doctor committed suicide with his own drugs, while others died of malnutrition and disease. In America, the government vetoed one attempt to launch a rescue party, but eventually relented. At the camp, spring had arrived so the survivors could eat lichen and flowers, but by the time relief arrived there were only seven of 25 left, and one of those died on the way home. Greely's book on the trip, with its harrowing detail, sold in large numbers, especially to the British: perhaps it offered some comfort to know that other nations could be just as bad at Arctic travel.

The *Jeanette*, already considered when dealing with the Russian Arctic above, was the next American attempt at the Pole, and the last before one of the most famous ventures, and one of the audacious of all Arctic expeditions.

## Nansen and the Fram

In Norway, Prof. Mohn's lecture on the *Jeanette* relics was read by the 23-year-old Fridtjof Nansen. He realised that such a current might take a trapped ship over, or very close to, the North Pole, provided the ship could survive entrapment it would be released near the coast of Greenland. What was needed was a ship strong enough to resist ice pressure and a team of men willing to spend perhaps five years on board. Nansen's ship was designed by Colin Archer, son of a Scottish immigrant to Norway, and was a work of genius. Named *Fram* – Forward – her cost was borne by the Norwegian government as an expression of national pride. She was large, 34.5m long on the waterline and grossing more than 400 tons. The hull was a half-egg in cross-section with a minimal keel and a removable rudder to avoid ice pressure pulling her down. The cross-beams and stern were huge and of well-seasoned oak to withstand ice pressure. The ship had both a steam engine

An exquisite drawing from *Farthest North*, Nansen's book on the *Fram* expedition. With the men's hats, the dogs and the rolls of equipment at the back, the illustration looks rather better suited to a child's book than one dealing with a life-and-death struggle in the Arctic. The harsh reality was, of course, that the dogs, here taking it easy as the kayaks (converted to a catamaran) took the strain, were all slaughtered and eaten.

and sails. She also had a wind turbine, which generated electricity for lighting.

*Fram*, with her crew of 12 plus Nansen, left Norway in June 1893 and sailed north to the New Siberian Islands to enter the pack. On 5 October 1893 the rudder was raised: *Fram* was frozen in. The ship drifted north through the winter, and then the summer of 1894 and into a second winter. By now it was clear that her direction was north-west rather than north. On 12 December 1894 she passed the record northing for a ship (set by Nares' *Alert*), but Nansen had realised she would never reach the pole and announced his intention of heading north with one companion and the dogs which had been taken. With Hjalmar Johansen he departed on 26 February 1895 with six sledges, 28 dogs and 1,100kgs of equipment, then again on 28th after returning to repair a broken sledge.

Nansen had thought that reaching the Pole would be easy, but the pressure ridges and biting cold slowed progress at 86°14'N, a new record by almost 3°; the pair were forced to turn south and headed for Franz Josef Land. Having failed to wind their watches, the men were now unsure of their longitude and were afraid of missing

land. Eating meat from slaughtered dogs, they sledged on. Eventually food ran short, but open water allowed the two kayaks they had taken to be lashed together to form a catamaran. They shot a seal, then a polar bear, and finally reached Adelaide's Island in north-eastern Franz Josef on 10 August. The two men continued to Jackson Island, but with winter now fast approaching they were forced to build a hut from stones, roofed with walrus hide.

On 19 May 1896 the two men refloated their catamaran and headed west again, but were still unsure of their exact whereabouts. In fact, they were close to Cape Flora on Northbrook Island which the Englishman Frederick Jackson was using as a base. Told that a man was approaching, Jackson went out of his hut and met the two Norwegians. The pair sailed south with Jackson, reaching Vardø, Norway on the 13 August to a hero's welcome. On the 20th they heard that the *Fram* had also arrived home: it had reached 85°56'N in November 1895, but then drifted south again. She was released from the ice in August 1896. After calling at Svalbard to see if there was any news Sverdrup took the ship on to Norway, arriving on 20 August just a week after Nansen and Johansen.

The site of Nansen and Johansen's winter camp at Kapp
Norvegia, Jackson Island, Franz Josef Land.

Fredrick Jackson, whose team Nansen and Johansen
had the good fortune to meet, was himself making
a (rather unconvincing) attempt on the North Pole.
Jackson took both dogs and ponies on his expedition,
noting that the latter suffered from the cold and
frequently fell belly-deep in show because of their small
hooves. Nevertheless he concluded that the ponies
were 'an unqualified success', a summary thought
to have influenced both Robert Scott and Ernest
Shackleton.

## Andrée and the Eagle

Almost as famous, though for tragic rather than heroic reasons, was the Pole attempt of another Scandinavian, the Swede Salomon August Andrée. After taking an interest in balloon flying, and discovering that by using guide ropes trailing on the sea to slow the voyage, and sails to influence direction, Andrée thought that he could use a balloon to reach the Pole. The endpoint of such a flight could not, of course, be fixed, but Andrée assumed that given enough time in the air the balloon was bound to reach land somewhere.

In 1897, with a crew of three – himself, Knut Frœnkel and Nils Strindberg – Andrée sailed to Svalbard and on 11 July the balloon, named *Eagle*, was launched. There was an immediate problem, the guide ropes meant to keep the balloon close to earth detached and it rose to 600m. The balloon headed north-east and, an hour later, disappeared. In the days prior to radio, Andrée was reliant on pigeons and buoys (rather like messages in bottles) to dispatch news of his expedition. Only one pigeon was recovered: its message read 'July 13 at 12.30pm. Lat 82°02'N, Long 15°E, good speed towards east 10° south. All well. This is the third pigeon post. Andrée'. Five buoys were eventually recovered, but by then all hope for the expedition was long gone. Searches were made, but it was not until August 1930 when a Norwegian sealer anchored off the south-western tip of Kvitøya, the rarely visited island to the north-east of Svalbard's Nordaustlandet, that the truth was established. Men from the ship found a camp and the remains of the three balloonists. A later search revealed the diaries of the crew and, most remarkably, about 20 photographic negatives that could still be printed. On 12 July rain had caused the balloon to descend. One short guide rope became trapped in the ice for 13 hours, then broke free. The balloon rose, but ice formed on it and the gondola was soon being dragged across the ice again. By 7am on 14 July the flight was over: the balloon had travelled 830km in 65½ hours and had landed at 82°56'N.

A marvellous cartoon – surely the best ever produced. Swedish in origin (of course!) it shows Andrée's balloon being reeled down to the North Pole by a friendly Inuk while Nansen, in *Fram*, looks on in dismay. The men's respective countrymen are suitably appalled (the Norwegians) or delighted (the Swedes). Exactly where the Inuk and his cog wheels were going to appear from or indeed, how Andrée was actually going to get to the North Pole, or back if he succeeded in doing so, has been ignored.

One of the most poignant photographs in Arctic history from the exposed negatives found on Kvitøya. It was taken immediately after the *Eagle* had landed on 14 July.

The diaries and photos tell of the three men hauling three sledges, on one of which was a boat. They headed for Franz Josef Land, but changed plan when they realised they were moving too slowly. They then became sick with stomach cramps and severe diarrhoea. On 5 October they reached Kvitøya, but now winter had arrived. Diary entries stopped in mid-October. The cause of death of the men has been the subject of speculation ever since the discovery. From the photos and diaries it is clear Polar Bears were shot for food. Analysis of meat samples discovered at the camp showed the presence of parasitic nematodes called *Trichinella*. If the meat had been eaten raw or poorly cooked the men could have developed trichinosis (which would explain the severe diarrhoea), a condition that can be fatal without treatment.

## Peary and Cook

The next attempts to reach the Pole were from Franz Josef Land. The American Walter Wellman led a team to 81°N in 1894, then an Italian-Norwegian team led by the Duke of the Abruzzi, tried in 1899. After overwintering the attempt, a team of four led by the Duke's deputy, Umberto Cagni, supported by three supply teams travelled 1,200km and reached 86°34'N, a new record. On his return, Cagni claimed that the journey over the sea-ice was too difficult and that future attempts should be made from Greenland. But two further expeditions financed by New Yorker William Zeigler, tried, neither getting past 82°N

We have already met Robert Edwin Peary (1856-1920) while exploring the history of Greenland. In letters to his mother after those trips he writes 'I <u>must</u> have fame' and 'Fame, money, and revenge goad me forward till sometimes I can hardly sleep with anxiety lest something happen to interfere with my plans', views which should be borne in my mind when considering his later claims and behaviour.

In the long expedition of 1898–1902 Peary had twice attempted to reach the North Pole from Ellesmere Island, his northernmost point in 1902 (84°17'N) being a record in the western Arctic. The attempts seem curiously tentative, at odds with his obsession with fame and the pole; though ice conditions and the cold in 1902 had contributed to the limited achievement, the attempt had lasted only 16 days. His 1901 journey seems lacking in conviction, an opinion apparently shared by his backers who sent a Dr Frederick Cook to examine him. There is doubt as to whether Cook did examine him, though he later claimed to have suspected pernicious anaemia and, looking at his feet had told him 'you are through as a traveller on snow on foot'. Peary was then 46. It would not be the last time that Cook and Peary would clash.

Whatever the truth, Peary set out again in 1905. He sailed to northern Ellesmere and from there early the following year started to lay supply dumps to the north. But by early April it was clear that ice conditions and bad weather would thwart a concerted effort on the Pole. Peary therefore tried a dash north and declared he had reached 87°06'N, bettering Cagni's record, though his daily travel distance claims are now viewed with

suspicion. From his northernmost point he turned south towards Greenland. Only his excellence in sledging and his knowledge of Greenland allowed the expedition to return safely.

In September 1909 an astonished world was informed that the North Pole had finally been reached – not once, but twice. On 2 September Dr Frederick Cook (1865–1940) announced by way of a telegram office in the Shetland Islands (where the Danish supply ship taking him from Greenland to Denmark stopped briefly) that he had stood at the pole on 21 April 1908. Then, on 6 September Robert Peary used a similar office in Indian Harbour, Labrador, to say that he had reached the pole on 6 April 1909. Each had friends in high places, and within days the *New York Herald*,

Photographer Per Michelsen, one of the first to visit the site in modern times, inspects the Andrée crew's camp site.

which had backed Cook, and the *New York Times* and National Geographic Society, Peary's backers, had declared war. It was a dirty war, one in which the reputations of both men were tarnished beyond redemption, and one which, a century on, shows no signs – nor has much chance – of ending in a truce, honourable or otherwise.

Cook's claim was the more remarkable as he claimed to have pioneered a new route. After overwintering in NW Greenland Cook and two Inuit, Ahwelah and Etukishook, two sledges and 26 dogs, headed west to Axel Heiberg Island, then turned north. They reached the Pole on 21 April, having travelled about 800km in 34 days. On the return journey, ice drift pushed them west and persistent fog and poor weather prevented them from calculating their position. They overwintered again, reaching their starting village in Greenland after 14 months away.

Peary's final Pole attempt did not start in 1907 as he had hoped, his ship needing a refit. He started out in 1908, and after overwintering on northern Ellesemere, set out for the Pole on 28 February 1909 with 24 men, 19 sledges and 133 dogs. A huge lead stopped him for six days, but progress was otherwise steady. On 1 April Peary sent the last support team back (from 87°45'N). Peary's team now comprised Matthew Henson, his black personal assistant and four Inuit, with five sledges and 40 dogs. This team reached the pole at about 1pm on 6 April. On those last five days they had averaged almost 50km/day (straight line distance). On the first 31 days they had averaged about 17km/day (straight line distance again). Peary remained at the pole for about 30 hours, then raced back to his base, arriving on 27 April, three days after his last supply team which had travelled at least 490km less. On 17 August Peary heard that Cook was claiming to have beaten him to the North Pole, but on Labrador went ahead with his announcement.

Cook was feted on his arrival in Copenhagen, but things rapidly turned sour for him. On his journey south Peary spoke with the two Inuit who had accompanied Cook and claimed they told him they had never been out of sight of land. But in the United States Peary's vitriolic attacks on Cook, in telegrams and to the press, had the opposite affect to that intended, rapidly drawing sympathy for Cook. In several polls

The Pole photographs of Cook (above) and Peary (below). The Peary photograph was hand-tinted after his arrival back in the US rather than being a colour shot.

public opinion was 80% in Cook's favour, often higher. Outside the States the less heated atmosphere allowed more sober judgements, and these tended to favour Peary. Cook's position was made much worse when his claimed ascent of Mount McKinley (the highest mountain in the USA) was declared fraudulent. In the end Peary won over the majority, often grudgingly as Cook was a much more amiable man, a complete contrast to the blustering Peary who lacked Cook's social skills. After his imprisonment (for oil stocks fraud: Cook served seven years of a 14 year sentence), Cook's reputation as a liar and fraud remained, despite the fact that he was pardoned in 1940. But to the end he maintained the validity of his claim to have been the first to the pole. Recording a tape for posterity, his final words were 'I state emphatically that I, Frederick A. Cook, discovered the North Pole'.

## Summarising the evidence

Millions of words have been written on Cook and Peary. At the time the dispute was simple – which of the two was first to reach the North Pole? Now the question is different – did either of them? Peary's claim is based on extraordinary rates of travel during the days after his last support was sent back (ironically, Peary's supporters used Cook's claimed rate of travel of 25km/day as evidence of his fraud when compared to rates achieved by earlier travellers, an argument they hastily dropped when Peary's account, with claimed rates of 50km/day, was published). Most of those who have sledged to the pole consider Peary's rates are unfeasibly (even ludicrously) high. In a carefully considered book Wally Herbert judged that Peary did not reach the pole, his claim being immediately rebuffed by a 'scientific study' commissioned by the National Geographic Society, Peary's staunchest supporter.

But much more important than his claimed travel rates was the fact that by his own admission, Peary made no measurements of his longitude, took no measurements of magnetic variation and made no allowance for ice drift. This gave even his staunchest supporters pause for thought, particularly those who had experience on ice. Peary's claim to have gone north along the 70°W meridian is at odds with all experience of ice movements and, unsupported by longitude readings, stretches credibility. If he really did travel by dead reckoning as he claims, then at his final camp he had no idea where he was. The strange omissions from his diary on such a momentous trip are also curious.

But if Peary's claim lacks credibility the situation is no less problematic for Cook's supporters. There is the curiosity of cropped photographs that appear to show land where none should be. There is the contradictory testimony of his Inuit companions; when interviewed by others, Cook's Inuit companions at first backed-up his claim to have reached the 'Great Navel' as the Inuit call the pole, but they are later said to have stated that they had never been out of sight of land. Cook's supporters make much of the earlier statements, claiming that the later ones were made under duress – the Inuit were often accused of telling the white man what he wanted to hear. Cook also claimed to have told the two Inuit that they were never far from land to calm them as they feared being far out on the sea ice. Yet it seems that Cook's two Inuit told a consistent tale throughout the rest of their lives, a tale that, though ambiguous, did suggest that they believed they really had not gone far from land.

At this remove in time the truth of the two claims can no longer be ascertained. While in general the polar environment does a remarkable job of preserving objects left either deliberately or casually, the nature of the Arctic Ocean precludes such survivals. No new evidence is likely either from the north or from the diaries and logs of the two men and their expeditions. No one will ever know for certain which, if either, was telling the truth. Overall it is probable that Peary got close to the pole (probably within 150km), but did not reach it, defeated by his own navigational naiveté and incompetence. It is likely, too, that Peary knew he had failed. Cook's claim is more intriguing. There are seemingly compelling reasons for discounting it, not least the fact that by his own admission Cook was a novice navigator – how could someone incapable of plotting longitude and having difficulty with latitude possibly know where he was on the shifting ice of the Arctic Ocean? Yet equally compelling is the evidence that suggests he did indeed travel a long way out across the ocean towards the pole and so, perhaps, might have reached it. Intriguingly, in their early retelling of the journey before Peary's supporters interviewed them, Cook's two Inuit companions gave a very accurate description of how the sun moved at the pole, something they had never seen before. That curious movement only occurs at (or very close to) the pole.

## Airships and aeroplanes

Throughout the period 1906–1909, voyagers continued to attempt to reach the pole by air. Walter Wellman, used an airship from Spitsbergen several times without success. The 1914–18 War halted further efforts, and by the time the world had recovered technology had moved on so when Roald Amundsen tried in 1925, with the help and finance of Lincoln Ellsworth, they could use planes – seaplanes so they could land on both sea and land. After a near disaster in 1925 the two tried again in 1926 but in an airship. However, while they were preparing at Ny Ålesund, Spitsbergen, Richard Byrd of the US Navy arrived with his plane, a Fokker tri-motor named the *Josephine Ford*, and his pilot Floyd Bennett. The pair took off in the early hours 10 May: they were back 15½ hours later claiming to have reached the pole. However, in 1960 a Swedish meteorologist checked timings and concluded that they were unlikely to have gotten past 88°N. It also emerged that Bennett had told a friend that the plane had developed an oil leak early in the flight so the pair had just flown around for many hours before returning to Ny Ålesund.

At 1am on 11 May Amundsen's airship, the *Norge*, took off. The crew included the Italian Umberto Nobile, and Oscar Wisting who had been with Amundsen at the South Pole. At 1.10am on 12 May the two men became the first to have definitely seen the North Pole, and the first to have seen both poles. The *Norge* continued to Teller, Alaska. The flight had been a major success, but its aftermath was ugly. To their annoyance, Amundsen and Ellsworth discovered that Nobile had, in secret, persuaded the Norwegian backers of the expedition to add his name as co-leader. In Italy, Mussolini promoted Nobile to general and ordered him to lecture to the 'Italian colonies' in the United States. A large crowd of Italian-Americans gathered at Seattle when the expedition team arrived; Nobile, in military uniform, made the fascist salute and was feted, while Amundsen and Ellsworth were virtually ignored. Nobile's lecture tour – in which he claimed to have both masterminded the expedition and piloted the airship, neither of which was correct – creamed off much of the available audience (and their entrance money), leaving both Amundsen and Ellsworth short of cash.

Ironically, Nobile eventually believed his own propaganda about being a major explorer, a delusion that would lead to tragedy and humiliation with the later, *Italia*, expedition. The *Italia* on an exploration flight from Spitsbergen crashed on the ice, detaching its gondola: the ship rose rapidly taking six men to their doom. Of the 10 in the gondola, one was killed. Rescue attempts involved 18 ships, 22 planes and 1,500 men of six nations. The first plane to land at the gondola site retrieved only the injured Nobile and his dog. The news spread around the world rapidly – a fascist general had saved himself and his dog before his companions. A furious Mussolini demoted Nobile. Of the remainder on the ice, one died in suspicious circumstances and another died of his injuries. Most tragic of all, Roald Amundsen, anxious to be involved in the rescue effort despite being 56 (and looking older), and despite his feud with Nobile offered his services. The offer was declined by both Italian and Norwegian governments. The decisions embittered Amundsen. He felt his honour was at stake, so when the French offered him a Latham 47 seaplane for a private mission he immediately accepted. With five others he took off on 18 June: wreckage from the aircraft was found weeks later.

## Later trips to the Pole

On 23 April 1948 a Soviet aircraft landed at the pole. The team led by Mikhail Somov – the others were Pavel Sen'ko, Mikhail Ostrekin and Pavel Gordienko – became the first men to be confirmed as having stood there; it seems that Gordienko was the first man out of the plane. Then on 4 August 1958 the US Navy submarine USS *Nautilus* reached the pole on a sub-surface crossing of the Arctic Ocean. The following year, on 17 March, the submarine USS *Skate*, surfaced at the pole. Not until 1977 did a surface ship reach the pole, the nuclear-powered Russian icebreaker *Arktika* arriving on 18 August.

The first overground expedition to indisputably reach the pole was in April 1968, when a Canadian/US team led by Ralph Plaisted used snowmobiles to travel from Ward Hunt Island, off Ellesmere Island's northern coast. Four men reached the pole on 19 April, and were then taken out by aircraft. That same year the British TransArctic Expedition, a team of four led by Wally Herbert set out from Barrow, Alaska with 40 dogs and four sledges. After wintering on the ice they reached Svalbard in May 1969 after having crossed by way of the pole. Later trips to the pole filled in the some of the gaps in these tales of human endeavour. On 5 March 1978 the Japanese Naomi Uemura set off alone (apart from a dog team) from Ellesmere Island, reaching the pole on 29 April. In 1986 the Frenchman Jean-Louis Etienne made a solo ski journey to the pole, with air resupply every 10 days. The first unsupported journey was made in 1986 when a team of eight (one of whom was evacuated when his ribs were broken by a sledge), led by Will Steger and Paul Schurke, used dog teams hauling three tons of equipment. Then in 1994 Børge Ousland made a solo, unsupported journey from Cape Arktichesky at the northern end of Severnaya Zemlya. In each case return was by aircraft. Also in 1994 the Russian icebreaker *Yamal* made its first trip to the pole. Since then it, and other, icebreakers have taken thousands there. An out-and-back, unsupported journey was not completed until 1995, when it was achieved by Canadian Richard Weber and the Russian Mikhail Malakhov. Weber rightly noted that the early pioneers had not had the advantage of air evacuation from the pole, and that the pair's journey would therefore be closer in spirit to them. More firsts followed, each significant in its own way, particularly to those involved. Such trips will continue as the lure of the North Pole encourages the brave and the reckless northwards.

Polar Eskimo Mammarut Kristiansen preparing for a narwhal hunt near Thule, north-west Greenland.

# 11 Indigenous peoples in the modern world

The interaction of Europeans and the peoples of the Arctic followed the same general pattern of all such meetings between a group that was technologically advanced and saw itself as civilised and one that it considered 'primitive'. Ownership also caused a problem, northern peoples being baffled when they discovered that one of its consequences was citizenship, which meant that people you had never seen, who knew nothing of your land and lived far away could exercise authority over you and how you lived. Most dramatically, this ownership meant that those far-off people could grant the rights to exploit your land to others without consulting you. At the time this seemed fair: as John Quincy Adams, the sixth President of the United States once said 'What is the right of a huntsman to the forest of a thousand miles over which he has accidentally ranged in quest of prey?' To modern eyes this seems a monstrous injustice, but it was a common idea then and, it must be said, variations of the theme are still current.

One of the more tragic consequences of ownership was the forcible relocation of native people of Big Diomede to the mainland. The international border between the USA and the USSR ran between the Diomede islands, which are just 4km apart. Soviet perception that espionage between family members was a threat to their security resulted in separation for some families that lasted throughout the Cold War.

## Canadian native peoples

For the purposes of understanding how the treatment of northern native peoples developed, it is convenient to consider five nation groups – Canada, Denmark (Greenland), Russia, Scandinavia (Norway, Sweden and Finland) and the USA (Alaska).

In Canada, decisions over the Inuit population were complicated by the lifestyle of many Inuit families and an initial reluctance to accept responsibility for them. Only when it became clear that a watching world would be unimpressed if famines or epidemics were ignored did this attitude change. But the Canadian Arctic is vast, the Inuit population small. The only realistic solution was to concentrate the people in settlements where welfare could be administered, a solution that brought its own problems: local wildlife was hunted to virtual extinction

Thule Air Base, north-west Greenland. When the American air base expanded in 1953, the Greenlanders of Pitffik, the local village, were moved 250km north to Qannaq to make way.

The Rasmussen Museum, Illulissat, Greenland. The juxtaposition of the old vicarage in which Knud Rasmussen grew up (his father was the town's pastor) and the modern flats says much about modern Greenland where tradition and modernity live side by side, sometimes uneasily.

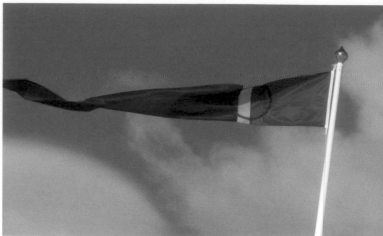

The Sápmi flag flying in northern Sweden.

and the desire for the accumulation of southern material goods – TVs, cars, etc. – without the work to support payment created dependency. Successive Canadian governments also had to contend with an increasing desire by the Inuit to control their own destinies and to have a greater say in the administration of their ancestral lands. In 1993 the Nunavut Land Claim transferred a vast area of land and substantial funds to an Inuit corporation. This was followed in 1999 by the creation of Nunavut, an essentially self-governing territory covering more than 2,000,000km² and constituting about 20% of Canada's land area.

## Greenland

Following the agreement on home rule in 1979 Greenland has effectively become an independent nation with a population of some 50,000, though Denmark retains control of international affairs.

## Scandinavia

In Scandinavia, the early relationship between southerners and the Sámi had been one of imposed assimilation, with enforced conversion to Christianity, the banning of Sámi languages and the execution of 'rebels' who resisted. Assimilation was not really questioned until the late 1960s and remained

government policy until a decade or so later. Not until the 1978 protests over the proposal to dam the Kautakeino/Alta River was the Norwegian government shocked into accepting that the Sámi had legitimate rights over their homeland. Even then it took another 11 years for this acceptance to evolve into the creation of the *Sámediggi*, a parliament with control over many aspects of governance of Sámi land. Similar parliaments were formed in Sweden in 1993, in Finland in 1996, but not until 2010 in Russia. In 2001 a Sámi Parliamentary Conference and Council, with representatives from the three Scandinavian parliaments, was formed. It will be interesting to see how the situation develops as the Council has shown enthusiasm for the creation of a single Sámi nation known as Sápmi and frequently refers to it. More recently (early 2016) in Sweden a district court returned the right to control local hunting and fishing to the Sámi village of Girjas, overturning a 1993 government decision that government alone had the right to determine such matters, independent of the views of an indigenous people. The government's opinion in court was that 'Sweden had no international obligation to recognise special rights of the Sámi people, whether they are indigenous or not', a view which was considered to be the 'rhetoric of race biology' by opponents. The government is considering appealing the decision.

## Russia

Under the Tsar, peoples whose primary means of life was Reindeer hunting were required to spend time hunting fur-bearing animals, the time lost causing hardship that was added to the abuses perpetrated by the incomers. Years of abuse and exploitation meant that by the end of the 19th century it seemed some of the native communities might become extinct. Communism initially aided the northerners by offering tax concessions and exemptions from military service. But collectivisation of reindeer herding with sales allowed only through the state office, which also sold life's essentials meant the herders were effectively tied to the system. Formerly nomadic peoples were forced into settlements as the new ideology abhorred nomadic and semi-nomadic lifestyles. Children were often removed from their parents and sent away to be educated. At distant schools they were forbidden to speak their own languages and taught an ideology intended to result in their integration into mainstream Soviet life. Later in the communist era the wholesale exile of 'dissidents' and criminals to Siberia made matters worse for the native peoples. The incomers often arrived stripped of all possessions; they felled trees to build cabins and grow crops. By the end of the 20th century the native peoples of Siberia constituted, on average, less than 5% of the population of their previous homelands.

During the late 1980s the USSR was in a state of near collapse and the peoples of Siberia, far removed from Moscow, were often left to fend for themselves as the country went into economic decline. But now, 70 years or more from their traditional way of life and without the safety net that socialism had offered, the people were often unable to adjust. Some of those who had moved to the north to aid the expanding communities, such as doctors and other professionals, moved out, making matters worse. In 1990, just a few months before the break-up of the USSR, the Russian Association of Minority Peoples of the North (RAIPON) was founded with the aim of addressing these problems. RAIPON was shut down by the Russian government in 2012

Inuit hunting camp on the sea ice of Baffin Bay. The Inuit were hunting Narwhal at the floe edge. In the background are the mountains of northern Baffin Island.

because of an alleged 'lack of correspondence between the association's statutes and federal law.' It was reopened in 2013 after statutes had been adjusted, and continues to function.

## Alaska

Following the purchase of Alaska from the Russians, the US government all but ignored the state, selling fur-trading rights to outsiders and then, during the Gold Rush, allowing a general lawlessness that further aggravated the condition of the native peoples. The treatment of the native Aleuts was arguably the worst meted out to any Arctic native population, and the change of ownership of Alaska did not put an end to the suffering, in part because the Americans made them 'wards' of the government rather than citizens of the United States, a status that was not corrected until 1971

despite Alaska becoming the 49th State of the Union in 1959. During the 1939–45 War the Japanese bombed Dutch Harbour and occupied the western islands of Attu and Kiska. Fearing an attack on mainland USA by way of the island chain, the Americans evacuated the remaining Aleuts to inadequate housing on the Alaska Peninsula while they fortified the islands. The Aleuts in south-east Alaska had a miserable time, with as many as 10% dying of malnutrition and disease. Some never returned to their island homes, but those that did found their houses had been looted and burned by the occupying American troops. Churches and houses had also been used for artillery target practice. The final settlement for these abuses was not paid until 1988; by that time there were only some 400 living survivors of the evacuation. Despite this tragic history, the Aleut population as identified in US census returns is now 24,000: estimates of the population before the Russian

Caribou spring migration, Courageous Lake, Canada's Barren Lands.

discovery of the islands is 10,000–20,000 implying either a remarkable rise in such a short time, or a number of unjustified claims.

In 1971, faced with demands from indigenous folk for restitution of their lands or significant compensation for the loss, and from settlers for the right to continue to exploit lands for the benefit of the state and the country as a whole, the US Government passed the Alaska Native Claims Settlement Act. This rejected all claims to original ownership of land by the indigenous population in exchange for a sum that totalled almost one billion dollars, and packages of land that amounted to 180,000km². No future claims on land ownership were allowed.

## The Arctic Council

While the treatment of the native peoples of the Arctic by all the Arctic nations has been, at best, poor, the setting up of the Arctic Council in 1996, largely at the behest of the Canadian government, has acknowledged their rights to a say in the development of what was once their land. The Council includes representatives of the eight Arctic nations, together with six indigenous organisations – the Aleut International Association, the Arctic Athabaskan Council, the Gwich'in Council International, the Inuit Circumpolar Council, the Sámi Council, and the Russian Arctic Indigenous Peoples of the North – as 'Permanent Participants'. While this is obviously a step on the road to recognising past injustices, the Council is purely an advisory body, so it is difficult to be optimistic over the influence it may have over government actions if there is an ownership scramble for the Arctic when global warming aids recovery of the area's resources.

Returning Sun. Winter in Sassenfjord, Spitsbergen, Svalbard.

# 12 After the ice

During the first decades of the 20th century the Serbian mathematician Milutin Milanković noticed an apparent periodicity in the climate of the Earth, and developed a theory based on the cyclical nature of the planet's orbital movements. Three variations affect the Earth's position relative to the Sun and so having the potential to affect insolation (the solar energy reaching the Earth) and, hence, climate (Figure 12.1). The first is the Earth's path around the Sun, which varies over time bringing our planet closer to, then further from, the Sun, the difference resulting in a variation of insolation with a period of about 96,000 years. The second is in the obliquity of the Earth's axis of spin (i.e. the angle between the spin axis and the plane of the ecliptic). This varies from about 21.4° to 24.4°, moving the poles closer to, and further from, the Sun with a period of about 41,000 years. The third cycle results from the Earth's precession (wobble) around its spin axis which also moves the poles towards and away from the Sun, with a periodicity of about 20,000 years.

It was initially assumed that these cycles, together with differences in atmospheric gas make-up, explained the gross changes in the Earth's climate, including the periodic Ice Ages, but more recently doubts have been expressed. The periodicity of obliquity and precession can be observed in the climatic record, but have a more limited effect on the total input of solar radiation than was first assumed. Orbital eccentricity is even more problematic, as it can be observed over the last million years or so, but not earlier, which is curious. Equally strange is that climatically, orbital eccentricity should be the weakest of Milanković's three effects, but it produces the strongest signal during the Quaternary era. Overall, the climate record does not readily accord with any of Milanković's effects, so it now seems that they are only part of the story, and other mechanisms have been proposed to account for cyclical climatic variations.

About 1,000 million years ago the appearance and evolution of photosynthesising organisms caused a

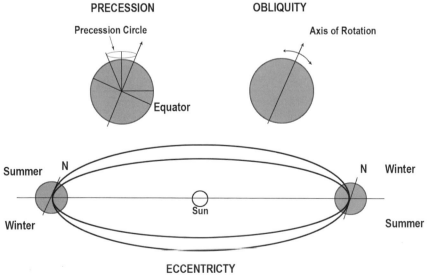

Figure 12.1 Variations in the Earth's orbit.

dramatic decrease in atmospheric $CO_2$ and a massive rise in $O_2$ (which killed off competing organisms). As $CO_2$ is a major contributor to the trapping of heat within the atmosphere (without the present concentration of $CO_2$ in the atmosphere the Earth's average temperature would fall from a balmy +15°C to a much more chilling -20°C: this is the 'greenhouse' effect, which is the subject of much current debate on climate change – see Chapter 18), this reduction caused a major cooling of the Earth. Eons later, a period of intense tectonic activity led to the emission of vast quantities of $CO_2$ and a rise in the planet's temperature. This cycle of increasing and decreasing levels of $CO_2$ in the atmosphere has repeated throughout Earth's history. When other effects are included – chemical processes during mountain building and erosion, thermohaline ocean currents and the distribution of land across the earth's surface – it is easy to see why no fully comprehensive theory of Ice Age periodicity has yet been established, even if the existence of Ice Ages is beyond doubt from evidence in cores taken from the Greenland and Antarctic ice sheets.

These cores offer climatic information because of the make-up of carbon and oxygen atoms. Just as carbon exists in different isotopic forms, aiding radiocarbon dating, oxygen also has different isotopes, and in an ice core the ratio of two these, O16 and O18 – the most abundant forms – depends on the volume of terrestrial ice. Water molecules in which the oxygen atom is O16, being fractionally lighter, evaporate more easily from the oceans. In interglacial periods the evaporated water falls as rain or snow, returning the lighter oxygen isotope to the oceans. However, in glacial periods the volume of ice on land increases and the amount of O16 locked in the ice rises: at the same time, since the O16 is not being returned to the sea, the concentration of the lighter oxygen isotope in the ocean falls (Figure 12.2). The ratio of the two oxygen isotopes in ice cores from the two ice sheets, and the ratio in alkenones (chemicals produced by marine phytoplankton) in ocean sediments, indicate that the temperature of the Earth began to fall some 2½ million years ago.

The ice cores also indicate a sustained period of cold from about 250,000 years BP, with interglacial periods when the temperature rose. There was a significant glacial maximum about 150,000 years ago, and another that ended about 18,000 years ago. Since that last maximum, the Earth's mean temperature has been more or less stable, with temperature fluctuations within a range of about 4°C. A rise in temperature led to a warmer period in the 9th and 10th centuries, when the Vikings settled Iceland and Greenland, while a cooling caused the 'Little Ice Age' of the 17th–early 19th century when, in Britain, the River Thames froze in almost all winters, the ice being thick enough for Frost Fairs, attended by thousands, to be held on it.

Ice core samples also reveal two events that may indicate what the future holds for northern Europe if the present increase in the Earth's temperature continues. The samples indicate that about 13,000 years BP the

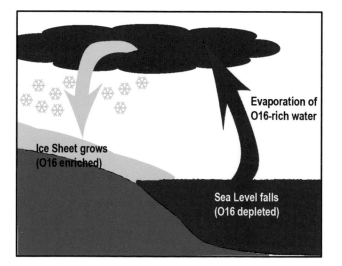
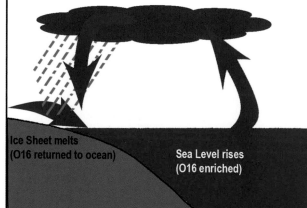

Figure 12.2 Oxygen isotopes and sea level changes

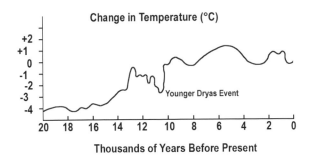

**Change in Temperature (°C)**

Younger Dryas Event

**Thousands of Years Before Present**

Figure 12.3 The change in the Earth's temperature over the last 20,000 years, as derived from ice cores. The mean temperature (i.e. the 0 point) is that for the first half of the 20th century.

climate warmed, then abruptly cooled again, but about 11,500 years BP warmed dramatically (Figure 12.3), the temperature in the Arctic rising by about 7°C in only 50 years (implying a mean temperature rise across the Earth of 4°C: temperature fluctuations are more pronounced in the polar regions, as we shall see later). This cooling and rapid warming is known as the Younger Dryas event (named after the spread of Mountain Avens – *Dryas octopetala*) and is thought to have resulted from a switching off and on of the Atlantic conveyor which drives the North Atlantic Drift.

It is believed that the initial warming phase induced copious glacial meltwater, which diluted the Arctic Ocean's surface water leading to an interruption of thermohaline circulation. The North Atlantic Drift shut down (though the Gulf Stream would have remained as an Atlantic gyre) and the northern hemisphere cooled. The cooling reduced the meltwater flow and, eventually, the conveyor switched on again, causing a rapid temperature rise. The warmer climate resulted in increased precipitation and a thickening of the Greenlandic ice sheet. It would seem that warming would again increase melting, but if local temperatures remained close to freezing then melting might not have been appreciable, preventing another switch-off of the conveyor. Following the Younger Dryas event, further cold troughs occurred about 8,000 years and 3,000 years BP, suggesting that a switch-off, or reduction of, the Atlantic conveyor is not as difficult a process as might be hoped. These cold troughs could imply that if the present increase in the Earth's temperature continues, the conveyor might again be reduced or switched-off, with disastrous climatic effects for north-west Europe.

## The last Ice Age

The last glacial maximum occurred about 18,000 years ago. At that time ice covered about 30% of the Earth's land surface, though it was not distributed evenly across the northern hemisphere, its extent differing between North America and Eurasia, and within those continental landmasses.

## The Ice Age in North America

At the glacial maximum there were three ice sheets, the enormous Laurentide ice sheet, the smaller Cordilleran sheet on the Pacific coast, and the much smaller Innuitian sheet on Ellesmere and Axel Heiberg islands. The three coalesced at their borders, effectively giving continuous ice coverage over most of the continent south to latitude 39°N.

The Innuitian ice sheet also coalesced with the ice sheet on Greenland. However, although there was an ice cap on Alaska's Brooks Range, much of central, western and the extreme north of Alaska was ice-free.

The ice sheet was up to 4,000m thick near Hudson Bay, an ice mass so heavy that it compressed the land beneath it, creating corresponding land bulges at the sheet's periphery. When the ice retreated, the compressed land rebounded, the process – known as isostatic uplift – raising King William Island above the sea and lifting the land beneath Hudson Bay, to the north of James Bay, by 120m in the subsequent 7,000 years. The shorelines of Hudson Bay have risen by about 80m: the creation of raised beaches by uplift can provide valuable information on an area's history. Rebound can now be measured with great accuracy using GPS systems so the continuation of the uplift can be confirmed: by the time rebound ceases, the shape of Hudson Bay will be very different from that of today. James Bay will not exist, Southampton Island will become part of the mainland, and the Bay's area will reduce to some 35% of its present extent. The long term history of Hudson Bay is a fascinating insight into the way the maximum glaciation of the last Ice Age affected, and continues to affect, modern geography.

As well as isostasy – the rise and fall of the land relative to the sea – there is eustasy, the rise and fall of the sea relative to the land. In the area now occupied by Hudson Bay a much larger sea – the Tyrrell Sea – formed as the

Laurentide ice sheet retreated, its size governed by the opposing processes of isostasy and eustasy. At first the eustatic rise of the sea dominated because the melting of ice was faster than isostatic rebound, and the Tyrrell Sea grew. Ultimately, as the volume of ice shrank, the situation reversed and the Tyrrell Sea began to shrink: it became today's Hudson Bay and so will eventually be an even smaller sea.

After the glacial maximum, ice retreated as the Earth warmed. Initially warming would have chiefly affected the thickness of the ice sheets, as the calving of glaciers into the northern oceans would have maintained low water temperatures. Only as the ice sheets retreated from the sea would calving have ceased and the oceans warmed: onshore winds would then have enhanced melting. The major effect of melting was to produce proglacial lakes, water masses formed at glacial snouts, the lakes dammed on one side by naturally higher ground or by land undergoing isostatic uplift and on the other by the ice sheet. The waters of proglacial lakes aided ice retreat, with wave action gnawing away at the sheet and calving icebergs, which drifted south to melt in water warmed by winds blowing north across the continental landmasses. Among the proglacial lakes formed by the retreat of the Laurentide ice sheet in North America were precursors of the Great Lakes. Large though these are, they would have been dwarfed by Lake Agassiz, which formed south of the present south-western corner of Hudson Bay, reached a maximum size of over 350,000km² and lasted about 2,000 years. It is likely that the lake drained into Hudson Bay during a violent event as ice retreat allowed an outflow about 8,000 years ago: it is thought that perhaps 150,000km³ of fresh water discharged, raising the global sea level by an estimated 20–40cm over a period as short as two days. The discharge would also have radically affected the surface salinity of the North Atlantic, and it is speculated that the cooling of

*Dryas octopetala* in the Barren Lands of Nunavut, Canada, near the Hood River.

the Arctic that occurred at the time of the discharge was caused by a corresponding change in the North Atlantic Drift. Today Lake Winnipeg is set at the heart of what was Lake Agassiz, the remainder of the old lake having become prairie. To the north, the proglacial Lake McConnell, another enormous body of water, was the precursor of the Great Bear, Great Slave and Athabasca lakes.

## The Ice Age in Eurasia

In the Palearctic during the last glacial maximum, much of Britain, Scandinavia and Denmark, together with parts of Germany, Poland, and Russia, were ice-covered. But, as with the Nearctic, the coverage was not complete, being surprisingly limited in northern Asian Russia. There were ice caps on Franz Josef Land, the northern island of Noyava Zemlya and Severnaya Zemlya (much of which is still glaciated), but there was little ice on the mainland, and Wrangel was ice-free. As in North America there were several ice sheets, these again coalescing at their borders. The principal sheets covered the British Isles, Scandinavia, and Eurasia, the latter often referred to as the Barents Ice Sheet: an ice sheet would also have covered much, probably all, of Iceland, ice coverage increasing the number of *jökulhlaups* that would have occurred. It is also probable that Jan Mayen was completely ice covered. The ice sheet reached a depth of 3,000m at the head of the Gulf of Bothnia, where the isostatic rebound has been approximately 100m to date, with peripheral lowering of about 10m in northern Germany.

Finally, on Greenland it is probable that the entire island was covered with ice, either as a continuation of the present ice sheet, or by glaciers. The weight of the present ice sheet has complicated the isostatic rebound of the island, but considerable uplift has been detected under the sea off the east coast near Scoresbysund, off the west coast south of Disco Bay and off the north-east coast north of Qannaq. The present isostatic depression of Greenland by the overtopping ice sheet has created a saucer-shaped landmass beneath the ice. On its eastern edge Greenland rises to an average of more than 2,000m above sea level. On the west the average is 1000m lower. However, over most of central Greenland the land beneath the ice is at, or below,

sea level. Occasionally, elevated parts of the landmass protrude through the overlying ice: such protrusions, known as nunataks, which occur or formerly have existed in other Arctic areas as well as Greenland, can form either as a consequence of ice retreat or ice depth reduction, or may always have existed as they were never inundated by the ice. The latter may have acted as small refugia for plant life and so aided the recolonisation of the Arctic fringe when the ice receded. Though conditions on most nunataks were so harsh that most plants could not survive, the plant life of today's Arctic includes species whose habitat suggests that they could indeed have survived (if not flourished) during the glacial maximum in such refugia. In particular, the Woolly Lousewort *Pedicularis lanata* has been discovered in well-separated places – northern Canadian islands, northern Greenland, Svalbard and north-east Siberia – suggesting they survived on nunataks when a formerly more extensive

Old sea bed at Badlanddalen, north-east Greenland. The bed is now 10 km from the sea.

Jan Mayen Island. It is likely that the island was completely ice-covered during the last Ice Age. Recolonisation has therefore occurred in about the last 20,000 years. Colonising plants have had to contend with the island's volcanic evolution.

circumpolar distribution was largely inundated by ice. However, caution is necessary when making such suggestions. Grasses have been discovered growing on Canada's Arctic islands well to the north of the rest of their range: these are assumed to have been taken there in the kamiks of Inuit, who are known to have used grass as an insulator in their boots. Similarly, some plant species in south-eastern Greenland are believed to have arrived with Norse settlers (they also brought insect species that have survived in the areas the Norse settled), and scattered outcrops of Ross's Avens *Geum rossii* in the Canadian and Greenland High Arctic are best explained by the transport of seeds on the feet or plumage of geese.

However, it is worth noting that plant refugia were not necessary for the rapid recolonisation of the land as the ice retreated. Seeds from winter caches have been found in lemming burrows dating back some 10,000 years: the subfossil skeletons of the lemmings that collected them were also found. When given ideal growing conditions, the seeds germinated rapidly and grew into healthy flowering plants. In places where the soil was not entirely scraped away by the ice, seeds may therefore have survived the glacial maximum and flourished as soon as more equable conditions returned.

## The development of Arctic habitats

As the ice of the last glacial maximum retreated it left behind bare ground. Not until plants had colonised this bare ground could birds and animals thrive. The position in freshwater lakes was analogous, with the establishment of photosynthetic organisms such as diatoms required to form the base of any potential food chain. Freshwater systems differ, for although some aquatic organisms can become established by means other than direct inflow from streams or rivers, larger aquatic organisms such as fish cannot. For lakes to acquire fish there must have been inflowing rivers, implying either the existence of freshwater refugia in ice-free areas or changes in local hydrology caused by the retreating ice. But freshwater systems represent a special case: elsewhere it is the return of plants that established habitats.

The retreating ice left behind bare rock and glacial till, much of the latter as a silt of ground-up rock debris. The thickness of the ice coverage (even during ice retreat) probably generated fierce katabatic winds, which would have redistributed the silt to form extensive deposits of loess (as such wind-driven silt is known), which would have been augmented by soil blown in from unglaciated areas to the south by the persistent anticyclonic winds that swept the ice: these latter winds would have brought seeds and spores. The large proglacial lakes formed at the ice sheet front would have induced a local 'maritime' climate so that rain dampened the fertile loess to create an excellent habitat for plants. Analysis of pollen from lake and pond sediments allows such colonisation to be studied. However, the results of pollen capture must be treated with caution because several factors must be taken into account. The speed of ice retreat and wind speed are obvious ones, but colonisation of ground exposed by retreating ice was also not straightforward. For a range of plants – flowering plants, conifers and shrubs – to establish, factors other than available, ice-free soil were necessary. One was the depth below the surface of the permafrost, with a shallow active layer restricting root growth. Another was that nitrogen-fixing plants had to become established before other plants could flourish: the decomposition of nitrogen-fixers acts as an effective nitrogen fertilizer. Yet another was the presence in the soil of mycorrhizal fungi, which form relationships with many plants, invading their roots. The fungi benefit by obtaining carbon (in sugars) from the plant, in exchange for providing the plant with a greater area for absorption of minerals from their hyphae (the fungal filaments). This association, termed mutualism, is particularly beneficial for trees attempting to establish themselves in soils that are poor in nutrients or poorly drained, precisely the conditions encountered by trees moving north in the wake of the retreating ice. Although deciduous trees form such relationships conifers are, not surprisingly, particularly rich with them. For trees and many other plants to become well-established, the spores of mycorrhizal fungi (e.g. basidiomycetes) must therefore be present in the soil.

The rate of colonisation of newly uncovered ground was initially dependent on the speed at which seeds arrived, plants with winged seeds preceding those which required dispersal by birds, with seeds with large wing areas relative to weight arriving before those with smaller wing-to-weight ratios. Later, the rate became dependent on the extent of available bare soil after the first plants had taken root. It is easy to assume that the plant cover we see today has existed throughout the time since the ice retreat. In fact, the coverage has changed due to fluctuations in climate. Pollen analysis shows that towards the northern limit of the forest there were once oaks and elms among the spruce: there are none now, indicating that the winter temperature (the limiting factor for these groups) has become colder.

## Avoiding freezing

In general, the effect of temperature on biological function is straightforward. Organisms depend on enzymes to allow the chemical reactions that drive life processes to occur, and these proteins have optimal temperatures at which they work best. As the temperature of an organism increases the function rate of the enzyme increases until this optimum temperature is reached: at higher temperatures enzymes denature (i.e. undergo structural change) and stop working, while at low temperatures they cease to function. Both extremes lead to death of the organism. At high temperatures damage to the cell structure may also occur, so the organism dies even though the enzyme function may not have ceased. In the absence of systems to combat changes in temperature the organism therefore has a limited temperature range over which it can survive (ignoring spectacular survivals such as those of nematodes, water bears and the eggs of some insects that can survive, in an inactive state, after a bath in liquid helium at just a few degrees above absolute zero).

In the Arctic it is, of course, low temperature that presents a significant threat to the survival of an organism. It might seem that the problem of reduced function rate would apply solely to ectotherms (those organisms whose temperature depends largely on the ambient temperature – fish, insects and amphibians), since endotherms (those organisms that can regulate their own temperatures – birds and mammals) can in principle increase their body insulation by adding blubber, fur or feathers, or increase their metabolic rate, assuming sufficient food is available. But it is worth noting that while the distinction between ectotherms and endotherms appears straightforward, organisms of each group can utilise some of the techniques of the other to their advantage. Small birds and rodents often reduce their metabolic rates – this may be seasonal or at night – while mammals and birds 'sunbathe', the heat absorbed reducing the level of internal heat generation required. Bumblebees rapidly vibrate their wing muscles to generate body heat to raise their flight muscles to working temperature, and have insulatory body hair: some other insects also use this shivering technique. However, despite this minor blurring of the distinction between cold- and warm-blooded organisms, it is the ectotherms that require very specific techniques to avoid death as ambient temperature falls.

If the temperature of individual cells in an organism falls enough for cellular water to freeze, the ice crystals created can grow rapidly and damage membranes and other cell structures. To survive at low temperatures ectotherms have therefore had to evolve strategies to combat these damaging and fatal effects. There are two strategies, freeze-tolerance and freeze-avoidance, though these are not mutually exclusive, some species using both: some beetle larvae switch strategies. Neither strategy is foolproof: each has a lower temperature limit below which cold injury or death will occur regardless, and within a species there is variation in effectiveness, so some individuals survive while others succumb.

Freeze-tolerance utilises the fact that ice crystals must have a nucleus around which they can form. Such particles inevitably exist, but hydrophilic proteins are pumped out of the cells into the extracellular fluid. There they act as ice-nucleators, so the freezing occurs outside the cell. The proteins also order the water molecules so that crystal formation is slow, limiting local damage. The cellular fluids remain ice-free: it is within the cell that ice crystals do the damage that causes the organism's injury or death. Solutes concentrated in the cell lower the freezing point of the cell fluids: glycerol is the most frequently found, and sugars are added to aid the prevention of damage to membranes and to maintain cell function. Freeze-tolerance is found in many invertebrates, including many marine species and some insects. It is rare in vertebrates, but does occur in some amphibians. Freeze-tolerance is effective in some species to temperatures as low as

-70°C, though temperature limits of -25°C to -40°C are more common.

Freeze-avoidance utilises the fact that very pure fluids, ones from which nuclei for crystal formation have been excluded, can be supercooled, that is cooled well below the normal freezing temperature. In this way water can be supercooled to about -40°C without ice formation. If solutes are added to the fluid so that the freezing temperature is further reduced, the strategy becomes even more effective. The added solutes are known, not surprisingly, as antifreeze compounds: glycerol and related compounds are again the most commonly found (and are also the basis of the antifreeze compounds which are routinely added to vehicle cooling systems). In some Arctic insects, antifreeze compounds can amount to 25% of body weight. To reduce the number of particles that can aid ice crystal formation, organisms have efficient cleansing methods so that, for instance, they can empty their gut of food as residual particles can act as nuclei. Freeze-avoidance is practised by insects and many spiders, and by more vertebrates – particularly polar fish – than practise freeze-tolerance. Though more energy-costly, as it requires the manufacture of antifreeze compounds, freeze-avoidance has the advantage of being more rapid in terms of switching from an inactive to an active state, and reduces the water loss that can arise through using freeze-tolerance. However, in general the lowest tolerable temperature is higher (usually in the range -5°C to -20°C) and the strategy is riskier: if intracellular freezing occurs it is very rapid and results in the swift death of the organism.

## Plant adaptations

Plants exhibit a combination of freeze-tolerance and freeze-avoidance as well as a variety of insulation adaptations to survive low Arctic temperatures. These adaptations allow broad-leaved evergreen trees to survive at temperatures of about -15°C. Broad-leaved deciduous trees push this survival temperature down towards -40°C, but at lower temperatures only conifers can survive.

Though low temperatures are, of course, a problem, one aspect of the temperate climate that young, tender Arctic plant parts have to deal with less frequently is

Polar willow, Cape Tyrol, Wiener Neustadt Island, Franz Josef Land.

frost. In temperate regions the cooling of the soil at night, particularly if the sky is clear, causes the temperature at ground level to fall rapidly, trapping cold air below warmer, slightly higher, air. This temperature inversion allows ground frosts even on nights when the ambient temperature is above freezing. The 24-hour Arctic day reduces – indeed, may eliminate – frosts so that plant growth is not retarded as it often is in temperate areas.

Arctic plants grow close to the ground, but this low form can be either a genetically inherited morphology or one fashioned by the local environment. A classic example of the former is Arctic Willow *Salix arctica*, a dwarf willow species that grows as such even if transplanted to a less hostile environment. In species where form is determined by the environment a dwarf will grow taller. The low form of Arctic plants offers protection from the wind as well as allowing a more hospitable microclimate to be created. Wind desiccates the plant, a particular problem in winter when the frozen ground prevents water uptake. Because of friction, wind speeds are lower closer to the ground than they are higher up. Lower wind speeds also mean that abrasion from dust particles and snow is reduced. The depth of the active layer of permafrost also affects plant height; if the active layer is shallow (and so uniformly cold) and root growth is restricted, plant growth is slowed. Arctic travellers will notice that plants prefer to grow in sheltered places, but they may also be surprised to come across plants in isolated and exposed positions. The tenacity of life, here as elsewhere, is remarkable.

Air temperature is usually higher closer to the ground, and the matted or cushion form of most far-north plants allows air to be trapped. Held close to the (relatively) warm ground, the warmer air helps to create a local microclimate of higher temperature. Many species have this form, an example of convergent evolution, the most notable examples being Moss Campion *Silene acaulis* and Purple Saxifrage *Saxifraga oppositifolia*, both of which are often claimed to be the most northerly flowering plant in the world (the claim is difficult to prove, but both must be strong contenders). Moss Campion in particular can, in certain circumstances, be easily confused with a true moss, so tightly packed are its stems and leaves. Studies in the field have shown that

Moss Campion, Bear Island, Norway.

in some cushion plants the internal temperature of the cushion can be as much as 15°C above ambient.

In winter the low form of Arctic plants also allows a covering of snow, which insulates the plant against plunging ambient temperatures. This is of value for species that are evergreen. Evergreen plants have the advantage of being able to photosynthesise in winter (if the sun rises and it is a bright day) and of being able to gain time at the start of spring by not having to wait until new leaves have grown. The new year's leaf growth is also protected during the early stages by the

Woolly Lousewort, Hold-with-Hope, North-east Greenland.

## Other adaptations

There are many other adaptations. Leaves and stems, and the branches of woodier forms, are darker for greater heat absorption. Many species have hairs on stems, leaves and even flowers. The hairs on willow catkins not only trap air for insulation purposes, but the still air also reduces water loss. The parabolic shape of many flowers mimics that of solar furnaces, directing sunlight towards the plants reproductive structures to speed development. The 'sun trap' that this induces within the flower also provides a warm, welcoming microhabitat that attracts pollinating insects. One adaptive characteristic of the trees at the edge of the boreal forest will be obvious to the traveller: while the trees of temperate forests take the traditional cone shape, the lower branches extending beyond those above to collect maximum light, the low angle of the sun at the Arctic fringe negates this form, so the trees instead have branches of approximately equal length along the trunk. This shape also helps reduce the weight of accumulated snow, which might otherwise cause branches to break. Boreal forest trees are usually evergreen, to allow photosynthesis to begin as soon as the sun returns. But the needles are thin and wax-coated, with the stomata (porcs through which gas exchange takes place) set deep to reduce desiccation.

Arctic flowering plants exhibit other adaptations in addition to these physical characteristics. Most are perennials: annuals are few, because even with overwintering buds and evergreen leaves, the Arctic summer is often too short for the plant to go through its entire life-cycle, particularly in the far north. However, some species are annual/biennial, being annuals if the polar summer is long enough, but taking two years for the life-cycle if not.

overtopping older leaves. While dead leaves at the top of the plant add insulation, those at the base are trapped close to the plant and ultimately add nutrients to the soil. Arctic soil is poor, and plants need to take advantage of any available nutrient resource: there is often relatively luxuriant plant growth beneath the breeding sites of birds where the ground is fertilised by droppings, or near an animal carcass. Some species also overwinter with well-developed flower buds to save development time in the short Arctic spring and summer.

Many Arctic flowers are highly phototropic, tracking the sun throughout the 24 hours of the polar summer day. The flower does not follow the sun by rotating – a recipe for disaster as following the midnight sun for day after day would twist the flower stem. The movement is created by the stem growing continuously, but always at a slower rate on the side towards the sun so the flower head tilts. Though the polar day is long, Arctic plants are also able to photosynthesise at low light levels to take full advantage of available light.

The Arctic Poppy, photographed here at Hold-with-Hope, north-east Greenland, has parabolic flowers to concentrate sunlight. The plant is also phototropic.

Just as in more southerly plants, Arctic species disperse seeds using the wind, birds and mammals. Many Arctic plants are berry-producers, the berries being an especially rich food source for animal life. However, not all Arctic species are seed-producers. Some spread by producing rhizomes (horizontal, underground stems that take root at intervals), some by growing stolons (above surface stems that produce new plants at their tips), while others produce bulbils (or bulblets), buds that form in the place of some, or all, flowers. Rhizomes and stolons have the advantage that the new plant will grow in a suitable habitat. Bulbils, as with seeds, are at the mercy of the wind, and root only if they land in a suitable spot. Bulbil production has the advantage over seed formation of being asexual: the plant requires neither a pollinator nor another plant, each of which may be scarce. However, there is a disadvantage: unlike seeds, bulbils, being much more fragile, cannot survive for long periods if they do not implant.

Viviparous Knotweed, which reproduces by bulbil production, central west Greenland.

142

## The adaptations of terrestrial invertebrates

Though most Arctic invertebrates are aquatic (and chiefly marine), a surprising number inhabit both the tundra and the taiga edge. In terms of abundance the most numerous are worms (particularly nematodes and oligochaetes) and rotifers, with planarian worms being important within the taiga. There is also a diverse collection of freshwater copepods, and many insect species. Spiders are also important within the taiga. Almost all insect orders are represented in the Arctic (though there are very few beetles) and most are important as food sources for birds or as plant pollinators. Those species that are parasitic on Caribou have an extremely deleterious effect on an infected animal, while mosquitoes can be a great nuisance to Arctic species (and travellers), the resultant blood loss from them even being life-threatening to Caribou and other species.

In general, insects found in the Arctic are smaller, darker and hairier than their southern cousins. All three characteristics are adaptations to the northern environment. The smaller size is almost certainly because the insects are constrained by food resources (a smaller size also enables the insect to warm faster – but, of course, it also cools quicker). Arctic individuals also tend to be stockier, reducing their surface area-to-volume ratio, resulting in a relative reduction in both heat and moisture loss. The darker colour seems at first a contradiction of 'Gloger's Rule', which suggests that individuals tend to be white in the Arctic to aid camouflage. But that rule was formulated for vertebrates, and does not allow for the fact that in ectotherms the increased heat absorption (both direct radiation from the sun and reflected heat from the ground) of dark colours is generally more important than camouflage: a warmer insect can fly faster. The hairiness has arisen for similar reasons to the hairs of Arctic plants – the hairs trap air, reducing convective heat losses and providing insulation. Experiments have shown that if the hair is shaved from a caterpillar, the insect will lose heat by convection much faster than its unshaven siblings. It is difficult to know whether to marvel most at the ingenuity of the experiment or to pity the poor caterpillar.

## Wings and basking

Insect antennae and wing sizes are, in general, smaller than in southern species, to reduce heat loss. This is particularly noticeable in stoneflies, crane flies and some moths. In some species wings are entirely absent. Those insects that do have wings fly close to the ground where the air is warmer and wind speed lower. However, the wings of Arctic butterflies are not significantly smaller and are utilised as sun 'catchers'. Indeed, the basking strategies of butterflies are often useful in helping to identify species. Butterflies use one of three strategies – though individuals may not always limit themselves to just one. Dorsal basking involves flattening the wings so that the back and the upper wings are exposed to the sun: the butterfly also leans towards the sun. This strategy is favoured by members of the Pieridae, the white butterflies (chiefly *Colias* spp. in the Arctic). In lateral basking the wings are raised so that the upper surfaces touch: the body is then turned sideways to the sun so the flanks and lower wing surfaces are exposed to it. Most fritillaries use this strategy. The wings may also be held in a V-shape, the wings acting as both collector and reflector, efficiently 'trapping' sunlight. Members

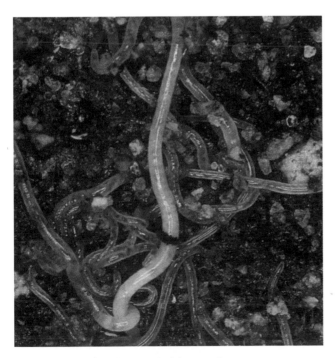

Oligochaetes (*Enchytraeides*) beneath a stone in south Greenland. In the lower part of the photo, a dark blue springtail is crawling over the palest worm.

of the Lycaenidae, the blue butterflies, often adopt this strategy.

Basking is not, of course, limited to butterflies, and travellers looking for insects should seek out sheltered areas that favour both the insects and the plants on which they feed. An alternative for camping travellers is to look at the sunny side of the tent. In order to maximise solar heat input all insects bask, and there are few better places than tent panels, the fabric warming quickly and so offering not only a conveniently flat platform, but also the advantage of heat on both sides. In the absence of a convenient tent, insects bask on bare ground, which offers similar 'all-round-heat'.

Insect flight muscles must achieve a minimum temperature before they can function, and those species that do not generate muscle temperature by shivering must bask. However, flying produces a lot of heat, particularly as wingbeat frequency can be incredibly high (up to 200Hz). So insects, even Arctic insects, have mechanisms to prevent overheating, as flight muscles must be maintained below a critical temperature to function efficiently. Normally this is accomplished by having an uninsulated section of the abdomen – usually the ventral side.

The search for extra energy from the sun is also seen in insect larvae. Basking is practised by some: the Woolly Bear caterpillar spends 60% of its active time basking, with just 20% spent eating. Mosquito larvae will move around a pool to track the movements of the sun, while the larvae of blowflies hatched from eggs laid on a carcass will be found in greater abundance and at a more advanced state of development on the southern side of the corpse.

Though these sun-seeking strategies are clearly effective, Arctic insects nonetheless complete their life cycles in ambient temperatures that are lower than those enjoyed by their southern cousins. They are therefore more active at lower temperatures. Mosquitoes in the far north go through their egg and larval stages in temperatures that may exceed 1°C only rarely, and Arctic travellers will (sadly) note that mosquitoes continue to fly and feed at temperatures close to freezing, though they become much more active as the temperature increases.

## Extended life-cycles

Yet despite being able to remain active at lower temperatures, and having life spans comparable to those

of southern cousins, many Arctic insects take much longer to go through their life-cycle than do related southern species. Some insects that range across both the temperate and Arctic zones have different lifespans in different latitudes. The springtail *Hypogastrura tullbergi*, for instance, has a lifespan as short as eight weeks in southern areas of its range, but in the far north these insects can live for five years. Though this difference may be extreme, there are many northern insects with lifespans of three or four years, and some far north midges may take seven years to pupate. The most extreme case of such an extended span is the Arctic Woolly Bear caterpillar (of the moth species *Gynaephora groenlandica*), which takes at least seven, and as many as 14, years to develop to the pupal stage. In sharp contrast, the adult moth completes the life-cycle within a few days. The male dies after mating, the female after egg-laying: neither adult feeds. The caterpillar is parasitised by wasps and bristle flies, which each take three or four years to develop. In almost all cases of extended spans, the insect may take advantage of favourable conditions by shortening the life-cycle.

Arctic Woolly Bear caterpillars, which display an extraordinarily extended life-cycle.

Life-cycle extensions invariably result from extended periods in the larval stage: few species overwinter as adults (although it is not unknown, with the adults seeking shelter beneath a stone or detritus and trusting to their freeze-avoidance or freeze-tolerance strategies). Most insects, however, overwinter as larvae or pupae: for some species this has clear advantages, young flies, for instance, being able to overwinter within the carcass or dungheap in which they hatched. However, mosquitoes overwinter as eggs. Female mosquitoes lay their eggs close to the water's edge at the southern edge of the chosen pond. Small ponds are usually selected as they thaw and warm more rapidly. Snow coverage insulates the eggs during the winter. In the spring the Arctic sun, rising in the south, melts the snow, this both triggering the eggs to hatch and providing a pathway to the pond. Eggs that undergo a prolonged period of freezing probably survive, but their development will be delayed. Larval mosquitoes can develop in four weeks in water temperatures down to 1°C. On first emerging as adults, mosquitoes feed on nectar and are important plant pollinators before the females become the misery that afflicts all Arctic animals and travellers.

The reproduction of Arctic insects is, in general, as for southern species, but there are notable exceptions. In some species males are rare and females reproduce parthenogenically (for example, some stoneflies, midges, black flies and caddisflies) Asexual reproduction allows spectacular and rapid population increases and so is favoured as it reduces the time for courtship, but it reduces genetic diversity. In the extreme case of the far northern black fly *Simulium arcticum*, the insects do not even go through the full life-cycle of egg, larva, pupa and adult, eggs developing inside the pupa, being released when the pupa dies, without an adult stage occurring at all.

## The adaptations of aquatic organisms

As fresh water freezes at 0°C and sea water at about -1.8°C, aquatic organisms – which are ectotherms apart from the marine mammals and birds, and some large fish such as tuna – need in principle employ only freeze-resistance techniques adequate for a range of temperatures close to those figures. However, this logic only holds if the organism is in liquid water; for both freshwater and marine organisms situations may occur whereby they experience much lower temperatures, low enough to require more sophisticated freeze-resistance methods in order to avoid certain death.

## Survival in freezing freshwater

Large lakes and most rivers do not freeze entirely (surface ice acting as an insulator), with fish and aquatic invertebrates surviving by employing freeze-resistance techniques and avoiding contact with the ice undersurface. Some river species may migrate to large lakes to reduce the risk of freezing. Invertebrates move to depths to escape the advancing ice and may burrow into the lake bed, though some insect larvae freeze into gravel beds close to the pond edges, using freeze-avoidance or tolerance to overwinter without ill effect. It is not, however, only the cold that must be endured. With ice coverage preventing the absorption of oxygen at the surface, and the halting of photosynthesis, decomposition may cause an oxygen deficiency. Some species can switch from aerobic to anaerobic respiration (metabolising without the need for oxygen) to survive, but for many – both plants and animals – oxygen starvation results in death.

## Survival in freezing seawater

Though tuna and other large, fast-swimming fish exhibit what is termed regional endothermy, in which metabolic output allows certain muscles to operate at temperatures above that of the remainder of the body, all marine creatures, with the exceptions of the cetaceans, pinnipeds and sirenians (together with penguins and certain other marine creatures, such as the largest turtles), are ectotherms, with body temperatures only marginally above that of the water. They are also stenothermic, i.e. they can tolerate only minor deviations from normal body temperature. Most marine ectotherms use freeze-tolerance or freeze-avoidance techniques, but also rely, to a lesser or greater extent, on the fact that the freezing point of seawater declines as depth increases, so that migration to greater depth is an effective precaution against freezing.

Shellfish of the intertidal zone, Disko Bay, west Greenland.

## Arctic amphibians and reptiles

As would be expected, there are no truly Arctic amphibians or reptiles, but several species are found north of the Arctic Circle in the Palearctic. Only one species has been recorded north of the Circle in the Nearctic: the Wood Frog *Rana sylvatica* breeds in Alaska as far north as Bettles, and in north-west Canada, where it has been found in the Mackenzie delta. The frog can survive short periods (up to about two weeks) at temperatures of -4°C.

In Fennoscandia and Russia as far east as the Urals, the Common Frog *Rana temporaria* is found to the northern coast and remains active to +2°C. The frog's range then becomes more southerly. The Moor Frog *R. arvalis* is even more cold-resistant, being found in Fennoscandia and east to the Urals. In eastern Russia, the Siberian Wood Frog *R. amurensis* breeds to the Arctic Circle around the Lena River and around the northern shores of the Sea of Okhotsk. The Marsh Frog *R. ridibunda* has also been found in Kamchatka, though its status there is not clear.

The Common Toad *Bufo bufo* has been found north of the Arctic Circle in Scandinavia. The Great Crested Newt *Triturus cristatus* occurs north of the Circle in Sweden. However, the most astonishing of all these amphibians is the Siberian Newt *Salamandrella keyserlingii*. This dark-spotted, brown or olive newt is 120–160mm long and is the most widespread of all 'Arctic' amphibians, including the southern Taimyr Peninsula and central Chukotka (to Anadyr). Adult newts can survive freezing to -40°C for extended periods and are active (though obviously sluggish) to +1°C. Most remarkable of all, newts excavated from depths of 14m in the permafrost have revived without apparent trauma, while eggs have been known to survive short periods of freezing within an ice matrix. The species is very long-lived (probably up to 100 years): this may be a response to periods of prolonged freezing.

All these amphibians hibernate during the winter, choosing burrows or piles of rotting vegetation where they stay either singly or in groups. The Siberian Newt may hibernate for up to 8 months.

Two species of terrestrial reptile, the Adder *Vipera berus* and Common Lizard *Lacerta vivipara*, have been regularly recorded north of the Arctic Circle in Fennoscandia and western Russia. The Common Lizard is the more northerly of the two species, having been recorded to the northern coast. There are also records of the Grass Snake *Natrix natrix* above the Circle, though these are fewer and from more southerly latitudes. At sea, the Loggerhead Turtle *Caretta caretta* has been recorded in the Barents Sea, and the Leatherback Turtle *Dermochelys coriacea* in the Bering, Chukchi and Labrador seas, and in waters around Iceland.

For intertidal animals (e.g. shellfish that graze algae at the tide level) the situation is very different as they may be exposed to ambient temperatures that are much lower than the freezing point of seawater. Such animals are invariably freeze-tolerant, able to survive if as much as 90% of their body water freezes. However, even for these remarkable survivors there is a problem if they remain above the sea ice, as prolonged exposure to the low ambient temperatures of the Arctic winter would almost certainly result in death. These animals must also avoid being caught where the sea is freezing, as during this time the ice is still in motion: moving ice is highly abrasive, with fast ice destroying most local life-forms during its formation. Being entombed within the sea ice as it forms would be equally deadly. Intertidal animals therefore migrate downwards as winter approaches. In rock pools sea ice formation results in salt leaching into the underlying water, raising its salinity and hence depressing its freezing point. Animals within these pools can therefore safely remain below the ice, often buried in the bottom sediment.

Fertilised by droppings from the seabird nesting cliffs, the plant growth below Alkhornet, Isfjorden, Svalbard, is luxuriant.

## 13 Arctic habitats

The Arctic Ocean can paradoxically be described as among both the least and the most productive seas on Earth. The central ocean, where the thick multi-year ice restricts sunlight transmission to the open water beneath, is an area of very low productivity, but the shallow seas above the extensive continental shelves that surround the ocean are, seasonally, highly productive. The reduced sunlight of winter, coupled with seasonal sea ice, restricts the growth of the photosynthetic organisms that are at the base of the Arctic Ocean food web, but with the annual thaw summer's increased hours of sunlight, coupled with the nutrients that flow in from the huge Asian and North American rivers, increases productivity dramatically.

## Life under the ice

Even though the sea ice restricts light transmission and so limits sub-ice productivity, it should not be assumed that either the seas below the ice or, indeed, the ice itself is devoid of life. On the surface of the ice there may be meltwater pools, particularly during the summer thaw, within which micro-organisms deposited either by overtopping waves, the inflow of river water or from the feet of seabirds can flourish. As sea-ice forms it traps organisms. Many of these will die, but some survive, living within the ice matrix, while other organisms live on the lower surface of the ice. Within the ice, phytoplankton occur – diatoms and other types of algae, such as flagellates. Diatoms are single-celled organisms that reproduce by dividing into two: more than 200 species have been identified so far living within Arctic ice. They live in the brine channels of the ice matrix, a home that demands not only an ability to withstand very low temperatures, but requires adaptations to cope with osmotic pressures created by high salinity. Reproduction of the diatoms depends on local temperature: although they can survive at low temperatures their growth rate is slowed – a diatom that might divide every day at 0°C, would perhaps take three days at -4°C, and 50 days or more at -8°C.

Diatoms stain the ice brown: as this darker colour absorbs more heat, the local temperature within the ice increases. This causes local melting, and the multiplying diatoms spread into the honeycomb structure this creates. Larger organisms feed on the diatoms and other phytoplankton: these also live within the ice matrix – the juvenile stages of many crustaceans are found

Pack ice stained brown by diatom growth in brine channels.

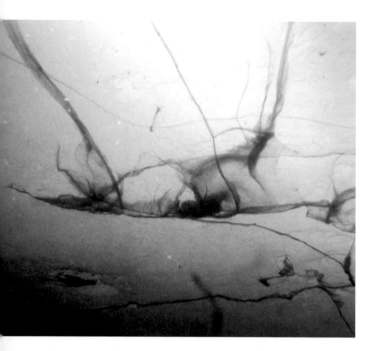

Diatom strands hanging below an Arctic ice floe. The strands wave in any passing current.

within or on the underside of the ice. These creatures – amphipods, copepods and euphausiids – are themselves part of a food chain that includes fish, marine mammals and seabirds at higher trophic levels.

Phytoplankton also live on the underside of the ice: one diatom, *Melosira arctica*, forms filaments and even sheets that can grow up to 15m long. These filamentary structures act as nets, trapping nutrients from the water. Under-surface phytoplankton is grazed directly by some fishes.

As noted in Chapter 3, polynyas occur in the ice-bound Arctic, and these form an important habitat. The cliffs close to the largest, North Water in Baffin Bay, hold some of the densest seabird colonies in the Arctic, is a significant wintering site for Bowhead Whales and the local seal population has double the density of other sites. By way of comparison, the polynya-free Beaufort Sea has a much lower overall productivity, and many of the animals at the top of the food chain have relatively fewer offspring, with those offspring maturing later: female Polar Bears resident on the Beaufort Sea, for example, breed one year later on average than bears to the east.

It is now believed that in some Arctic waters bacteria and viruses may be a significant biomass, though the exact role of both in the production processes of all the

Earth's oceans is still debated. It seems possible that they are a major source of organic material, caused by cell breakdown, on which zooplankton, the lowest trophic level of the marine ecosystem, feed. Phytoplankton, and some bacteria, photosynthesise.

## Photosynthetic organisms

Where the sea ice melts, photosynthesis occurs at all depths to which light penetrates. As light is transmitted through water, the red wavelengths are preferentially absorbed: in clear oceanic water, the energy input from red light at a depth of 10m is only about 1% of that falling on the surface. By contrast, blue light penetrates much further, only reaching 1% of the surface energy at *c*.150m. Both depths reduce in turbid waters. The net effect of this differential absorption is, of course, to make the water appear blue. But the fact that light can be transmitted to 150m creates a euphotic zone – a zone in which photosynthesis is possible – of corresponding depth. The euphotic zone is the habitat of photosynthetic phytoplankton, the organisms varying their light-sensitive pigmentation to allow for changes in light wavelength. During the winter months beneath the ice, photosynthesising phytoplankton enter a state that can be compared to hibernation, with cell metabolism depressed to a level merely capable of sustaining life.

The most regularly encountered marine photo-synthetic organisms are the seaweeds. Seaweeds grow surprisingly far north, and some intertidal species can survive temperatures as low as -60°C. Equally surprising is the fact that some species can begin to grow during the late stages of the Arctic winter, i.e. in darkness, using starches stored during the previous summer. As elsewhere, kelp 'forests' are an important marine habitat for invertebrates and fish, which feed on, and hide among, the swaying blades. The most important Arctic species are the brown seaweeds *Laminaria saccharina* and *L. solidunula*.

Seaweeds form part of the diet of Snow Geese during the late summer (a dietary change from tundra vegetation that makes late summer geese unattractive to the Inuit who claim seaweed spoils the taste). Eelgrass *Zostera marina*, a flowering plant that thrives in the shallow water of estuaries and tidal lagoons, also provides a rich microhabitat. Crustaceans and small fish live on and within the plants, while geese and other wildfowl feed on it. When it dies, the decomposing plant is home to myriad crustaceans, which are taken by shorebirds and even by foxes and bears.

## Marine invertebrates

When the ice melts and sunlight increases, the phytoplankton blooms as do the herbivorous zooplankton that feeds on them. The most abundant of these are the Calanus copepods (particularly *Calanus glacialis* and *C. finmarchicus*) and amphipods (particularly *Apherusa glacialis* and *Gammarus wilkitzkii*). These creatures are the Arctic equivalent of the Southern Ocean's Krill *Euphausia superba*: Krill is a euphausiid, a group not found in significant numbers in the Arctic. In certain circumstances the dominant grazers on the phytoplankton are nematode worms, which are common in Arctic waters but virtually absent in the Antarctic. The reason for this (and the fact that this imbalance also occurs in rotifers) is not known.

Most adult crustaceans do not penetrate the ice matrix, due in part to their inability to squeeze into the brine channels, but also because they are unable to cope with the high salinity. Copepods (particularly *Halectinosoma*, *Harpacticus* and *Cyclopoida*) are the crustaceans most likely to be found within the ice. Those that do not enter the ice feed on diatoms that fall from the matrix. Many of the smaller crustaceans are essentially planktonic; they have cilia or flagella for local locomotion, but generally drift with ocean currents. Larger copepods are less at the mercy of ocean currents and migrate to depths below about 300m during the winter. When the phytoplankton blooms, they rise to begin feeding. They tend to rise during the night to reduce their chances of being eaten.

## Arctic marine fish

The herbivorous zooplankton sustains the nekton, the generic name given to marine organisms capable of moving independently of currents. Nektonic organisms comprise small fish, the larger fish that feed on them and other marine animals up through the food chain. A little over 100 species of fish have been identified to date in Arctic waters. Of these two of the most important are the Arctic Cod *Boreogadus saida* and Glacial Cod

The amphipod *Gammarus wilkitzkii*, a predator 2–3cm long. It sits head down in a brine channel in the sea ice waiting for prey.

*Branchiomma* polychaetes.

Sea lily *Heliometra glacialis* whose arms grow to 20cm.

Sea slug *Coryphella fusca*, from the Bering Sea.

Arctic Cod *Boreogadus saida*.

*Arctogadus glacialis* (smaller cousins of the Atlantic Cod *Gadus morhua* and Pacific Cod *G. macrocephalus*) which are important prey species of pinnipeds and whales. The Arctic Cod (which can occur in huge shoals: one studied in Canada was believed to comprise almost 1,000 million fish) has an unusual mouth, the lower jaw being elongated beyond the upper so the fish can graze on the underside of sea ice. The fish have been known to use cracks within the sea ice as feeding, resting and hiding places, implying an ability to survive across a range of salinities. In more southerly waters halibut (the Greenland Halibut *Reinhardtius hippoglossoides* and the Pacific Halibut *Hippoglossus stenolepis*) are commercially important. Halibut have an extraordinary life story, starting out as 'conventional' fish. The left eye then migrates across the top of the head until it nears the right eye. The fish turns on its left side, which becomes white, while the right (now upper) side becomes mottled, an excellent camouflage for its voracious bottom-feeding habits. Other southerly species include the Capelin *Mallotis villosus* (which has north Atlantic and north Pacific subspecies), herring *Clupea* spp. and pollock (particularly the Alaskan or Walleye Pollock *Theragra chalcogramma*). Neither the Arctic cod species nor capelin have been extensively fished commercially, unlike the Atlantic Cod which has been fished almost to extinction: the increase in fishing for Walleye Pollock in the Bering Sea is raising fears of over-exploitation.

As well as cod, capelin and pollock, and several eel species (particularly ammodytids or sandeels – called sandlances in North America – which are the regular prey of auks), a major food resource for Arctic marine mammals are members of the Salmonidae family. The salmonids are chiefly anadromous, the adults being pelagic, but returning to freshwater streams to mate and lay eggs, young salmonids reversing their parental swim to reach the ocean. However, certain subspecies may be resident in rivers or estuaries. The spawning journey of the salmon is one of the most remarkable in the natural world: some King Salmon *Oncorhychus tshawytscha* travel almost 2,000km to reach their spawning grounds in the Yukon River. A young salmon is still attached to a large yolk sac when it hatches. It remains in or close to the redd (nest) prepared by its mother (the adult female salmon makes a shallow depression in the gravel of the stream bed with its tail). Once the yolk sac has been absorbed the young fish (fry), feed on copepods and larval insects. The amount of time they spend in freshwater before heading downstream to the ocean depends on the species, and varies from weeks to several years.

The salmon runs of spawning adults are a famous feature of the Pacific Arctic fringe and are an important food resource, particularly for Brown Bears: the bears that feed on the annual run are the largest of their species. There are seven salmon species that spawn in the rivers of Alaska and north-eastern Russia within the Arctic. Of these seven, five are widely distributed – Pink Salmon *Oncorhychus gorbuscha*, the smallest species, Chum or Dog Salmon *O. keta*, Coho or Silver Salmon *O. kisutch*, Sockeye Salmon *O. nerka* and King Salmon, the largest, with specimens reaching more than 50kg. A further species, Steelhead Salmon *O. mykiss*, breeds in the rivers of the Alaska Peninsula, in south-east Alaska and in western Kamchatka, while a seventh species, Masu Salmon *O. masou* breeds in the rivers of western Kamchatka, plus Japan and the adjacent Asian mainland. Steelheads and Masu are the least Arctic of the seven species, and have resident (i.e. non-anadromous) subspecies (though the taxonomy of the Masu is still debated, with some experts considering the resident fish to be a different species, *O. rhodurus*, the Amago Salmon).

All Pacific salmon undergo remarkable changes in their appearance prior to spawning (after which they die). Sockeyes change from the normal silver-blue colour to bright red, while the male Pink Salmon develops a humped back, hooked jaw and enlarged teeth. The purpose of these strange changes is not understood, and nothing similar is seen in the Atlantic Salmon *Salmo salar*. Atlantic salmon also differ in frequently surviving spawning to return to the sea. All salmon are fished commercially, over-fishing having eliminated some local populations. Concerns have been expressed over the effect of over-fishing on the Atlantic Salmon, and on the effect salmon farming might have on wild stocks if diseases and genetically modified fish were able to escape.

The Arctic Char is another salmonid: strictly the char is *Salvelinus alpinus*, but there are many closely related fish, the taxonomy of which has kept experts busy for years and now includes around 20 recognised subspecies. The char complex has a circumpolar

Sockeye Salmon, Alaska.

Pink Salmon.

Spawned Salmon.

distribution, breeding throughout North America including the Aleutians and islands of the Canadian Arctic, in Greenland, Iceland, mainland Scandinavia and across Russia to Kamchatka, including the more southerly Russian Arctic islands. Some char subspecies are anadromous, but others are purely freshwater fish, some populations having presumably been isolated in lakes by the effects of glaciation.

One particularly diverse family of fish found in the Arctic are the Cottidae or sculpins, some species of which are circumpolar. These benthic fish inhabit the shallow seas above the continental shelves, often being found close to the shore and even in less salty parts of coastal estuaries. Sculpins are known to the Inuit as sea scorpions, and were sought beneath stones in the shallows.

The shallow Arctic seas above the continental shelves are, because of their silt burden, home to rich benthic communities – jellyfish, sea anemones, sea urchins, sponges, starfish and worms. The largest of the world's jellyfish, *Cyanea arctica*, which has tentacles up to 25m long, lives in Arctic waters, while some of the Arctic sea's unsegmented worms can reach 10m in length. To the surprise of many, corals are also found: Coral Harbour on Southampton Island is named after the fossil corals found there, and although these undoubtedly lived in much warmer waters, cold-water corals are found, in small numbers, in the seas close to the Aleutian Islands and southern Alaska. Significant numbers of molluscs, such as clams, mussels, oysters and scallops, occur in Arctic waters. One gastropod worthy of mention is *Clione limacine*, often called 'sea angels' because of their angelic appearance, which live throughout Arctic waters. Growing up to 3cm, these gastropods have a larval stage which feed on phytoplankton, but an adult stage which feeds almost exclusively on Limacina 'sea butterflies', a form of sea snail. Equipped with far from angelic hook-like mouth parts the gastropod extracts the unfortunate snail from its shell and consumes it whole.

Scallops are fished commercially in the Barents Sea and near Iceland, as are whelks in the Sea of Okhotsk, while in the north Pacific king crabs are fished commercially. The largest of these, the Red King Crab *Paralithodes camtschaticus,* has a leg span of up to 2m, though the carapace is rarely more than 25cm across. King crabs are interesting creatures, migrating

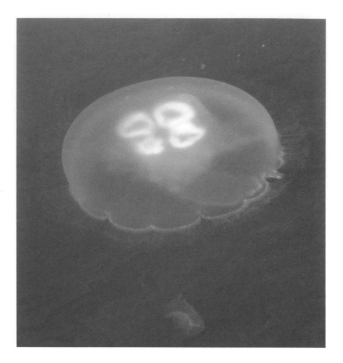

The Moon Jelly (*Aurelia aurita*) is a commoner jellyfish. Resurrection Bay, Alaska.

annually to and from shallow waters, where they breed, and deeper waters down to 100m, where they feed, often travelling more than 150km between the two. The life-cycle of the crab is complex, with an egg stage (the number of eggs per female can be as high as 500,000, an astonishing number given that she incubates them between her abdomen and cephalothorax), and four larval stages before tiny crabs with carapaces little more than 2mm across emerge. Travellers to Alaska will often see king crab on restaurant menus, along with locally caught salmon and halibut. As well as the Red King Crab, Blue Kings *P. platypus*, which are smaller, and Golden (or Brown) Kings *Lithodes aequispina,* which are smaller again, are also fished. Blues are chiefly found around the Bering Sea islands, while the Golden Kings are fished close to the Aleutians. Tanner Crabs *Chionoecetes bairdii* and Snow (or Queen) Crabs *C. opilio*, which are smaller than the king crabs, are also caught for the table in the Bering Sea.

The northern deepwater shrimps of the Atlantic and Pacific are also fished commercially. *Pandalus borealis* is found in Atlantic waters from eastern Canada to the Barents Sea and is fished in Canada, Greenland (where it is a major export), Iceland, Norway and Russia (and by other, non-Arctic nations). In the northern Pacific

*P. goniurus* is found in the Bering and Chuckchi seas where it is fished by the Canadians, Americans and Russians. The sand shrimp *Crangon crangon* is fished commercially in Russia's White and Barents seas.

## Freshwater rivers and lakes

Freshwater in the form of rivers, lakes and ponds is a very important habitat on the landmasses that surround the Arctic Ocean. In Canada close to the Great Slave Lake and southern Hudson Bay, in European Russia and in parts of Siberia, land coverage of fresh water exceeds 30%, and over much of the rest of mainland Arctic Canada and Siberia it exceeds 15%. At Canada's Arctic fringe there are two huge lakes, the Great Bear Lake extending over 30,000km$^2$ and the Great Slave Lake being over 28,000km$^2$ in surface area: each is more than 400m deep. No other Arctic lakes can compare to

these, though Baffin Island's Lake Netilling and Russia's Lake Taimyr are close to 5,000km$^2$. To these statistics must be added those of the rivers that discharge into the Arctic Ocean. These rivers are among the biggest in the world and form important habitats at their coastal boundaries. The Yenisey and Ob river systems of northern Russia are the fifth and sixth longest in the world, each at more than 5,000km: the catchment areas of both rivers exceed 2,500,000km$^2$. The annual discharge of the Yenisey (whose name derives from the Evenki *Ioanessi* – 'great river') exceeds 600km$^3$, while that of the Ob exceeds 400km$^3$. The Lena, which also drains continental Russia into the Arctic Ocean, is more than 4,000km long, as is Canada's Mackenzie River. The annual discharge of the Lena is over 500km$^3$, that of the Mackenzie exceeding 300km$^3$. But though these numbers are enormous, they must be viewed in context; the total annual inflow of river water into the

## An Arctic shark

The Greenland Shark *Somniosus microcephalus* is the only shark, and the largest fish, found in Arctic waters. It is also one of the world's largest sharks, reaching up to 7m, and one of the largest deep-water fish known. It has been recorded at depths of more than 2,000m, but is frequently seen close to the surface. The shark, the skin of which is uniformly covered in tooth-like denticles, has a varied diet, feeding on fish and other marine creatures (such as crabs and jellyfish), but it will also take birds and seals, and is claimed to have taken swimming Reindeer. Mammalian prey must be ambushed as the shark is a slow, somewhat ponderous fish that would be unlikely to catch a seal by pursuit. Greenland Sharks are most numerous in the waters of the eastern Canadian Arctic and west Greenland, being found as far north as Ellesmere Island, but they are also seen off Iceland and Svalbard and in Russia's White Sea.

Research in 2016 suggested that the sharks were probably the longest-lived creatures on Earth, perhaps living to 400 years. But if longevity seems an enviable prospect it is worth noting the presence on almost all females of a parasitic copepod, *Ommatokoita elongata*, which attaches itself to the fish's cornea. These parasites, which can grow to 3cm in length and hang like pale pink worms from the eyes, feed on cells of the corneal surface and ultimately cause lesions that cloud the fish's eyesight. Some scientists have suggested that the parasite and the shark are an example of a mutualism, the copepod being highly visible and perhaps even luminescent and so attracting prey to the fish. As the shark is thought to hunt mainly by smell in the murky waters of the deep Arctic seas, the partial loss of vision may be a relatively minor handicap.

Fig.5. The eye of the Greenland Shark.

Fig.4. UNDER SIDE VIEW of the GREENLAND SHARK

Fig.3. GREENLAND SHARK, 12½ Feet in Length.

Arctic is estimated to be about 2,800km³, which is a little over 1% of the annual water transfer (i.e. warm water in, cold water out) of the Fram Strait, between Greenland and Svalbard.

These huge inflowing rivers, together with other smaller but still sizeable rivers – the Yukon, Pechora, Kolyma and Nelson – form large deltas due to sediment build-up (the Lena, for example, deposits more than 11 million tonnes of sediment into the Laptev Sea annually, the silt creating a dark plume that extends for up to 100km offshore). The Lena's delta covers about 30,000km² and comprises more than 6,000 channels and a collection of about 30,000 lakes. In the Mackenzie River delta, beginning just north of Tsigehtchik, the river divides into three main channels, but has hundreds of lesser watercourses that wind through a vast maze of islands, ponds and lakes to the ocean, for some 200km, the entire delta covering about 10,000km². Ice damming of these great rivers results in a seasonal back-up of water as the spring thaw begins in the higher reaches of the river before it occurs at the coast. Periodic flooding by water and ice as these ice dams break drastically changes the deltas, forming networks of channels interspersed with flooded plains. It was the nightmare maze of the Lena delta that doomed one group of sailors who had escaped the sinking *Jeanette* in 1881.

Deltas are a good habitat for wildfowl and shorebirds, but the shifting nature of the channels, ice abrasion of the channel banks and the new layer of silt that flooding leaves behind makes much of the area too unstable for plant growth. Silt and sand bars at the river's edge shift too often for anything to grow successfully, though horsetails and sedges can take hold a little further back, with trees on higher ground where flooding is less frequent.

Arctic lakes and ponds are usually limited in vegetation – with horsetails *Equisetum* spp. and pondweeds dominating – because of the unpredictability of ice depth and duration. Some aquatic invertebrates can arrive as eggs, carried on the wind or on the feet of birds, as can the seeds of aquatic plants, both eggs and seeds being unharmed by periods of drying. Many adult aquatic insects are poor fliers (dragonflies are an obvious exception), and the distribution of these, and particularly that of fish, which may use a water route to become

established in a new area, aids an understanding of the Earth's periods of glaciation. The present distribution of mayflies and stoneflies in central Canada confirms the existence of Lake Agassiz, in which their aquatic larvae would have matured, while the distribution of fish in the Alaska/Yukon area helped unravel the sequence of events as the Laurentide ice sheet retreated and the Bering land bridge became submerged. The Lake Whitefish *Coregonis clupeaformis* is found in both the Mackenzie and Yukon rivers, even though the two are separated by a substantial mass of high land. In this case, it seems that two proglacial lakes formed as the ice sheet retreated north-eastwards. The lakes were connected by a channel that allowed the fish free passage, but as the water levels in the lakes fell (when iceberg calving ceased), the lake populations were isolated. The linked lakes had drained into the Yukon River, and the western lake continued to do so. But as the ice retreated across the valley through which the Mackenzie now runs, the more easterly lake drained that way, taking the fish with it. The Blackfish *Dallia pectoralis*, a form of mud-minnow, has an even more curious distribution, being found in the rivers of Chukotka, western Alaska and on islands of the northern Bering Sea (eg. St Lawrence Island). These islands were once hills rising above the plateau of Beringia, the plateau becoming the seabed as the ice melted. But by then the fish had populated all Beringia's rivers.

Of the fish identified in Arctic freshwaters the most northerly is the Arctic Char, a population of which lives in Ellesmere Island's Lake Hazen at 82.5°N. In the cold waters of that lake, and other Arctic lakes and rivers, the Char grows slowly, a fact that has given rise to concerns over the survival of some populations as it is also a popular sport fish. Another sport fish is the brilliantly coloured Arctic Grayling *Thymallus arcticus*. The Grayling is circumpolar in distribution, as are Burbot *Lota lota* and Northern Pike *Esox lucius*, though in all cases the inevitable isolation of discrete rivers and lakes means that subspecies with marked differences have arisen. Other lake and river fish include trout, minnow, carp and perch species. During the Arctic winter, river fish can migrate towards the sea, ensuring that they reach a section of river that remains unfrozen. Fish in large lakes survive below the ice, though those in smaller bodies of water or in small

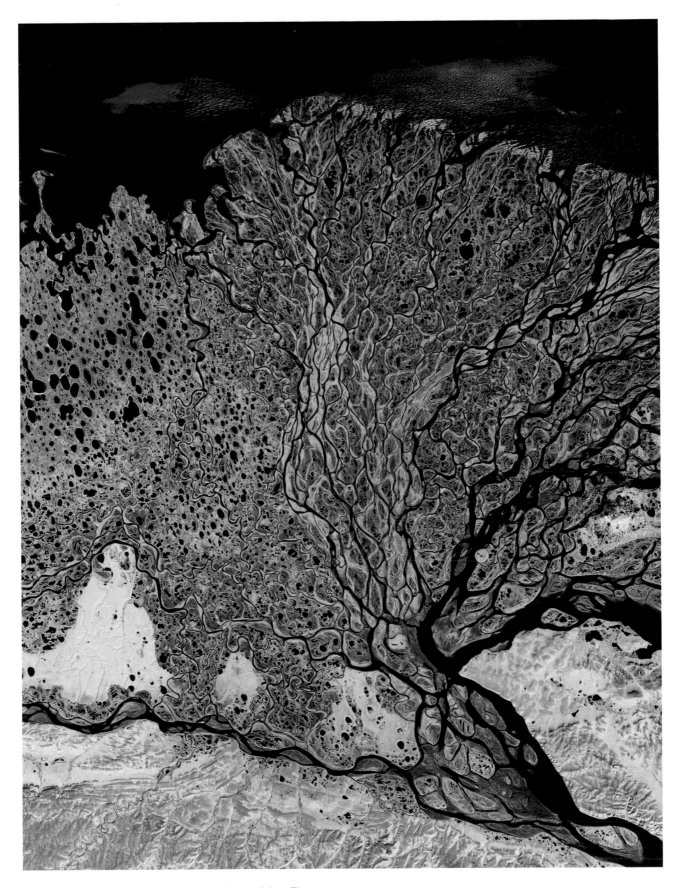

Colour-enchanced satellite image of Lena delta. The
myriad of channels is well illustrated.

Arctic Char.

streams that freeze entirely cannot survive. It is often claimed that Blackfish can survive freezing, but this is not so. They can survive very low temperatures and short periods of freezing, during which they may be partially frozen, but complete freezing is fatal.

There are numerous species of zooplankton, as well as molluscs (including gastropods and bivalves) in Arctic freshwaters. There are also some freshwater mites. There are few Arctic aquatic insects, though many insects with terrestrial adult forms have aquatic young – e.g. caddisflies, black flies, mayflies, stoneflies, midges and mosquitoes. The larval stages of these, which can reach astonishing densities, are important food resources for fish and birds: the curious 'spinning' of swimming phalaropes is a way of stirring up insect larvae, which are then pecked from the water surface.

## The Greenland ice sheet

Perhaps the most amazing of all Arctic habitats are cryoconite holes, the 'oases' of micro-organisms found on the Greenlandic ice sheet. Dust finds its way on to the ice sheet, both wind-blown towards its edge and, remarkably, cosmic dust collected by the Earth as it sweeps through space. In each case the dust is dark and so absorbs heat, forming water-filled holes in which the temperature can be up to 5°C higher than the surrounding ice. Within these miniature ponds, which can vary from a few millimetres to several centimetres in diameter and can be as much as a metre deep, bacteria

and algae find a home and are grazed by nematodes, rotifers and water bears.

## Terrestrial habitats

The total number of vascular plants identified within the Arctic is around 1,000, that number rising towards 2,000 when subspecies are taken into consideration. There are also many hundreds of embryophites (mosses and liverworts), more than 1,000 lichens and a surprising number of fungi. Most Arctic vascular plants are tundra-specific, though some are also found in the taiga and to the south. Many species are circumpolar (though with subspecies) while others show the influence of Beringia by having a trans-Beringian distribution. Yet others show a distinct trans-Atlantic distribution. The Hairy Lousewort *Pedicularis hirsuta*, which is familiar to visitors to Svalbard, Greenland and Baffin Island, is a good example of a trans-Atlantic species.

Given the number of tundra plants and the complexity of the relationship between the various species and subspecies, no detailed description is possible here. Instead, only a general introduction to the species likely to be seen is given, together with details on some of the more common species. About 60% of tundra vascular plants are circumpolar (with subspecies), this number rising to about 90% for the polar desert. Non-vascular plants show similar percentages, but with a greater number of species.

Lichens are particularly important in the Arctic, providing a valuable winter food source for Musk Oxen, rodents, hares and, especially, Reindeer: Rock Tripes *Umbilicara* spp. helped save the first of Franklin's overland expeditions from starvation. Lichens also add splashes of colour in otherwise uniform landscapes, something which all travellers to the area appreciate. Lichens were long thought to be dual organisms, a combination of fungus and alga, the algae lying a little way below the surface of the fungal thallus. Lichens were therefore an example of a facultative mutualism, the fungus providing water and minerals to the algae in exchange for the products of photosynthesis: free forms of lichen fungi exist, but they grow more slowly. However, recent work has shown that many lichens are more complex, many including a third organism, a yeast, so that the individual lichen can be considered an entire eco-system. They are also incredibly robust, lichens having survived outside spacecraft for up to 18 months, despite being in a vacuum and high radiation field. Most lichen reproduction is vegetative, but some associated fungi produce spores, these only forming lichens if they capture algae cells.

Lichens are of three forms – crustose (crusty), foliose (leaf-like) and fruticose (shrub-like), and they will grow on virtually any substrate – soil, rock or tree bark. In the far north grey and orange crustose forms that colonise rocks are frequently seen, sometimes in places where no other vegetation is visible. One of these, *Rhizocarpon geographicum* (occasionally called Map Lichen because of its irregularly shaped black patches) grows slowly but at a defined rate, and can be used to measure the time since the retreat of ice from an area: it is estimated that some specimens of the lichen are at least 9,000 years old. Of the fruticose forms the most famous is the inaccurately named Reindeer Moss (*Cladonia* spp., but particularly *Cladonia stellaris* and *C. rangiferina*), which occasionally forms extensive and dense patches in open areas at the edge of the timberline. These lichens represent as much as 90% of Reindeer winter diet and a good fraction of the summer diet as well.

Lichens derive much of their nutrient uptake from the air and so accumulate pollutants and, as they are long-lived, represent a long-term indicator of local pollution. Lichens were used to check radioactive fall-out levels during early nuclear weapon testing, and following the Chernobyl accident. A significant amount of the fall-out from Chernobyl came to ground in Fennoscandia, where the Reindeer consuming lichens became contaminated, as did the Sámi and other reindeer-herding groups who ate the meat.

The importance of mycorrhizal fungi in aiding plant growth and, therefore, the development of Arctic habitats has already been discussed (see Chapter 12): other fungi are much more conspicuous. Indeed, the number of fungi in the Arctic is very high, probably greater than the number of vascular plants, though few are seen in the far north. With the invertebrates that are the primary decomposers of the temperate world relatively scarce, fungi are the major agents of decomposition in the Arctic, and many of the fungal families familiar in more southerly latitudes have representatives north of the treeline.

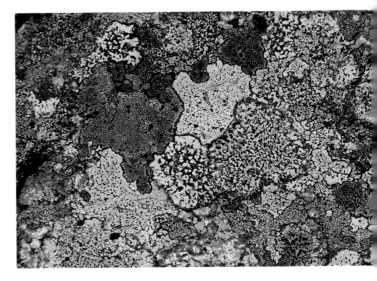

A specimen of *Rhizocarpon geographicum*, north Norway.

*Cladonia stellaris.*

Northern forest carpeted with Reindeer Moss, Sweden. Reindeer Moss is actually several species of *Cladonia* lichen.

Yellow wall lichen.

Puffball (*Leccinum* spp.), Victoria Island, Canada. Fungi are also a feature of southerly Arctic habitats.

A thick carpet of *Racomitrium* mosses, Eldhraun, Iceland.

## The tundra

The Finnish word *tunturia*, describing a treeless plain, has evolved into *tundra*, which describes the circumpolar treeless belt that lies between the Arctic Ocean and the treeline. In North America, the northern section of the Canadian mainland between the Smoking Hills and Hudson Bay is often called the Barren Grounds or Barren Lands. This was the last section of the Canadian Shield to emerge from the ice after the last Ice Age, its lack of vegetation reflecting that limited ice-free history. The name is used to describe forms of Caribou and Brown Bear, and will occasionally be heard across North America to describe the northern lands, though in general 'tundra' is now the accepted label worldwide.

Tundra is characterised by low temperature, low precipitation (particularly to the north) and a short growing season. But despite these characteristics applying in large part to the entire land belt between the treeline and the ocean, the tundra is not homogeneous: it has sub-divisions based on latitude and the characteristics of the landscape, each sub-division having its own vegetation. Beneath the tundra lies permafrost, the annual thaw of which provides water to compensate for the area's low precipitation. The active layer created by the annual thaw allows plants to become established, but though it provides a growing medium, the active layer is a harsh environment. It may have a negative thermocline, i.e. the temperature decreases through the layer, and it may also be waterlogged, as the underlying permafrost inhibits drainage. The depth of the active layer also defines the depth of the root structure of plants, as permafrost is as impenetrable to roots as it is to water. The summer thaw of the active layer is slowed by plant growth, leaf coverage preventing soil heating: in areas where vegetation is absent, the active layer may be two or three times deeper than that beneath local vegetation cover.

Attempts have been made to divide the tundra into zones of vegetation, and while this has been reasonably successful, the boundaries between zones are flexible since local conditions – shelter from the wind, differences in snow accumulation, presence or absence of streams, etc. – can mean that oases of plant life occur in otherwise unsuitable locations. I attempt here to define the differing types of tundra. In the main these are also latitude-based, but they do allow for the influence of microclimates.

## Polar desert

This is the most northerly region of the tundra, covering the Arctic islands of Eurasia (but see below regarding Wrangel Island) and most of those of Canada (though the southern parts of the southerly islands have more extensive plant coverage), and the shield area of the Canadian mainland (the Barren Lands). The polar desert is dark, cold, arid and windswept. Several factors reduce vegetation cover: the long polar night and low sun angle of the polar summer shorten the growing season and reduce the sunlight available for photosynthesis; low temperatures and lack of water inhibit plant growth; the wind scours exposed places, and piles snow into depressions where long melt times mean that the advantages of insulation are outweighed by a further shortening of the growing season for emerging plants. Consequently, there are only scattered patches of vascular plants, the ground cover often being less than 5%. This lack of ground cover indicates a further problem – a form of 'negative feedback': fewer plants means that an individual plant is offered less protection against the elements by its neighbours and so finds it increasingly difficult to survive.

The polar desert is a place for only the hardiest of plants, such as Saxifraga, Papaver, Cerastium, Dryas and Draba. 'Saxifrage' derives from the Latin 'stone-breaker', which has led to the occasional suggestion that the plants aid the production of soil in the stony alpine terrain or tundra they prefer: this is not correct, since the name actually refers to the similarity of its reproductive buds to kidney stones, a similarity that once led to the use of the plant as a remedy for dispersing the stones. The saxifrages include Purple Saxifrage *S. oppositifolia*, which grows to 83°N in northern Greenland, making it probably the most northerly flowering plant. Competition for this title comes from Moss Campion *Silene acaulis* and perhaps from Mountain Avens *Dryas integrifolia/D. octopetala*. The hesitation over the scientific name for the latter species arises from there being two closely-related species whose ranges overlap in eastern Siberia, western North America, and eastern Greenland. In North America both plants are called Mountain Avens, with the two differentiated by being given a second name (White Dryas and Eight-petalled Dryas). In limited areas of North America and in north-west and north-east Greenland hybrids of the two species have been identified. Just to add further confusion, the Eight-petalled Dryas occasionally has seven or nine petals.

The Papaver poppies are among the most delightful of the northern flowers, the long stems making them look particularly fragile. Arctic Poppy *P. radicatum* can be sulphur-yellow, but also white or even pale pink. As with other northern species, although the plant

Marsh Saxifrage *Saxifraga hirculus*, Myggbukta, north-east Greenland.

Tufted Saxifrage *Saxifraga cespitso*, Spitsbergen, Svalbard.

is essential circumpolar, closely related species are recognised in certain areas (for instance the Svalbard Poppy *P. dahlianum*). Many of the other polar desert plants are also circumpolar, but with subspecies. Draba, the white and yellow Whitlow-grasses, and Cerastium (mouse-ears and the related chickweeds) are particularly difficult to differentiate for the non-expert. Other plant genera represented in the tundra of the extreme north include Ranunculus, which includes the Snow Buttercup *R. nivalis*, Lapland Buttercup *R. lapponicus* and Arctic Buttercup *R. hyperboreus*, each of which is circumpolar. As with the northern poppies, the long-stemmed, fragile appearance of the buttercups is in sharp contrast to their actual hardiness. Potentilla (cinquefoils) and Minuartia (sandworts) may also be seen.

In sheltered spots the northern dwarf willows occur – Arctic Willow *Salix arctica*, a Chukotka, North American and Greenland shrub, and Polar Willow *S. polaris*, which is found in the European Arctic. The difference between the two is marginal and they may well be subspecies. Arctic Bell-Heather *Cassiope tetragona* may also be found in these places. In more exposed areas the willow is more of a creeping woody plant, often only 2 or 3cm high. Yet it is a true tree, its leaves changing to a beautiful red in autumn. Arctic Bell-Heather, a circumpolar species called Arctic White Heather in North America, is common on Svalbard and Greenland: in Greenland it was important as a fuel for the local Inuit.

Miniature grasses occur both inland (*Poa* spp. and *Festuca* spp.) and close to the coast (*Puccinellia* spp), and despite the region's aridity there are also sedges, particularly the drought-tolerant Cushion Sedge *Carex nardina* and Rock Sedge *C. rupestris*. In wetter areas such as stream valleys there may be Arctic Sedge *C. stans*.

Occasionally within the polar desert there are areas of exceptional plant vitality, akin to the oases in the more familiar hot deserts. These exist where local topography allows a good, well drained soil to develop in a spot where those plants that manage to take root are protected from the wind. Such an area is the valley of Ellesmere's Lake Hazen, at about 82.5°N, where more than 100 flowering plant species have been identified. Similar oases may also be discovered on a micro-scale: small areas where there is protection from the wind,

Arctic Poppy *Papaver* spp., Fosheim Peninsula, Ellesmere Island, Canada.

Wild Iris *Iris setosa*, Kamchatka, Russia.

Pasque-flower *Pulsailla ludoviciana* near Inuvik, NWT, Canada.

where an animal has died and so fertilised the soil, or beneath a look-out perch frequently used by a raptor, will all show greater productivity, either in terms of species diversity or an abundance of growth.

## The southern tundra

To the south of the polar desert, tundra vegetation covers a greater percentage of the land, though initially the list of species is much the same. Ultimately more species appear – close to the treeline there are about four times the number of vascular plants as seen in the far north. Some of these are shrubby plants, a fact that has led to the suggestion that in addition to a treeline there is also a shrubline. The height of the shrubs, and of some other plants, also increases as the climate becomes, relatively, more benign. But within this graded approach to the treeline there are specific forms of tundra in which differing species dominate.

### Dry tundra or fell fields

Both names are frequently used for this tundra type, the latter deriving from the Scandinavian *fjell*, mountain. Dry tundra is an area of poor soil, a rocky or stony habitat, often exposed and so with a limited number of vascular plants, most of which maintain a low form to avoid dessication. Dry tundra is common in the southern areas of Canada's southerly Arctic islands, Russia's Taimyr Peninsula, and other upland Arctic areas. Because dry tundra tends to be windswept and so has limited snow cover, it is an important winter feeding area for Musk Ox. The plant species are similar to those of the polar deserts, but sometimes with a more extensive coverage. There are often species of Oxytropis (the oxytropes and crazyweeds of North America, and the milk-vetches of Eurasia). In more southerly areas of dry tundra, patches of the matted Alpine Azalea *Loiseleurisa procumbens* add a splash of colour, the gentle pink-and-white bell-shaped flowers contrasting with the deep green of the leaf mat, while the Black Bearberry *Arctostaphylos alpina* adds a valuable berry to the diet of Arctic wildlife. Lapland Diapensia *Diapensia lapponica*, another species that forms low cushions, also occurs here.

Dry tundra areas are also seen in more southerly locations, particularly those with limited snowfall. In such locations, the number of berry-producing heath species increases – Vaccinium species such as Northern Bilberry (or Blueberry – *V. uliginosum*), Cranberry (*V. oxycoccos*) and Rock Cranberry (*V. vitis-idaea*), of which there are subspecies throughout the Arctic, and Crowberry *Empetrum nigrum*, a circumpolar species, while other heath species such as Labrador Tea *Ledum palustre* and some heathers flourish.

Crowberry *Emperium nigrum*, west Greenland.

Autumn colours, Muskusoksfjorden, north-east Greenland.

## Mesic tundra

Mesic tundra is an intermediate form between dry tundra and the wetter sedge and tussock forms. Watered by streams of melting snow yet adequately drained, mesic tundra is home to many varieties of grasses and other flowering plants. Mesic tundra is found in eastern Canada and across the European Arctic to the Taimyr Peninsula, though in the eastern Russian Arctic, and in Alaska and Yukon it is largely replaced by tussock tundra.

On mesic tundra Dwarf Birch *Betula nana* occurs among several varieties of willow – e.g. Woolly *Salix lanata* or Downy *S. lapponum*. On southern mesic areas alders (e.g. *Alnus crispa* and *A. fruticosa*) occur. Berry-producing plants and heaths also thrive, with additional circumpolar species such as Arctic Bramble or Dwarf Raspberry *Rubus arcticus* and Cloudberry *R. chamaemorus*: the latter is particularly common in northern Fennoscandia where it is much sought after. In damper areas of mesic tundra a wide range of mosses and sedges occur.

Some authorities recognise another form of tundra, dwarf-shrub tundra, as being intermediate between dry tundra and mesic tundra. They place this form on well-drained soils, usually close to rivers or in areas with limited snow fall. However, the vegetation species list for such areas is largely as that listed above – birch and willows, berry-producing plants and species such as Labrador Tea, Arctic Rhododendron *Rhododendron lapponicum* and Lapland Diapensia (with Matted Cassiope *Cassiope hypnoides* more common in eastern Canada, Greenland and Fennoscandia). Dwarf-shrub tundra is prevalent in western Alaska, Fennoscandia and eastern Chukotka. There it is an important feeding ground for Reindeer and for Snow Sheep in Chukotka, particularly as fruticose lichens often thrive among the shrubs.

Although it is one of Russia's Arctic islands, the dominant habitat on Wrangel Island is mesic tundra, with a remarkable collection of around 400 species of plant. The island was not covered by an ice sheet during the last Ice Age, and it is thought that the plant variety developed due not only to this lack of glaciation, which allowed an existing flora to flourish, but from occasional periods of attachment to Beringia that allowed the spread of southern species. Periods of isolation from

Labrador Tea *Ledum palustre*, Victoria Island, Canada.

Beringia allowed the development of endemics, of which the island has many.

## Wet tundra

Wet tundra covers around half of northern Siberia, large areas of the central Canadian mainland, much of northern Alaska and areas of Greenland. In many places in the southern Arctic it is the predominant form, and though it is an excellent habitat for waders and waterfowl it is rather less welcomed by the Arctic traveller. In wet tundra the dominant plant species are the Common Cottongrass *Eriophorum angustifolium*, a circumpolar species, together with the Harestail Cottongrass *E. vaginatum* and White Cottongrass *E. scheuchzeri*, and Arctic Sedge *Carex stans*, Mountain Bog Sedge *C. rariflora* and Water Sedge *C. aquatilis*. In western Siberia the dominant sedge is a particular subspecies, *C. ensifolia arctisibirica*, the absence of which signifies the transition to east Siberia for Russian scientists. As well as this broad switch, there are also more local changes. For example, on the southern island of Novaya Zemlya and on adjacent Vaygach Island, Shortleaf Hairgrass *Deschampsia brevifolia* dominates.

In general, either the cottongrasses or the sedges prevail in an area of wet tundra. Where cottongrass dominates, the white, fluffy seed heads create one of

Arctic Harebell *Campanula uniflora*, west Greenland.

the Arctic's most aesthetically pleasing sights. Mixed with the cottongrasses and sedges are grasses such as Arctic Marsh Grass *Arctophila fulva*, and mosses. Where the ground is continuously waterlogged there are sphagnum mosses, while on drier ridges dwarf birch, heathland vegetation and berry-producing shrubs occur.

One form of wet tundra is tussock tundra. This has a circumpolar distribution, but is most frequent in areas where the active layer of the permafrost is about 50cm deep. It is a feature of the Russian Arctic east of the Kolyma delta, particularly in Chukotka, and of the western North American Arctic from Alaska to the Mackenzie. Tussocks are an adaptation to keep the Harestail Cottongrass away from the permafrost. The plant stores nutrients in tubers within the tussock and so isolated from the frozen ground. The plant can also photosynthesise when the temperature of early spring is still below zero using a phenomenon known as 'snow greenhousing'.

## Forest or shrub tundra

Close to the treeline the shrubs, particularly the birch, willow and alder species, grow taller and further species – *Populus* spp. i.e. aspens and poplars – become established, creating an area of forest tundra. Interestingly, forest tundra often has fewer species than either the tundra to the north or the boreal forest

## Snow Greenhousing and Tussock Homes

Cavities created between the soil and the snow by local solar heating may be 10°C warmer than ambient air allowing photosynthesis. The tussock forms when the dead leaves of cottongrass take time to decompose in the cool and acidic waterlogged ground at the base of the plants. Dead material therefore builds up at the plant base. Eventually this material is converted to a soil that may be exploited by other plants, dead leaves from these being added to the base so the tussock height increases. Tussocks can be more than 100 years old and have significant heights. As a micro-habitat tussocks are superb for many species. The Siberian Brown Lemming finds a home in the tussock where the greenhousing allows winter breeding, extremely unusual behaviour which can result in spring

population peaks which in turn causes irruptions of Snowy Owls. Insects also make a home in tussocks, while waders and other birds may use them for feeding or nesting. But for the Arctic traveller tussocks can be a nightmare. An individual tussock wobbles and is unstable, making tussock-hopping a risky means of travel, and the ground between the tussocks may be waterlogged. The combination makes for slow, hazardous travel, the misery compounded by the fact that tussock tundra is the ideal breeding place for mosquitoes. Tussocks can burn, the dead, dry leaves at the tussock base making good tinder, but as the new buds of the cottongrasses and other plants are often buried deep inside the tussock they survive, so that the flames become a useful regenerator.

Arctic cottongrass, Badlanddalen, north-east Greenland.

to the south. In forest tundra berry-producing shrubs tend to grow taller and set more fruit, making the area particularly attractive to Reindeer. Rushes are found in the wetter areas. Forest tundra is a feature of eastern Russia, where there are large expanses between the tundra of the Taimyr Peninsula and the taiga, and significant expanses to the east of Taimyr, extending as far as the border with Chukotka.

Some flowers of these areas are impressive. Of the lily family, the Chocolate Lily *Fritillaria camschatcensis* of Kamchatka and the Aleutians), a magnificent chocolate brown flower, is perhaps the finest example. Among the orchids, Calypso *Calypso bulbosa* has a flower reminiscent of a masked carnival figure. It grows in northern Fennoscandia, across northern Russia and in north-western North America. Spotted (or Pink) Lady's Slipper *Cypripedium guttatum* has a distribution that includes the Mackenzie delta and the Aleutians, and Eurasia (though in the latter it is usually confined to areas south of 60°N). The purple-spotted white flowers are distinctive. Also in Fennoscandia and across Russia are the various marsh orchids, while Kamchatka and the Aleutians are home to several local species. One of these, the Bering Bog Orchid *Platanthera tipuloides*, which grows on Attu and more rarely on islands east to Unalaska, is the rarest North American orchid. It is tall (*c.*20cm) and has up to 20 tiny, golden-yellow flowers on a single stalk. The Pasque flowers are a collection of related species of circumpolar distribution, species including *Pulsatilla ludoviciana* of north-western North America and *P. pratensis* of Fennoscandia. Wrangel has its own endemic species *Pulsatilla nuttaliona*. Visitors to Alaska and the Yukon will see the Nootka Lupine *Lupinus nootkatensis*, which thrives on the Aleutians, the Pribilofs and in southern Alaska, and the Arctic Lupine *L. arcticus,* which occurs in northern areas of Alaska and nearby Canada. Finally, there are the gentians – the Alpine Gentian *Gentiana nivalis* of northern Eurasia and the Northern Gentian *G. acuta* of North America and Aleutians. On the latter, the Aleutian Gentian *G. aleutica* is an endemic species. The Aleutians is also home to Cow Parsnip *Heracleum lanatum*, a member of the hogweed family, which grows up to two metres tall in sheltered stream valleys.

Kamchatka Rhododendron *Rhododendron camschaticum*, Chukotka, Russia.

Chocolate Lily *Fritillaria camschatcensis*, Kamchatka, Russia.

## A natural calendar

Fireweed *Epilobium angustifolium*, a coloniser of burned or otherwise disturbed areas, acts as a makeshift calendar for the inhabitants of Alaska and Canada's Yukon Territory. The flowers open progressively up the tall stem; when the last flowers, at the stem tip, open, then winter is just around the corner. The progressive opening is a reproductive strategy, as bees visiting a plant always start collecting at the bottom and work their way upwards. The official flower of Yukon Territory, Fireweed is related to the willowherbs of Eurasia.

## Boreal forest or taiga

Boreal forest and taiga are names given to the forest belt that crosses North America and Eurasia, particularly the northern reaches which are characterised by a climate of long, dark, cold winters and short, cool summers. In general precipitation is low. There are, of course, variations: the Pacific coast of North America is warmer in winter and wetter overall, while inland Siberia has winters that are ferociously cold. The two titles are interchangeable: *boreal* derives from *Boreas*, the North Wind of Greek mythology, but *taiga* has a much less definite origin. Some authorities suggest a Turkic word meaning a dense coniferous forest area rich in wildlife, while others see origins in the language of indigenous Russian peoples, deriving from phrases for 'swamp-forest' or 'stick forest', a reference to the short, stunted form of trees at the northern forest edge, or from *tiy*, the name for the Reindeer.

Although the treeline is a useful construct it is not a clear-cut limit (see Figure 13.1). There are no trees north of Alaska's Brooks Range, while across the border in Canada, near the Mackenzie delta, trees grow to the shore of the Beaufort Sea. From there the treeline heads south to the southern shore of Hudson Bay, then north again into Quebec and Labrador. In Eurasia, the treeline is no better behaved, being well above the Arctic Circle in Fennoscandia (because of the influence of the North Atlantic Drift), then heading south to the Circle, before turning north yet again. In the far east of Russia the treeline is almost a north-south line, with minimal encroachment on to the tundra of Chukotka. In mountain areas, where elevation adds extra complexity, the treeline is even more difficult to draw.

At the northern edge of the taiga, trees become more widely spaced, creating an area of forest tundra as noted above. This zone is of variable extent, depending upon local topography. On Russia's Taimyr Peninsula and inland Labrador in Canada it is occasionally several hundred kilometres wide, yet is only a few kilometres wide in eastern Labrador and on mainland Scandinavia. But frequently there are stands of trees even further north. These indicate where the treeline used to be when the climate warmed after the retreat of the ice at the end of the last Ice Age: today's treeline is now several hundred kilometres south of its position during that period. These stands often survive by suckering. In North America Black Spruce *Picea mariana* sucker relatively easily, while White Spruce *P. glauca* and Tamarack (or American Larch) *Larix laricina* sucker less often. Suckering means that a whole stand of trees can be clones of an original tree or trees on a site and so could be viewed as being perhaps 5,000 years old. If fire destroys them they would not be replaced, but if climate improves then it is possible they will revert to sexual reproduction. In exposed places close to the edge of the taiga, trees can be deformed and twisted. Such expanses of misshapen trees are known as *krummholz* (from the German 'twisted wood') and can occasionally form areas that are almost impossible to penetrate.

Despite many species of birds and mammals having circumpolar distributions, the trees of the taiga differ in the Nearctic and Palearctic, though in each case the dominant species are coniferous, these, as we have seen, being better adapted to withstand the Arctic climate. As most of the taiga grows on permafrost, the distribution of both trees and species depends on the thickness of

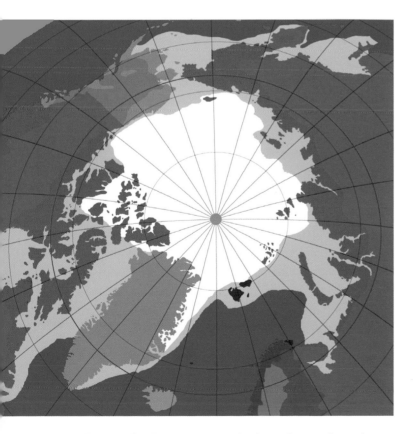

Figure 13.1 The position of the boreal forest (taiga) across the northern hemisphere is shown in green.

the annual active layer. Spruces and larch dominate where the active layer is thinnest, as they produce only shallow root systems. Larches shed their leaves in winter as a protection against the intense cold: they are the only deciduous conifers. The ground cover of larch forests is dominated by lichens, while that of spruce forests is mainly green mosses, though there are clear exceptions to these general rules.

The North American taiga is dominated by spruce, though Alder, a nitrogen-fixer that is intolerant of shade and so prefers open ground, probably preceded the first spruces. White Spruce favours well-drained land, but it can grow on inorganic soils and so was among the first to colonise ground uncovered by the retreating ice. Black Spruce prefers a damper environment and requires a soil richer in organic material, and so would have moved in later. Other prominent species are the Jack Pine *Pinus banksiana* and Tamarack, while broad-leaves Balsam Poplar or Cottonwood *Populus balsamifera* and Trembling Aspen *P. tremuloides* may also be seen. Firs *Abies* spp. are almost entirely absent except in northern parts of coastal Labrador, where the Balsam Fir *A. balsamea* is a dominant species right up to the treeline.

The Palearctic taiga is vast, particularly the section that extends across Russia. This Siberian forest covers over 500 million hectares, constituting about 20% of the Earth's entire forested areas, and more than 50% of the total coniferous forest. The forest covers a huge longitudinal range (from Scandinavia to the Chukotka border) and forms a closed, often dense forest across most of that great range, except in eastern Siberia where it gives way to open larch forests in the north (but with the southern taiga maintaining its denser form). South of the Taimyr Peninsula, in the valley of the Novaya River, are the world's most northerly stands of trees, with 'forest islands' of Daurian Larch *Larix gmelinii* at 72°30'N. One of these islands, Ary-Mas, is actually the world's northernmost forest, separated by more than 30km from the taiga. Ary-Mas extends to almost 60km² and is home to several species – plants, birds and rodents – which are not found on the surrounding tundra, only in the taiga to the south.

Though the form of the Eurasian forest remains more or less constant, the tree species change heading east. In mainland Scandinavia, the forest is chiefly of Norway Spruce *Picea abies* and Scots Pine *Pinus sylvestris*. In European Russia, Norway Spruce and Siberian Spruce *Picea obovata* dominate. Further east, the great forests of western Siberia are predominantly Siberian Spruce, Siberian Fir *Abies sibirica* and Siberian Stone Pine *Pinus sibirica*. Western Siberia also has swathes of wetlands (the largest wetland area in the world) because of its poor drainage.

In central Siberia, larch species dominate, particularly Siberian Larch *Larix sibirica*. To the east, Daurian Larch is the main species, together with *L. cajenderi*. To the south of the larch forests Siberian Spruce and Siberian Pine grow, and Scots Pine is found at the edge of the steppes. One characteristic of the central Siberian taiga is the presence of alases, treeless areas with a meadow-like vegetation. Alases form when thawing permafrost creates a lake. The soil near the lake collapses, halting tree growth so that the area surrounding the lake changes to meadow. Alases may vary from a hundred metres or so across to about 10km, and can form as much as 50% of the taiga in some areas.

East again, the severe Siberian climate limits the spread (and size) of Larix species, the forests of Daurian Larch becoming more open with the shrubby Dwarf Siberian Stone Pine *Pinus pumila* often dominating. Ayan Spruce *Picea ajanensis*, White-barked Fir *Abies nephrolepis* and Sakhalin Fir *A. sachalinensis* also occur. *Larix ochotensis* is endemic to the coast of the Sea of Okhotsk.

In western Eurasia alders, birches and willows are the main broadleaved trees. In western and central Siberia broadleaved species, particularly aspen and birch, form a narrow band at the southern edge of the conifer forest, separating the conifers from the wooded steppe to the south. In eastern Siberia this narrow band disappears, the conifers extending to the edge of the steppe, though there are scattered stands of broadleaves, principally the alder *Alnus fruticosa* and the birch *Betula middendorffii*. In Fennoscandia, in sheltered spots, a zone of Mountain Birch *Betula pubescens* is often found between the conifer forest and the tundra. The birch is not drought resistant and so requires shelter to avoid desiccation. The species can withstand winter temperatures to at least -30°C, low for a broadleaf, though the Stone Birch *B. erminii* of north-eastern Eurasia can apparently withstand temperatures even lower (to -45°C). Stone Birch is a feature of the

Sundew *Drosera rotundifolia*, Kamchatka, Russia.

coastal plains and river valleys of Kamchatka, Stone Pine and the alder *Alnus maximowiczii* being found on the mountain slopes.

Within the taiga a surprisingly high number of vascular plants flourish. Much less of a surprise is that shade- and cold-tolerant species dominate. In the Nearctic the main species are members of the Asteraceae (aster), Onagraceae (willowherb), Ranunculaceae (buttercup) and Rosaceae (rose) families, as well as berry-producing shrubs. In the Palearctic taiga the main species are similar, with the addition of the Brassicaceae (crucifers). There are also areas of bog, some of which are extremely large. The bogs of the Palearctic are known by a variety of names, most of them local in origin and with little standardisation, while in North America the most usual term is muskeg. To the specialist the use of muskeg to describe all northern bogs is wrong, though the difference between the various forms is academic to the casual traveller. Muskeg refers to the wetter bogs, created where meltwater saturates the ground, and is, technically, not a true bog. True bogs form by the infilling of lakes and ponds so that the bottom layer of peat is formed of pondweeds and other water plants, while muskeg forms in areas of poor drainage. Bogs are also formed by the process of paludification. Here tree litter such as leaves on the forest floor decomposes slowly because of the low temperature. Mosses flourish on the litter, retaining moisture and adding to the insulating properties of the litter itself. The underlying soil cools, the combination of wet litter and cold soil inhibiting the growth of seedlings. Over time a bog forms. In principle, all northern forests would give rise to bogs in this way, but fire can eradicate the sodden ground cover, giving seedlings a chance to establish. The northern bogs are home to marshland plants, including the insectivorous Sundew *Drosera rotundifolia* and Pitcher Plant *Sarracenia purpurea*. But as with tussock tundra to the north, the bogs are breeding grounds for mosquitoes and other biting insects, and a misery to cross.

Within the taiga fire is a hazard. Detritus on the forest floor builds up, acting as tinder when a lightning strike starts a blaze. The fire spreads rapidly, and consequently temperatures do not become sufficiently high for long enough to damage the soil. Taiga trees have evolved to deal with regular fires: their seeds are stimulated to mature and release by fire, with the mineral-rich ash soil being an ideal growing medium, while the open areas created by fire allow seedlings to develop without competition for light. Fire may also aid the introduction of species that specialise in colonising 'disturbed' ground. However, if the natural cycle is interfered with, particularly if fires are suppressed so that detritus builds up, then blazes can be catastrophic as local temperatures are elevated, allowing real damage to occur to both soil and trees. Regular fires are therefore both useful and important to the forest and its well-being. In general Moose benefit from the new growth, though Reindeer and Caribou do not as the flames destroy the lichens on the forest floor.

## Invertebrates of tundra and taiga

Although most Arctic invertebrates are aquatic (mainly marine), a surprising number of terrestrial species inhabit both the taiga edge and the tundra. In terms of abundance the most numerous are worms (particularly nematodes and oligochaetes) and rotifers, with planarian worms being important in the taiga. There is also a diverse collection of freshwater copepods and many insects. Though there are marine and freshwater molluscs, there are no Arctic slugs or snails.

Spiders occur to northern Ellesmere Island. Indeed, several species from that area are endemic, suggesting the presence of ice age refuges that allowed speciation to occur. Many of these spiders are very small, but on the tundra the traps of funnel web spiders can occasionally be seen, while larger wolf spiders are among the more likely specimens to be encountered. Spiders are also important on the taiga.

## Arctic insects

Insects are the most numerous of Arctic animals (yet the number of species represents only about 0.3% of Earth's known insect species), with densities sometimes reaching staggering proportions. On dry tundra in Svalbard the density of springtails was found to be almost 40,000/m²: on damp tundra this density rose to more than 240,000/m². Similar densities have been measured on the tundra close to Barrow, Alaska. For larger insects the densities decline sharply, but even for dipteran (fly) larvae a density of 50/m² has been measured on Svalbard tundra.

Almost all insect orders have Arctic representatives, though there are relatively few beetles. The most successful species are the non-biting midges or chironomids, which account for as much as 25% of the total insect population in areas of the far north. Many insect species are circumpolar – as many as 80% of mosquitoes and nymphalid butterflies – while many others show distinct trans-Atlantic or trans-Beringian ranges.

Many insects are nectar and/or pollen feeders, these species being important as pollinators and food for birds.

They can, however, also have a negative effect. Because sections of taiga can effectively be a monoculture, any outbreak of insects that feed on the trees can swiftly reach epidemic proportions. Trees may then be killed off by the insect horde (or by microorganisms that they introduce). As well as herbivorous insects, there are also those that feed on carcasses, dung feeders, several predatory species, and significant numbers of parasitoid wasps and flies. The hosts of these insects are often home to the larvae of two or more parasitoid species.

The far north has no mayflies, stoneflies or dragon-flies, though these occur in the southern Arctic, as do grasshoppers, though these are few and found only at the Arctic fringe. Beetles, as noted earlier, are few – they represent only some 10% of the total number of species (this figure reaches about 35% in temperate areas). There are two Arctic diving beetles (*Hydroporus polaris* and *H. morio*), some rove beetles and a single ground beetle, *Amara alpina*. Interestingly, subfossil *A. alpina* have been found in Greenland though the beetle is no longer extant there, having failed to survive the last glaciation. In the southern Arctic there are representatives of other coleopteran families, including ladybirds and leaf beetles.

There are about 15 species of caddisfly in the southern Arctic, but of these only *Apatana zonella* is also found in the far north. The aquatic larvae of these species share the same unusual habit as their southerly cousins, constructing protective tubes of twigs and small stones.

Taiga Bluet Damselfly *Coenagrion resolutum*, Potter Marsh, Alaska.

Of the hymenopterans, there are stingless ants on the tundra, and some Arctic bees, although only two bee species are seen in the far north. Each is large and uses shivering to raise body temperature. The muscle mass required for this process, together with dense insulating hair, explains their large size, which usually comes as a surprise to the first-time Arctic traveller. The queen of *Bombus polaris* overwinters, having been fertilised during the summer. In spring, she founds a new colony, laying two batches of eggs. The first may contain workers, though sometimes does not, but both the first and second batches contain fertile young – new queens and male drones. In some colonies the absence of workers means that the queen herself must forage in order to feed the colony in its early stages. The second, larger, species, *B. hyperboreus*, is a social parasite. A queen of this species overwinters as in *B. polaris*, but on emergence in spring seeks out a *B. polaris* nest site. She bypasses any workers (perhaps using chemical secretions), then finds and kills the *B. polaris* queen. She then lays her own eggs, which are coated with chemicals that trick the *B. polaris* workers into treating them and the larvae that hatch from them as their own siblings. All the *B. hyperboreus* larvae are queens or drones, with the queens ready for overwintering and a new year of cuckoo-like social parasitism with a twist.

## Flies

Dipterans represent about half of all insect species, and there are representatives of the order in the far north wherever their 'normal' larval habitats of dung, carrion and other detritus are available. Of particular interest to the Arctic traveller are mosquitoes and black flies, the females of which are facultative bloodsuckers, i.e. they will feed on blood if the opportunity presents itself. Male mosquitoes feed exclusively on nectar. Females feed on nectar to obtain the energy required to fly, but seek a blood meal to provide the nutrients for producing abundant, healthy eggs. Females who do not obtain a blood meal may also lay eggs, but these will be far fewer in number: they may also produce eggs autogenously, using food reserves accumulated while they were larvae.

Female mosquitoes that find a host may consume up to five times their own body weight in a single blood meal. The insect injects saliva into the host's blood to prevent clotting: it is this that causes the swelling and irritation. Female mosquitoes will feed on any warm-blooded animal – bird or mammal. The great herds of Caribou in the Nearctic are driven, it sometimes seems, to the point of madness by their attention. Although it is rare, cases are known of Caribou dying from blood loss due to mosquito bites. Birds have also been known to succumb: a team researching a Brünnich's Guillemot colony during a particularly warm spring noted the deaths of many birds resulting from blood loss, mosquitoes attacking the feet of the birds, which had blood vessels close to the surface to aid heat loss. It has been calculated that a naked human making no effort to protect himself from the attentions of mosquitoes would die from blood loss within a day. The main mosquito species in the region are *Aedes impiger* and *A. nigripes*, which are widespread with ranges that extend into the High Arctic. There are, however, about a dozen species of Arctic mosquito in total.

Mosquitoes are drawn to sources of carbon dioxide, and, when close to a victim, to body heat. The ability of the insects to sniff out victims is both amazing and infuriating. On one canoe trip to the Canadian Arctic I deliberately camped on an island situated over 1km from the banks: while it is not possible to state with certainty that there were no mosquitoes on the island when my canoe landed, none were seen for about 40 minutes, at

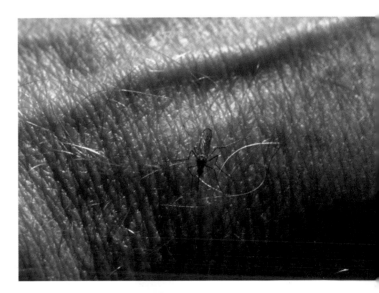

Mosquito attacking the author's hand, NWT, Canada.

which point they appeared in a swarm. At such times it is easy to understand the basis of the Inuit description for such swarms – *sordlo pujok*, 'like smoke' – and to recall one of the first descriptions of Arctic mosquitoes by a European. In Martin Frobisher's account of his journey to Baffin Island in 1576, he writes of meeting insects that were like 'a *small fly or gnat that stingeth and offendeth so fiercely that the place where they bite shortly after swelleth and itcheth very sore*'. It is a description that can hardly be bettered.

It is often claimed by some people that they are more prone to biting than are their companions, and some experiments with (non-Arctic) species, using a Y-shaped tube so that the insects could travel in one of two directions, showed that they do often preferentially chose one person rather than another. This has yet to be explained, but one plausible theory suggests that 'healthier' people are more likely to be bitten: 'unhealthy' people would have relatively fewer nutrients in their blood and therefore represent a poorer investment for the insect, though exactly how a mosquito decides on fitness by smell alone is not understood. It is also known that mosquitoes preferentially bite pregnant women.

Black flies are numerous and particularly ferocious in the taiga and at the taiga edge, where their blood-sucking habits produce a more damaging wound

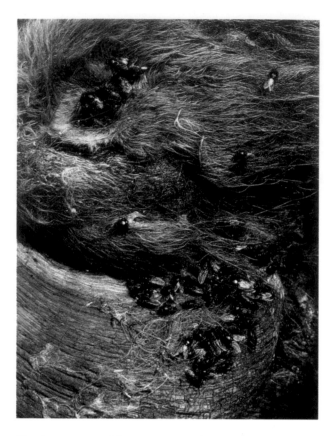

The Arctic is no different from any habitat in needing insects to aid the decomposotion of corpses. Here blowflies are laying eggs on a dead Musk Ox. The maggots will play a major role in disposing of the corpse, and may also overwinter in it.

## Parasitic insects of Caribou

Not only do Reindeer and Caribou suffer blood loss from the swarms of mosquitoes that fill the summer skies – they also have infestations of two highly specialised fly parasites to contend with. The Caribou Warble Fly *Hypoderma tarandi* lays a sticky egg on the legs or underside of Caribou. When the larvae hatch they burrow into the animal, then migrate subcutaneously to the back, close to the spine. There they excavate a breathing hole and begin to feed on the tissues of their host. Most Caribou are infested; some carry up to 2,000 larvae, their skins becoming useless to native hunters. The larvae overwinter in the animal then emerge through their breathing holes in spring, falling to the ground to pupate. Occasionally the breathing holes become infected, adding to the suffering of the host animal.

The Caribou Nose Bot Fly *Cephenomyia trompe* deposits live larvae (that hatch inside their mother) at the entrance to the host's nostrils. The larvae then migrate to the opening of the throat, where they cluster. The larval mass can be so large that it interferes with the host's breathing: the coughing often heard in groups of Caribou is usually caused by animals attempting to dislodge the mass of parasites. The Caribou eventually succeeds in expelling the mass, but only when the larvae are ready to pupate.

Both warble and bot flies are stronger fliers than mosquitoes and so are more difficult for the host animals to evade. The animals lower and shake their heads when a Caribou Nose Bot Fly is seen, and the apparently random jump and run of an individual animal is usually a sign that one or other fly has been spotted.

than that of a mosquito. Black flies are thankfully less common and less damaging in the far north, as there the females do not suck blood because their mouthparts (and those of their mates) do not fully form. Instead the eggs are produced solely using reserves built up during the larval stages. In the Nearctic taiga the famed *No-see-ums*, ceratopogonid biting midges, are another dipteran scourge, particularly as they seem able to penetrate the tightest mosquito netting.

## Arctic butterflies

Butterflies are one of the surprises and joys of the Arctic summer. They can be seen as far north as flowers bloom – the Greely expedition famously spent an idyllic summer at Fort Conger at the northern end of Ellesmere Island (at almost 82°N), with butterflies flitting among the flowers.

Although there are more moths than butterflies in the Arctic they represent a smaller fraction of the resident lepidopterans than is usual in more temperate areas. Of the 90 or so moth species so far identified, most are micro-moths from the families Tortricidae, Noctuidae, Geometridae and Lymantriidae. As noted earlier, the caterpillar of the Arctic Woolly Bear *Gynaephora groenlandica* may develop for up to 14 years before pupating.

Of the families of butterflies, the nymphalids (Nymphalidae) and whites (Pieridae) include the true Arctic dwellers, with blues (Lycaenidae) at lower latitudes, and skippers (Hesperiidae) and Swallowtails (Papilionidae) prominent in the southern Arctic. Species seen in the far north are invariably darker than their southern cousins, an adaptation to allow greater absorption of heat from sunlight. All Arctic butterflies overwinter as larvae or pupae, with some species requiring several years for the larvae to develop sufficiently to pupate.

## Circumpolar species

The most northerly of the butterflies are circumpolar. Of the nymphalids the Polar (or Polaris) Fritillary *Boloria polaris* is found on the tundra of the Arctic islands to northern Ellesmere, on north-west Greenland and across northern Eurasia. The larvae feed primarily on

Mountain Avens. The Arctic Fritillary *B. chariclea* has a similar distribution, though it is not found as far north on Canada's islands and is uncommon in Fennoscandia. The larvae feed on Arctic Willow. The taxonomy of some of the northern butterflies is still debated, as some species and areas are poorly studied. Consequently, there are species recognised by Russian lepidopterists that are not generally recognised elsewhere. In Chukotka and on Wrangel *B. butleri* is recognised and is also considered to be a west Alaska species, but it may well be conspecific with *B. chariclea*. Other *Boloria* species also occur in both Chukotka and Alaska (though problems occasionally occur with the nomenclature due to inconsistencies between Russian and American authorities).

Mating Polar Fritillaries (*Boloria polaris*), Victoria Island, Canada.

Of other nymphalids, the Dusky-winged Fritillary *Boloria improba* breeds on Novaya Zemlya and Canada's southern Arctic islands. Frejya's Fritillary *B. freija* has a similar distribution. Of larger butterflies the Camberwell Beauty *Nymphalis antiopa* (named Mourning Cloak in North America) is rare everywhere, but may be seen at the treeline throughout the Arctic. The Red Admiral *Vanessa atalanta* and Painted Lady *V. cardui* may be seen, but are uncommon north of the Arctic Circle in Eurasia: they also occur on the southern shores of Hudson Bay and eastwards to southern Labrador.

Of the pierids the Northern Clouded Yellow *Colias hecla* (known as the Hecla Sulphur in North America) is found to northern Ellesmere and in north-western Greenland as well as across Arctic Eurasia. The Pale Arctic Clouded Yellow *C. nastes* (Labrador Sulphur in North America) has a more southerly distribution on Canada's Arctic islands, but a similar range in Eurasia. Booth's Sulphur *C. tyche* has a similar distribution: some authorities consider this to be a hybrid of *C. nastes* and *C. hecla*.

Of the skippers the Northern Grizzled Skipper *Pyrgus centaureae* is found in bogs and damp heathland in more southerly areas of Fennoscandia, Siberia and Kamchatka, Alaska and the Yukon, while the Chequered Skipper *Carterocephalus palaemon* (Arctic Skipper) and Silver-spotted Skipper *Hesperia comma* (Common Branded Skipper) have similar ranges. The papilionid *Papilio machaon*, the (Old World) Swallowtail, may be seen to, but rarely beyond, the treeline.

## Trans-Beringian species

Species common to far eastern Siberia and western North America include some that are found to the northern coasts. Nymphalids include the Eskimo Alpine *Erebia occulta*, which is found on gravelly areas of rocky tundra, and Young's Alpine *E. dabanensis*. The latter is the subject of debate as there are several butterflies in Chukotka considered by some to be full species but which may be subspecies of *E. dabanensis*. Similar taxonomic confusion surrounds other alpines. Although in general the swallowtails of the region are restricted to the southern Arctic, Eversmann's Parnassian *Parnassius eversmanni* is seen on the open

tundra of northern Alaska and the Yukon. In Siberia its range extends west to the Altai Mountains. The Phoebus Parnassian *P. phoebus* ranges from the Urals to central Alaska, but is more southerly, being seen on rocky areas of open woodland and at the forest edge.

## Eurasian Arctic butterflies

Apart from the circumpolar species noted above, most Eurasian species are seen in southern Arctic areas. Nymphalids include fritillaries, ringlets and browns. The Small Tortoiseshell *Aglais urticae* has a wide range, while the Indian Red Admiral *Vanessa indica* breeds in Kamchatka. There are several whites of the family Pieridae, together with the lycaenids such as Green Hairstreak *Callophrys rubi* and a small number of other blues and coppers. The only skipper is the Alpine Grizzled Skipper *Pyrgus andromedae*, which may be seen on open moors, heathland and alpine grass of western Fennoscandia and the Urals. There are no Parnassus in the European Arctic, but *Parnassius tenedius* may be seen on the forest-tundra of Siberia from east of the Altai Mountains to western Chukotka.

The resident butterflies of Iceland, and many of the moths, are introduced species, arriving with imported garden plants etc. It is not known if the island has any 'indigenous' species as some moth species that appear to predate the recent introductions may themselves have arrived with the crop plants imported by Viking settlers. Vagrant butterflies – Monarch *Danaus plexippus*, Painted Lady and Red Admiral – and moths – both macro- and micro-moths, and including the Death's Head Hawk Moth *Acherontia atropos* – are seen regularly on the island. Eastern Greenland exhibits a similar lack of resident butterflies, and those on the west coast are limited to the north-west where just a few kilometres of water separates Greenland from Ellesmere.

## North American butterflies

North American Arctic nymphalids include fritillaries, commas, alpines and arctics. Some of these may occur as far north as the Mackenzie delta (being essentially treeline rather than tundra species), though the

Canadian Tiger Swallowtail *Papilio canadensis*, Potter Marsh, Alaska.

Polixenes Arctic *Oeneis polixenes* breeds not only throughout Alaska and on the northern Canadian mainland, but also on the southern Canadian Arctic islands, including Baffin Island. Pierids include whites, sulphurs and marbles. Lycaenids are few, while of the skippers only the Persius Duskywing *Erynnis persius*, which breeds in central and east central Alaska, the Yukon and North-West Territory, can be considered Arctic. It breeds to the coast at, and to the immediate east of, the Mackenzie delta. The delta is also home to the only truly Arctic papilionid, the exquisite Canadian Tiger Swallowtail *Papilio Canadensis*, which can also be seen in east central and south-east Alaska.

177

Polar Bear tracks beside the sea ice at Kvalvaagen, Svalbard.

# 14 The adaptations and biogeography of birds and mammals

The survival strategies of Arctic endotherms (warm-blooded animals) against the cold are very different from those exhibited by the region's ectotherms. The simplest of these is adopted by most Arctic breeding birds – they fly south, escaping the cold and reaching places where food is abundant. Several seabirds stay within the Arctic, feeding at the ice edge or in polynyas. The Snowy Owl, the *Lagopus* species – (Rock) Ptarmigan and Willow Grouse – Gyrfalcon, Common and Arctic Redpolls and the Raven are often said to be resident throughout the winter. In practice, all will move south if local food supplies fail: individuals that remain on their breeding grounds throughout the winter tend to be at the Arctic fringe and so do not contend with the full rigours of the Arctic winter. Nevertheless, these species must also have adaptations to allow them to survive periods of intense cold. Marine mammals are highly specialised, and can also move ahead of winter's ice. For terrestrial species the situation is very different: they, too, are highly specialised.

## Avian adaptations

All Arctic birds have an increased feather density, while the *Lagopus* species and the Snowy Owl have feathered feet. The Snowy Owl also has modified foot pads to reduce heat loss, a characteristic it shares with the Raven. The owl's insulation is superior to that of the *Lagopus* species, but the grouse can reduce their metabolic rate: they also excavate snow caves in which they spend most of their time, emerging only to feed (on a poor forage of willow twigs) something the owl does not do. The redpolls also excavate snow holes, the ability of these much smaller birds to increase their feather density being limited, as there is always a trade off between insulation and efficient function.

As well as having a greater density, the feathers of Arctic birds form a particularly smooth outer surface,

The Ptarmigan has feathered feet, the Hood River, Nunavut, Canada.

Female Svalbard Ptarmigan, Kongsfjorden, Spitsbergen.

Female Snowy Owl, Hudson Land, north-east Greenland. The owl is the only other Arctic bird to have feathered feet.

shedding the wind so that ruffling, with a consequent breakdown of the insulatory layer and increased heat loss, are reduced. Arctic species also have down feathers below their contour feathers, the down having a modified structure, being 'fluffier' to trap air, enhancing insulation properties. The down of the Common Eider became synonymous with warmth in the second half of the 20th century, when the down was harvested commercially, and the 'eiderdown' became the standard word for the down-filled covering on British beds in the days before central heating became widespread. The down of the Common Eider has the best insulating properties of any natural substance: weight for weight and thickness for thickness, no synthetic material can better it. The mammalian strategy of adding insulation in the form of sub-cutaneous fat, or blubber, is not readily available to birds. Penguins do this, but flight has been compromised by increasing weight relative to wing area: birds able to add sufficient blubber to survive an Arctic winter would probably be unable to fly: the Great Auk *Pinguinus impennis*, the largest member of the auk family which was driven to extinction in the mid-19th century was hunted for its 'oil' as much as for its meat and feathers and so clearly had a form of blubber insulation, and was indeed flightless.

Common Eider nest, northern Iceland. Weight for weight, eider down is the best natural insulator in the world, and superior to man-made synthetics.

In very cold conditions birds seek shelter from the wind and stand motionless, occasionally with one foot tucked into the body to reduce heat loss, or sitting so that both feet are covered. Foraging for food requires energy and if food is scarce and local conditions hostile, doing nothing may be the most energy-efficient strategy (though if the temperature is low enough so the bird must shiver to maintain warmth then the strategy may fail, more energy being required to shiver than to forage). As in marine mammals, the arteries of Arctic birds are enveloped by veins in the legs and feet so that venal blood is warmed by arterial blood, allowing heat to 'bypass' the feet, reducing heat loss. The foot temperature of some Arctic species may be 30°C lower than the body temperature. Systems like this are known as counter-current heat exchangers, and are common in Arctic animals.

## Mammalian adaptations

For the terrestrial Arctic mammals, the avian option of migration is, at best, limited. For almost all, the gaining and holding of a territory is also critical to breeding success and so cannot be lightly rejected. The energetic requirements of long journeys, and the physical impossibility of such journeys for small animals such as rodents, also preclude migration. However, both Reindeer and Musk Oxen move south in search of herbage. In neither case are the animals territorial; reproductive success in these species is decided by harem possession (won by trials of strength).

## Torpor

One common strategy in terrestrial mammals that overwinter in the Arctic is torpor, the state in which an endotherm reduces its body temperature to a new, lower norm (which may be close to the ambient temperature), with metabolic processes slowing, usually to about 5% of the rate at normal body temperature. Just as the normal body temperature is critical in maintaining body functions (meaning the animal must increase metabolic rate and take positive steps to avoid hypothermia, or suffer potentially lethal effects), the new, lower temperature is also critical, so that if the body temperature falls below it then the same response is observed. It is, therefore, not true to say that mammals that pass the winter in a state of profound torpor are independent of their surroundings – if they do not respond to a significant fall in temperature they become hypothermic and die, just as any non-hibernating mammal would.

True torpor is confined to small mammals (rodents and insectivores: Arctic marmots seem to be at the upper weight limit for true torpor, with body weights of about 8kg) and a limited number of birds (some hummingbirds, poorwills and swifts, none of which are Arctic species), and is often practised not only during the winter but also at night, to reduce the call on fat reserves. Before entering torpor fat reserves are laid down, but these may not be sufficient to survive the long Arctic winter and Arctic rodents prepare food caches, waking at intervals to feed. Once torpid, heart and breathing rates decline and the carbon dioxide level in the blood increases, reducing the metabolic rate. The animal falls into torpor relatively slowly, but arouses from it much more quickly. During early arousal the body temperature is raised without shivering, but eventually shivering does occur, this causing a rapid increase in body temperature. Field Voles, from the fringe of the European Arctic, huddle together during winter, a strategy that may also be used by less well-studied Arctic species. However, not all Arctic rodents enter a state of winter torpor. Lemmings, for instance, excavate burrow systems beneath the snow which, when lined with grass, allow communal living (with huddling augmenting the insulating value of snow cover) while they remain active, feeding on sub-snow vegetation. This system is not without its drawbacks. Owls can hear the rodents even when they are moving

'Rodent roads', the name given by Norweigians to the paths made through vegetation by lemming and other rodents which are revealed by the spring snow melt. These roads were photographed beside Varangerfjord in north Norway.

below a substantial thickness of snow and will dive through to catch them.

The winter state of Arctic Brown and Black Bears, and female Polar Bears, differs from true torpor, and might more correctly be termed a state of winter sleep or dormancy (and is usually called hibernation). In these bears (and other large mammals such as beavers and skunks) the fall in body temperature is limited to perhaps only 2–4°C. However, the heart rate falls significantly (by about 80%) and the metabolic rate falls to about half that of the waking state. This would not be a sufficient fall to prevent hypothermia unless the animal accumulated very large fat reserves before dormancy, and prior to the onset of winter the bears gorge on a high-calorie diet. During winter they lose some 25% of their body weight. The animals avoid dehydration during dormancy by neither urinating nor defecating.

Female bears of the three species also give birth during winter dormancy. The cubs are very small at

Female Polar Bear and cubs, sea ice off east Spitsbergen, Svalbard. The family has only recently emerged from the maternity den.

Brown Bear in the 'No-Man's-Land' between Finland and Russia.

birth. As a proportion of the weight of the mother, bear cubs, at 0.4%, are the smallest of any mammal. By comparison, a human baby is around 5% of the mother's weight: even when it emerges from the birthing den, a bear cub usually weighs much less than 5% of its mother's weight. Although female Polar Bears sleep the winter away, males do not, though they may excavate a den in the snow to shelter during periods of bad weather.

## Other mammalian adaptations

Beside torpor, the other principal mammalian strategy is to increase insulation, either sub-cutaneously by using layers of fat (blubber), or externally through having a dense pelt. Blubber is the preferred option of most marine mammals, with fur being used mainly by terrestrial mammals. The Polar Bear, which spends significant parts of the year on land or sea ice and in the sea, uses a combination of the two, as do some other species, while beavers and otters use fur only. Beavers and otters have a dense underfur that traps air as an insulator, and a coarser outer fur. This outer fur becomes wet and must be shaken to remove water when the animal emerges. Wet fur is only about 2% as efficient an insulator as dry fur, but these animals spend only relatively short times in water and so maintain a dry underfur. The Sea Otter, which spends virtually its entire life in sea water, also uses fur as an insulator. They have the densest fur of any mammal, with the hairs of the underfur reaching $c.125,000/cm^2$, giving a total of $c.800$ million on the entire body: otter pelts were so prized for their luxurious fur that the animals were hunted ruthlessly. Sea Otters spend about 20% of their time grooming. Mother otters also frequently groom their young, the cub lying across the mother's chest, clear of the water. Sea Otters also blow into their fur or create water bubbles to enhance the air layer at the base of the pelt. This creates a layer of still air close to the skin. Still air is a marvellous insulator (about seven times better than the rubber of a wet suit) and the best furs can create insulating layers that are about 60% as good. The animals also maintain a very high metabolic rate to stay warm, and must eat up to a third of their body weight each day.

The reduced efficiency of wet fur and the need for grooming mean that fur is not a comfortable option for pinnipeds and is out of the question for cetaceans.

Nevertheless, pinnipeds do employ a combination of fur and blubber, though only for pups and the fur seals fur is a significant aid to insulation. An adult Northern Fur Seal has a hair density of 40,000–60,000 hairs/cm², about 35–50% of the density of the Sea Otter, but sufficiently high for the animals to have been hunted for their pelts.

Seal pups are born with a coat known as lanugo, this coat being shed when the pup has accumulated a blubber layer by feeding on milk that is super-rich in fat. Lanugo is thick as the pups rely upon it for insulation, and is usually white for camouflage as most Arctic seal pups are born on ice. Interestingly, the fur seals and seal pups do not curl as an aid to staying warm (by reducing exposed surface area) in the same way as terrestrial mammals (the Arctic Fox gains a further advantage by wrapping itself in its luxurious tail). Seals need to be relatively inflexible to allow efficient use of their hind flippers, while their body shape, optimised for hydrodynamic efficiency, also acts against curling. Their blubber insulation is also so efficient that overheating is the more usual problem for a hauled-out seal.

In principle, making fur thicker also increases its insulation properties. However, there is a limit to how thick fur can be. The polar bear can grow fur up to 6cm thick, a length that would not be feasible on a small rodent. Because of the limit on fur thickness,

The pelts of the Sea Otter (below, sleeping in Resurrection Bay, Alaska) and the Northern Fur Seal (above, sleeping on St Paul, Pribilofs, Alaska) are so luxurious that hunting once almost drove both species towards extinction.

## Polar Bear fur

In common with other species, Polar Bears have guard hairs, longer and thicker hairs that add little to the insulation properties of the fur. The hairs of the underfur are wavy, which allows them to interlock when overlapped, trapping air, with the guard hairs protecting the integrity of the system when the animal is underwater. When the bear emerges from the water, the springiness of the guard hairs allows the fur to 'bounce' back and so encourages air entrapment. As the skin of Polar Bears is black, it was thought that these guard hairs 'piped' sunlight to the skin where its heat was absorbed. While the conditions that would have made this feasible certainly exist, further research has shown that it is, at best, a minor effect.

smaller animals must compensate by increasing their metabolic rate and by burrowing beneath the snow. It is therefore not winter but autumn that presents the greatest danger to small animals, as it brings low temperatures without snow cover. For larger animals that remain active, snow is a very real hazard. Deep snow makes movement difficult and energy-intensive, and covers food. Moose and woodland Reindeer create 'yards', relatively small areas of significant resources where they maintain tracks in the snow to minimise the problem.

The insulation properties of fur are so good that Arctic species can maintain their metabolic rate over a wide range of ambient temperatures. This range, known as the thermoneutral zone (TNZ), is a measure of the adaptation of Arctic animals. The TNZ of tropical species is narrow, as they have evolved in an environment with a limited range of ambient temperatures. It widens for temperate species, and is massively enhanced for Arctic species. For example, a tropical mammal might have a TNZ of no more than 3 or 4°C, centred around 30°C: if the temperature falls below about 26°C the mammal must increase its metabolic rate to avoid hypothermia. For the Arctic Fox, the TNZ is very wide, about 60°C, extending to -40°C when it is clothed in its thicker winter fur, which extends over the paws, both top and bottom. When the temperature falls below -40°C, the fox's metabolic

rate increases, but much more slowly than that of a tropical mammal as its fur continues to provide good insulation. (As a digression, humans evolved in the tropics and are essentially a tropical species and have a limited TNZ, so that if the temperature falls below about 27°C human metabolism must increase to maintain a core temperature of 37°C. If the temperature decreases further, clothing must be added to prevent hypothermia. Although the Inuit, and other northern dwellers, do have a greater ability to resist cold than southern peoples, their TNZ is the same.)

Most arctic Foxes moult to a white pelage in winter, though the so-called Blue Foxes do not. The large and very luxurious tail is wrapped around the animal when it sleeps. The photograph was taken at Kongsfjorden, Spitsbergen, Svalbard.

## The colour of fur and feathers

The winter pelage of the Arctic Fox is not only thicker than its summer coat, but it is also white. This appears to be counter-productive as white is highly reflective, while a black pelage would absorb more of the sun's radiation, allowing the animal to gain 'free' energy. But white camouflages the animal against winter's snow, a fact that has led to the assumption that all white coloration serves the same purpose. Animals that change colour tend to be prey species (even if, as in the case of the Arctic Fox, for example, the species is also a hunter) and so camouflage is beneficial. But it is dangerous to make too many assumptions about animal coloration based solely on what we see as humans. The UV vision of many animals has not been studied, but in those cases where it has, particularly in birds, UV sensitivity has often been found and in UV a white animal may appear dark. Some Arctic species are also white all the time – Polar Bears, the Snowy Owl and

Gyrfalcon. A case could be made for this being to camouflage the animal against detection by its prey, the bears against the snow, the birds against the cloudy sky, but again there are contradictions. In the far north wolves are white throughout the year and so are highly visible against snow-free tundra. Their main prey, the Arctic Hare, is also white throughout the year at the same latitudes. In these cases it seems that because the summer is short, and moulting to a cryptic summer pelage is energetically costly (since it requires pigment production and hair growth), it makes more sense for these animals to forego the effort of moulting into a summer coat altogether. To further complicate matters, the Raven, a truly Arctic bird, is always black. In this case it seems that camouflage is not required – the bird is big enough to look after itself – and, as it is an omnivorous, opportunistic feeder, camouflage is of much less value.

*Staying white in summer makes the Arctic Hares of Ellesmere Island, Canada very conspicuous (left), but the wolves which prey on them also stay white and so are equally conspicuous.*

## Blubber

Sub-cutaneous blubber is important for insulation in marine mammals, including the Polar Bear (though not, as noted above, by the essentially marine Sea Otter). Blubber has advantages for marine mammals as weight is much less of an issue for these species, and not an issue at all for cetaceans. Indeed, for marine mammals bulk is an aid to keeping warm. Heat loss is proportional to body surface area, but thermal inertia (i.e. heat capacity) is proportional to body volume. Animals with small surface area-to-volume ratios can therefore maintain body temperature more easily. A straightforward example of this is the cooling time of dead animals. A dead shrew will cool to ambient temperature in a matter of minutes, while the body temperature of a dead adult Blue Whale reduces by just 1–2°C over a 24-hour period.

Blubber comprises lipids (fatty substances), collagen fibres and other connective tissues, and water. Cetacean blubber is rich in collagen fibres which prevent the non-rigid mass from hanging, bag-like, from the animal, adversely influencing streamlining. The lipid content of blubber varies from about 60 to 80%. It varies both from species to species and within an individual animal depending on age, position on the body and the time of year. In general, marine mammals have more blubber than is necessary purely for insulation purposes, the excess being an energy reserve. The insulating properties of blubber are dependent on thickness, lipid content and blood flow. Centimetre for centimetre it is a poorer insulator than fur, but it has the advantage of being essentially maintenance-free. But though not as effective as fur, blubber is so efficient at keeping pinnipeds and cetaceans warm in water that the animals need a method of keeping cool. This is particularly true for pinnipeds, which haul out of the water into an environment in which heat loss is usually much reduced (since water is highly conductive in comparison to air). To allow for this, pinniped flippers have a blood supply system in which a central artery is surrounded by a network of veins, another example of a counter-current heat exchange. When in the water the venal blood acts as an insulator for the inflowing arterial blood, absorbing heat in the process and so reducing overall heat loss. Then, when the animal is hauled out, venal blood flow can be increased, so increasing heat loss and helping the animal keep cool. As we have seen, similar systems are widespread in Arctic animals: for example, in the legs of Arctic birds and mammals, in the flippers of cetaceans, in the tails of rodents (such as beavers) and in the horns of ungulates. A particularly efficient counter-current system allows the tongues of baleen whales (particularly the Grey Whale and the right whales), which are large and uninsulated and therefore represent a potentially huge heat-loss surface, to remain cool even though they are richly supplied with blood.

Counter-current heat exchange systems result in the extremities being kept constantly cold relative to the core body temperature: the feet of an Arctic seabird may be below 5°C when the bird's core temperature is around 40°C. Similarly, the paws of an Arctic wolf might be only just above freezing and its nose at about 5°C, with a core temperature of 38°C. Counter-current exchangers may be used in conjunction with vasodilation of extremities, with the restriction of blood flow helping to reduce temperature. However, an animal must guard against frostbite; an occasional pulse of warm blood can be sent to the extremity if the ambient temperature is below 0°C.

Counter-current heat exchangers must also work in reverse in cetaceans. Because the testes of male cetaceans are held within the body cavity to improve streamlining, the temperature of the organs would be too high for successful sperm production. Consequently, cold blood from the flippers and fins is diverted to cool the spermatic arteries to maintain a cooler environment. A similar system is employed to cool the uterus of female cetaceans, as the metabolism of the foetus is about double that of its mother and so overheating would otherwise be a problem.

One significant adaptation of Arctic mammals, both seals and terrestrial species, concerns the newborn. These experience a large temperature differential at birth, with as much as an 80°C change from the uterus to the outside world. Despite insulation this could lead to thermal shock. The youngster will shiver to generate heat – shivering as a means of heat production is also used by almost all adult endotherms – but will also employ 'non-shivering thermogenesis', a process unique to young mammals. Here heat is produced by metabolism of brown adipose tissue (BAT or brown fat), which is found in body cavities and around major organs. The colour of this tissue type derives from the mass of capillaries within

Walrus, Appolonoff Island, Franz-Josef-Land, Russia.

it. Heat production by brown fat metabolism is about ten times higher than that of normal shivering. As the young animal develops, its brown fat deposits diminish, though a small amount is maintained in the adults of some species, the Musk Ox, for example.

## Size and shape

In addition to these specific adaptive strategies, the body shapes of northern species conform, in general, to a couple of 'rules', each first propounded in the mid-19th century. Bergmann's Rule states that within a species, as latitude increases or ambient temperature decreases, there is a tendency for a larger body size to reduce the surface area-to-volume ratio, and so reduce relative heat loss. Though recent studies have questioned the validity of the rule, a large body size is helpful if periods of torpor are a winter strategy, allowing the storage of the fat necessary to survive periods of starvation, and helps females to give birth to higher numbers of young, a useful investment as the Arctic summer is unpredictable: periods of harsh weather can spell disaster for small litters. The latter argument is supported by birds, which (again only generally) also obey the Clutch-size Rule which states that northerly species tend to have larger clutch sizes. In general, northern species of birds also hatch after shorter incubation periods and fledge faster than their more southerly counterparts, each a response to the short Arctic summer.

Another rule, Allen's Rule, states that endotherms such as mammals tend to have smaller appendages in cold regions, to reduce heat loss. A good example of this would be the ears of the Arctic Fox compared to those of, for instance, the African Bat-eared Fox *Otocyon megalotis*, which has large ears through which heat can be lost. However, this rule would imply smaller feet and this is certainly not the case in many terrestrial Arctic mammals, as large feet allow easier travel over soft snow. The Arctic Hare has much larger feet than its southern cousins and the Snowshoe Hare is named after this feature. Wolves not only have large feet but long legs, the better to move over soft snow.

## The biogeography of Arctic mammals

The current distribution, or biogeography, of birds and mammals in the Arctic depends not only on adaptation, but also on the distributions that preceded the last Ice Age. In general, northern species moved south as the ice advanced, then moved back north once the ice had retreated and acceptable habitats had been recreated. A good example of this is the Reindeer. Subfossil evidence shows that this tundra species occurred periodically in England (and as far south as northern Spain and Italy) throughout the Pleistocene: the latest subfossils found so far date from 10,000BP, while the species clung on in Scotland until as recently as 8,000BP.

The effect of the ice retreat can also be seen by observing the distribution of more localised species. One good example is a bird of the Arctic fringe, the Ring Ouzel *Turdus torquatus*. This is a bird of cold climates. At the height of the last Ice Age its range probably extended across Europe and western Asia. When the ice retreated and the climate warmed, Ring Ouzels sought out the remaining colder areas for breeding (while spending the winter further south): today they are found in the uplands of southern Europe, on the Caucasus, on the uplands of Britain and Norway, and in northern Fennoscandia.

## Refugia

In addition to the general northward movement of species after the last Ice Age, there were ice-free areas in which species would have been able to survive the glaciation. Nunataks have already been mentioned as potential 'micro-refugia'. Much larger refugia also existed: these formed in north-west Siberia (the Angaron refuge), in north-east Greenland (the Peary Land refuge) and on some Arctic Canadian islands (e.g. the Banksian refuge). The most important refuge was Beringia, an area that included ice-free areas in Alaska, Yukon and north-east Siberia and the negotiable land bridge between the two continents that now lies beneath the waters of the Bering Strait. These refugia are the basis of the three most important tundra areas of today's Arctic in terms of species numbers, emphasising their importance to Arctic biogeography: Russia's Taimyr Peninsula (with 43 species of bird), Canada's Arctic islands (42 species) and the Beringia region of east Siberia and west Alaska (47 species). Each of these areas has endemic species, Beringia having the greatest number. For bird species, the Bering land bridge may have been useful, but even today the narrow Bering Strait is little impediment to birds, and several have extended their ranges across

it since the last time Beringia was above the sea. The Arctic Warbler has spread east from Siberia to Alaska, while the Grey-cheeked Thrush *Catharus minimus* and Yellow-rumped Warbler have travelled the other way.

Beringia was important not only as a refuge but as a bridge to mammal species, including humans, between Asia and the Americas. However, the exact role Beringia played in the movement of mammals is debatable, in large part because there is no consensus on the land bridge's vegetation, evidence from the pollen analysis of lake sediments in Alaska being at odds with studies of surface remains of the permafrost from Seward Peninsula maar lakes. The pollen analysis suggests that Beringia was a polar desert (implied by the density of pollen rather than the number of species), unable to sustain herds of ungulates spreading east. However, ejected material from maar explosions, a plug of permafrost topped with the actual surface vegetation of Beringia, suggests a richer vegetation, one capable of supporting herds. Further work is required to resolve the paradox, though what is clear is that Beringia did indeed represent a corridor for mammalian migration and a range of large-mammal groups of Eurasian origins were established in northern North America by the end of the last Ice Age. The species included mammoths and mastodons, horses (a group that actually evolved in North America, expanded into Eurasia and then re-invaded following extinction in the Americas), several species of bison, Stag-moose (*Cervalces*), Giant Beaver (*Castoroides* – the size of a Black Bear, with huge incisors), and carnivores such as sabre-toothed cats, American Lion (*Panthera atrox*), Dire Wolf (*Canis dirus* – a larger, more formidable version of the Timber Wolf), and the massive Short-faced Bear (*Arctodus* – moose-sized, but apparently a fast-moving animal). Many of these large mammals, some 40 in total, became extinct in a very short period (3,000 years or less) from about 12,000BP, an extinction that raises the question of why, particularly as a similar level of species extinction did not occur in Eurasia. There are several possible explanations. One suggests overhunting by humans, the logic being that people with sophisticated hunting techniques crossed the Beringia land bridge and came into contact with species that had never been hunted (by humans) and so had not evolved to become wary and elusive as they had in Africa and Eurasia. The theory suggests that humans would have concentrated on large herbivores and that some carnivores and scavengers become extinct due to loss of prey. But though humans are a ferocious and relentless predator, could people really have killed off entire species, reducing numbers below the limit of biological survival continent-wide?

Another theory is that the change of climate was responsible for the extinctions. But a changing climate affects all species, so why should North American species fare worse than those of Eurasia, particularly as the continent is vast, allowing species to move to more equable areas? And besides, such groups had survived the transition from glacial to interglacial period (and *vice versa*) many times before. Perhaps it was disease that accounted for the extinctions, pathogens brought from Asia – perhaps by humans or their commensals – killing off local animals, which had no time to develop immunity. Probably it was a combination of all these factors, perhaps with the addition of others as yet unidentified.

## Refugia and speciation

As well as the major refugia formed by the uneven distribution of ice during the last glacial maximum, refugia would also have been formed during the glacial and interglacial periods of the entire 250,000 years of the late Quaternary ice ages. Evidence from studies of mitochondrial (mt) DNA supports this, indicating that in some species identifiable changes occurred each time populations were separated by glaciation events, with genetic divergence resulting in the formation of distinct subspecies. A good example is the Dunlin, a wader with a circumpolar distribution. The Dunlin is polytypic, with subspecies distinguished only by subtle differences in plumage colour and pattern. mtDNA studies indicate that the oldest form of the bird is the subspecies that breeds in central Canada, and that this split from an ancestral form about 225,000 years ago, a time that coincides with the Holstein interglacial. The studies indicate that further splits occurred during the Emian interglacial (120,000 years ago), and during a glacial period about 75,000 years ago. These data suggest that during interglacial periods the Dunlin could expand its range, populations then being isolated by further glaciation. Splits during a glacial period would arise if

Dunlin (above, photographed on Kamchatka, Russia) and Dall's Sheep (opposite, photographed above the Turnagain Arm, southern Alaska) provide clear evidence of speciation.

significant ice tongues developed, isolating populations on ice-free tundra to either side.

Such subspecific differentiation is possible only if the ancestral form of the species had a significant distribution. If that was not the case then glaciation might isolate an entire population, the species only being able to expand its range when the ice retreated. mtDNA studies indicate that this has also occurred, some northern species being monotypic, while in others the differences between subspecies indicate that splits have occurred only in relatively recent times. The Ruddy Turnstone is a good example of the latter phenomenon, having few (and virtually identical) subspecies.

In the case of the Dunlin, glaciation has driven speciation – the development of new species or subspecies – east-west, as might be expected by an ice front that developed southwards, with ice tongues separating populations to either side. However, the ice could also work north-south, pushing one population of a species southwards while another remained in a northern refuge. This process may have allowed the evolution of the two forms of diver seen in North America. The Great Northern Diver and the White-billed Diver are so similar that some authorities consider them subspecies of a single species. In general, the Great Northern breeds to the south of the White-billed, suggesting that an ancestral form remained in a northern refuge, from where it spread west to Siberia, while another population moved south and evolved subtle differences. A similar separation appears to have occurred with an ancestral form of sheep. During one of the ice ages a population became isolated in a forerunner of Beringia, while another moved south ahead of the ice and occupied the mountains of the western United States. The northern species is Dall's Sheep, the southern one being the Bighorn Sheep (*O. Canadensis*).

An equally interesting example of north-south speciation is seen in the Gyrfalcon and Saker Falcon (*Falco cherrug*). The former, the world's largest falcon, is one of the Arctic's most magnificent and emblematic birds. The Saker is a more southerly bird, breeding on the steppes of southern Russia and Mongolia. The birds look very similar and interbreeding by falconers has shown that cross-bred chicks are fertile, indicating that the two species are very closely related. In this case, it seems that the separation of two populations of an ancestral form occurred when, after the ice had pushed one population south, the development of the taiga across Asia separated the two completely: the falcons, being birds of open country, did not cross the forest belt. A similar process is assumed to have given rise to the Polar Bear, when a population of Brown Bears was isolated to the north of the ice. As with the Gyr and Saker, Polar and Brown Bears can mate and produce fertile offspring.

Although the differences between Dunlin subspecies are minor, as are the differences between the two divers, there are more pronounced variations between the two falcons (though the casual observer might be confused if a pale Saker and darker Gyr were seen together), and Polar and Brown Bears are very different. Adaptations required for survival in the harshest of environments have clearly hastened the evolution of the Polar Bear.

## Merging populations

With the retreat of the ice, populations could expand their ranges, perhaps contacting populations from which they had been long isolated. When contact occurred, hybridisation might have taken place (populations being subspecies and so able to interbreed), or two populations might overlap without interbreeding (the populations

then being considered separate species). A third outcome is also possible, the two populations occupying slightly different ecological niches so that they appear in adjacent habitats or geographical areas. The gull species (or subspecies) of the Arctic – for example Iceland, Thayer's and Kumlien's Gulls – provide a good example of this. Many authorities consider Iceland and Thayer's to have derived from a single ancestral gull, populations of which were glacially separated. With the retreat of the ice the two populations met again, with Kumlien's being the hybrid form.

One of the most interesting examples of contact between populations is the merging of two forms of Snow Goose. Until the early years of the 20th century there were two distinct populations of geese, white geese in western North America and 'blue' geese to the east. So distinct is the colour of the two that it was long assumed that they were two different species, the Snow Goose and the Blue Goose, a view reinforced by the fact that the two geese did not share either breeding or wintering ranges. Changes in agricultural practices then allowed the geese to expand their populations and ranges. When the two forms met, they interbred, the young being fertile so population merging could occur. Not until the 1960s was it finally confirmed that the geese are colour morphs of a single species. As the blue morph is genetically dominant, the population is tending to become blue, though this is a slow process as the birds prefer mating with geese of their own colour. Almost all flocks of Snow Geese now comprise some birds that are white, some that are blue and some that are intermediate between the two.

To add further interest to the Snow Goose story and its value in understanding the way in which the ice ages have isolated populations, and what happens when the ice retreats, there are also two other goose types to be considered. Blue and white Snow Geese make up about 80% of the total population of the species. The other 20% comprises a larger form of the goose, sometimes called the Greater Snow Goose, which breeds in northern Nunavut and Greenland. These birds are larger and heavier, and white (though rare sightings of a blue form have been claimed). In the central Canadian Arctic there is another goose species, Ross's Goose, which is almost identical to the white Snow Goose but about half the size. Blue forms of Ross's are also known, but are very rare. The rare blue morphs of Greater Snow and Ross's Geese might have resulted from hybridisation with blue Snow Geese. There is also the possibility of 'egg dumping', as it is known that some female Snow

A flock of mixed blue and white Snow Geese,
Southampton Island, Canada.

Geese occasionally lay eggs in the nests of other birds if their own clutch is already as large as they can manage. Blue Snow goslings imprinting on Ross's Goose parents might then hybridise with Ross's Geese when they mature.

When the ice retreated, populations could expand their ranges northwards. Another broad biological rule, Rapoport's Rule, states that, in general, northern species have larger ranges than their southern cousins because of the relative scarcity of food. Combined with the shrinkage of the circumference of the Earth towards the North Pole and the fact that the Arctic is surrounded by a near-continuous continental chain, an expansion of population, and therefore range, means that the occurrence of circumpolar species increases with latitude. Many Arctic species are circumpolar, and while most of these are birds, some are mammals – Polar and Brown Bears, Reindeer, Arctic Fox, Musk Ox. There are, of course, exceptions to Rapoport's Rule (indeed, it really applies only to resident terrestrial vertebrates). One of the most pronounced is an Arctic fringe species, the Bristle-thighed Curlew. The species' breeding range in Alaska is much smaller than the available habitat. In this case population size is capped by the extent of the species' wintering quarters – small Pacific islands that are not capable of accommodating the population that might be expected from the size of the potential breeding range.

The seas between Greenland, Iceland and Eurasia were never crossed by land bridges, so mammalian traffic between Eurasia and North America during the Pleistocene must have been via Beringia. However, for birds these gaps were less important: on Greenland about 70% of the breeding bird species are circumpolar. As the island is thought to have been completely covered by ice until about 6,000 years ago, all breeding land birds have arrived since then. Immigration of species has occurred in both directions: Nearctic species breed on the west coast, and Palearctic species on the east coast. Although Eurasian species could reach North America most easily across Beringia, it was also possible for them to head for the Americas westwards, with Iceland and Greenland being stepping stones *en route* to the eastern coast of Canada. One species, the Northern Wheatear, spread to North America in both directions: it now breeds in eastern Canada and in Alaska and the Yukon

Territory, but not in the huge expanse of the Arctic that lies between. The winter migration of some populations continues to indicate their origins: some species that breed in North America migrate to Eurasia, using wintering grounds that have been visited for millennia, since long before the expansion of the breeding range. A classic example of this is seen in the Light-bellied Brent Goose population (*Branta bernicla hrota*) of eastern Canada, which winters in Ireland, crossing Greenland and Iceland on a journey that forms the longest migration flight of any Arctic goose.

Interestingly, although the Beringia land bridge assisted in the expansion of Eurasian species into North America, it may also have assisted speciation by preventing the Pacific and Atlantic populations of ancestral species from interbreeding. The distribution of northern auks is highly asymmetric, with far more species in the north Pacific than in the north Atlantic. It is assumed that the few which are circumpolar – three species of guillemot – and the one that shows similarities – the Atlantic Puffin, which is very similar to the two Pacific puffins   crossed from one ocean to the other via the Arctic Ocean (probably some time in the early Pleistocene), a route that was then closed to them by the presence of Beringia and the frozen sea

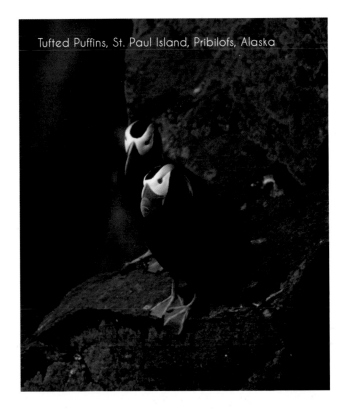

Tufted Puffins, St. Paul Island, Pribilofs, Alaska

Atlantic Puffin, Kjorholmane, Norway.

to the north. The evolution of the numerous species of auklets and murrelets in the Pacific and why there is only one corresponding form in the Atlantic (the Little Auk) is one of the puzzles of Arctic biogeography. The asymmetry in the number of auk species is mirrored in the distribution of pinnipeds that have a similar diet: the reasons for this are not clearly understood. Much less of a puzzle is the fact that in both the Atlantic and the Pacific there are more species (in general, rather than just of auks) on the eastern sides than on the western sides because of the direction of cold, productive currents.

A related issue concerns the distribution of penguins – why are there no penguins in the Arctic? The Arctic birds that most closely resemble the penguins in terms of diving ability are the larger auks, the smaller auklets and murrelets more closely resembling the diving petrels. Penguins are essentially confined to the southern polar region, but do extend northwards where cold water currents allow. It is thought that there are several reasons why penguins have not reached the northern hemisphere. Firstly, warmer waters are less productive. As the hunting method of penguins requires a large energy input, tropical seas are therefore unlikely to be able to support the birds. Secondly, the number of breeding sites free from terrestrial predators declines in equatorial areas, particularly in the Atlantic. Auks probably did not cross the equator southwards for similar reasons. While foot-swimmers such as grebes have colonised both hemispheres, wing-swimmers have not.

Wing-swimming is a high-energy technique, as is the flight of auks, which all have short, rapidly beating wings. Auks need to be volant because there are terrestrial predators in the north and the birds need to nest on cliffs the better to avoid them (though the flightless Great Auk survived until people with the capacity to reach its limited breeding sites arrived). Northern auks also breed preferentially on islands, a good strategy for avoiding terrestrial predators, except where annual sea ice allows predators to reach them. Of the auks, only the Atlantic Puffin is a truly ground-nesting bird (their nests being set at the ends of excavated burrows), the others being ledge nesters. However, Atlantic Puffins nest primarily on islands south of the sea ice limit. Where rats, cats or pigs have been introduced, Atlantic Puffins colonies have been rapidly depopulated, usually to the point where the island has been completely abandoned. The introduction of Arctic Foxes to some Aleutian Islands (for fur farming) led to similar destruction of those auks that nested in accessible positions.

## Human influences on biogeography

As noted above, the spread of the Snow Goose may have resulted from human changes to the environment. People have also aided other species, the population increase and range expansion of both the Northern Fulmar and Great Skua being attributed to the growth of the fishing industry, since fish processing on ships produces a great deal of waste. However, human influence has only rarely been beneficial. Humans may have caused or hastened the extinction of the mammoths and other megafauna: Arctic cetaceans have suffered from over-hunting; fur-bearing mammals have been threatened and, in some cases, almost extirpated; and some bird species have been drastically reduced by egg-collection and hunting. The Great Auk was driven to extinction by the mid-19th century (the last recorded birds were killed in 1844, the final phase of extirpation being by specimen collectors), and the once-abundant Eskimo Curlew was relentlessly hunted during its migration across the southern United States. This, together with loss of habitat and a consequent loss of essential food resources when the prairies were converted to agricultural land means that this species is now very probably extinct. The flightless Spectacled Cormorant (*Phalocrocorax perspicillatus*) was first described by Steller during the Bering expedition of 1741–42: it was extinct by the 1850s. Even more dramatic and appalling was the fate of Steller's Sea Cow (*Hydrodamalis gigas*), which was described by Steller on the same voyage. It was hunted to extinction within 27 years, its fate sealed by a lethal combination of docility, slowness and being good eating.

Evidence now suggests that humans are altering the Earth's climate. The warming of the Arctic will have a direct and disastrous effect on many Arctic species. But there will also be a subtler effect. Iceland has gained ten bird species since the early 20th century, probably in large part due to climatic changes as there has been little change in land usage on the island, Iceland's geology negating significant increases or changes in agriculture. Climatic changes and this northern movement of bird species has pushed other Arctic species further north – but there is a limit to how far north they can go, as birds need land sites for nesting. The same effect is also seen in the sea: as fish species move north, traditional hunting grounds for birds disappear. Nest sites are chosen to be close to those traditional sites, so chicks starve with population decreases adding to the problems of a shrinking range. If global warming continues this double-edged effect will, ultimately, result in a dramatic reduction in the number of Arctic species.

Arctic Tern, Seward Peninsula, Alaska.

# 15 Arctic birds

In the following sections wildlife that breeds beyond, or may be seen beyond, the Arctic boundary as defined in this book is explored. Some species with ranges whose northern extent barely overlaps the Arctic are also included, while for some areas – Fennoscandia, Kamchatka, and southern Alaska – a more pragmatic approach has been required as many non-Arctic species may be seen in those areas: some may also occasionally be found breeding. For those areas, therefore, the species lists that could be developed from the following sections are not complete, with species that are essentially non-Arctic having been omitted.

## Arctic Birds

Nomenclature generally follows that of the British Trust for Ornithology. Where it differs, the common name in North America is given in parentheses. In a book of this nature it is not possible to give rigorous descriptions of the various species. For such detailed descriptions the reader is pointed towards *A Complete Guide to Arctic Wildlife* (Christopher Helm) with text by Richard Sale and photographs by Per Michelsen and Richard Sale. Here, only brief details of where the species breeds, or may be seen are given.

## Divers

With their delicately patterned plumage – seemingly the work of a talented painter rather than comprising feathers – the divers are among the most attractive of all Arctic birds. For many, the haunting wailing or yodelling call of the Great Northern Diver is also redolent of the wilderness, and it is no surprise to discover that these birds have become entwined in the myths of northern dwellers. In North America, the call was thought to be the anguished cry of the dead calling for lost loves, an evocation that led to the belief that the birds guided the dead to the spirit world: sometimes the skull of a diver would accompany grave goods, carved ivory eyes replacing the originals, the better for it to find the correct pathway. The bird's diving abilities also led to the belief that it could see in the dark and, therefore, could restore sight to blind people, the bird diving with the blind person on its back: the prominent white markings on the backs of Great Northern and White-billed Divers were believed to be necklaces presented to the birds in thanks. In Siberia, native people incorporated the birds into creation myths, claiming that mud dredged from the bottom of the sea and brought to the surface on a diver's webbed foot began the process of building the land. Several Inuit dances included diver masks. Today, the Canadians have the Great Northern Diver (or Common Loon) on their one-dollar coin, as a reminder of the wilderness that holds a special place in the country's heart. Its appearance on the coin explains why, to the occasional puzzlement of first-time visitors from Europe, the coin is frequently called a 'loonie'.

Great Northern Diver (Common Loon) in the Barren Lands of North West Territories, Canada.

Yet despite the reverence, the birds were killed for both meat and clothing. The dense feathers, evolved for keeping out the chill of Arctic waters for this essential aquatic bird, allowed a diver carcass to fit snugly on the head: the clothing of 15th century mummies found in west Greenland included a parka hood made from the skins of two Red-throated Divers. The waterproof skins were also useful for carrying the means to prepare fire, with the bird being gutted to produce a bag. These are still occasionally seen in Nunavut where they are called loonie bags.

*Diver* is the British name for the five species of the genus *Gavia*, birds so highly specialized for swimming that they have considerable difficulty walking. The legs are positioned far back on the body, allowing the thrust from the feet, which are webbed between the three front toes, to be developed behind the body for maximum efficiency when diving. The birds are superb divers, reaching depths of 75m, though much shallower dives are more normal. Immersion is usually for about 45 seconds, though longer dives have been recorded. Underwater, the bird draws a transparent nictitating membrane across its eye as a form of 'contact lens' to retain excellent vision. The dagger-like bill is a highly efficient fishing tool.

On land the position of the legs makes walking difficult – the North American name, loon, for the birds, probably derives from the Old Norse word *lømr*, lame or clumsy – and divers rarely travel far from water. The larger species move by a series of inelegant hops or an equally inelegant shuffle, with the body held at an angle. Because of the difficulty of walking the birds take off from water, scurry across the water surface to obtain speed before lifting off. They then gain height slowly. Consequently divers look for large lakes that offer long stretches of water. If the lake is smaller and tree-surrounded, it is not unusual for the birds to have to circle, gaining height, until they are high enough to clear the trees. Yet despite these problems, divers are strong fliers, often travelling considerable distances to feed when rearing chicks. The position of the legs also prevents the birds from landing feet first as other water birds (e.g. swans) do, divers landing on their undersides, resembling seaplanes. In flight the feet extend beyond the tail, a diagnostic characteristic.

Divers eat a variety of fish species, though they will make do with just one kind if that is all their lake provides. They also eat amphibians and shellfish. The preferred nest site is an islet or floating mass of vegetation, but a marshy part of the lake shore will be used if these are not available. The nest, a pile of aquatic vegetation, is always close to the water because of the birds' poor walking abilities. There are usually two eggs, though one is not uncommon. The chicks can swim within 1 or 2 days, though they often rest on their parents' backs where they snuggle beneath the wings for warmth. The birds are migratory, but do not travel far from the breeding territories.

### Great Northern Diver (Common Loon) *Gavia immer*

Breeds across northern North America, including southern Baffin Island, on Greenland, Iceland, Jan Mayen and (irregularly) on Bear Island. In winter they may be seen off both coasts of the USA, and off the coasts of the British Isles and Norway.

### White-billed Diver (Yellow-billed Loon) *Gavia adamsii*

The largest diver. Breeds along the northern coast of Russia to the east of the Urals, and on the southern island of Novaya Zemlya. In North America breeds in northern Alaska and the central Canadian Arctic, including Banks and Victoria islands. In winter, they are seen off the north Norway coast, in the Bering Sea and the Great Slave Lake.

### Black-throated Diver (Arctic Loon) *Gavia arctica*

Similar to the Great Northern Diver. Breeds in Fennoscandia, across Arctic Russia and in Kamchatka, with a few pairs also in western Alaska. In winter, birds are seen off the Japanese coast, and in the North Sea and North Atlantic.

Black-throated Diver, Telemark, Norway.

Pacific Divers, NWT, Canada.

### Pacific Diver (Pacific Loon) *Gavia pacifica*

Very similar to Black-throated Diver in breeding and winter plumages, though the latter has a white flank patch, which is diagnostic. The two birds share habitats, habits and calls and for many years it was thought Pacific Divers were a subspecies of Black-throated.

Breeds across North America and on the southern Canadian Arctic islands, and also in north-eastern Chukotka. In winter, Pacific Divers are seen in the Bering Sea off Kamchatka and along the western US coast.

### Red-throated Diver (Red-throated Loon) *Gavia stellata*

The smallest diver and, with a red throat that develops for the breeding season, one of the most attractive. The call of Red-throated Diver is also distinct from the voices of the other divers, being more duck- or goose-like. Being smaller, Red-throated Divers can nest on smaller lakes. This often means that the local food supply is inadequate for chick-rearing, the birds having to fly to gather food.

Circumpolar breeders. In winter, they are seen in the North Atlantic, North Sea, Bering Sea and North Pacific.

Red-throated Diver, southern Kamchatka, Russia.

## Grebes

Though superficially similar to divers, grebes have some distinctly different characteristics, suggesting a very different evolutionary path. The feet are not webbed, the toes being lobed to provide the paddles necessary for pursuit of prey underwater. The paddle stroke is also different from that of divers, a surface paddling grebe using the feet one at a time, and twisting them so that they move parallel to the surface. As with the divers, the feet are placed far back on the body, making walking difficult, Grebes often falling over during their short journeys.

Grebes have conspicuous courtship ceremonies. A pair of birds will stand upright on the water, breast-to-breast, with the heads turning from side to side. They may then swim side by side, or even rush across the water side by side while remaining upright. In the 'weed ceremony' the birds dive together, each surfacing with its bill filled with weeds. They then stand facing each other, their heads moving sideways to display the weeds. The weeds may be used to build the nest, which comprises a heap of weed floating in the chosen pond or lake. The nest is anchored to aquatic vegetation and 3–5 eggs are usually laid. The downy chicks leave the nest soon after hatching and often ride on their parents' backs. They are fed the same diet as the parent birds – fish and aquatic invertebrates. The prey is collected in short dives, usually less than 30 seconds in duration, at moderate depths of up to 20m.

Grebes are poor fliers in comparison to divers, the wings beating so fast the birds appear panic-stricken. Consequently, they are rarely seen in flight at their breeding territories. Nevertheless, both northern breeding grebes are migratory, moving to southern

Slavonian Grebe pair, Great Slave Lake, North West Territories, Canada.

coastal waters in winter. On migration, they frequently fly at night. This has led to instances where in the early morning light exhausted birds have mistaken wet roads for streams and landed. They are then stranded, being unable to take off from land.

The luxurious nature of grebe plumage led to a European fashion fad for 'grebe fur', which was used to trim lady's clothing. This led to severe reductions in grebe numbers in Europe, though numbers have now increased. Some grebe species are endangered by habitat destruction.

### Slavonian Grebe (Horned Grebe) *Podiceps auritus*

A handsome bird when in its breeding plumage of chestnut, with yellow 'horns'. In winter, both are lost. Circumpolar breeders, but absent from Greenland. In winter the birds are seen around the British Isles, in the North and Baltic seas, and off both American coasts.

### Red-necked Grebe *Podiceps grisegena*

An Arctic fringe dweller. Circumpolar breeder, but absent from Greenland and Iceland, and the High Arctic. Wintering birds are found off both coasts of America, and in the North and Baltic seas.

## Tubenoses

The tubenoses, distinguished by their external, tube-like, nostrils comprise albatrosses, petrels and storm-petrels, though only one species is a true Arctic dweller.

### Northern Fulmar *Fulmarus glacialis*

Superficially similar to gulls in its lighter morph, the Fulmar is distinguished at close quarters by the prominent tubenose and the fact that the suture lines of the bill plates are visible, and at a distance by the stiff-winged flight, the wings remaining stiff even during bouts of vigorous flapping. The bird comes in two basic morphs, dark and light. Light-morphs have pearly-grey heads and upperparts, with whiter underparts. Dark-morphs are darker grey above, though often with a paler head and tail, and pale grey underparts.

Circumpolar breeders, nesting on sea cliffs. Pelagic in winter, moving ahead of the ice, but feeding as far north as the ice edge, either by surface feeding or making shallow dives.

Dark and light phase Fulmars, eastern Barents Sea, Russia.

## The foul gull

The name Fulmar is Icelandic, meaning 'foul gull'. This almost certainly derives from the way the birds regurgitate fish oil when approached at a nest site. Both adults and young do this. The oil is viscous and evil-smelling, and it is renowned among rock climbers and birdwatchers who have been hit for being almost impossible to fully remove from clothing. Fortunately, the bird's aim does not match its enthusiasm for regurgitation, and most oil jets miss their target. An alternative suggestion for the name is that the high oil content of the birds meant they could be rendered down and used as lamp oil. As the oil would have had much the same smell as the defensive jet, the name again seems appropriate.

Albatrosses are essentially birds of the southern oceans, most species being confined between 45°S and 70°S, though four species breed in the equatorial and northern Pacific. Of these two, wintering Black-footed Albatross (*Phoebastria nigripes*) and Short-tailed Albatross (*Phoebastria albatrus*) may occasionally be seen off southern Alaska and, very rarely, in the southern Bering Sea.

Of the shearwaters, the Short-tailed Shearwater (*Puffinus tenuirostris*) winters in the north Pacific and Bering Sea where huge flocks form. Flocks of more than 100,000 birds (or significantly more, perhaps to 500,000) have been reported as far north as St Lawrence Island. The Sooty Shearwater (*Puffinus griseus*) and Great Shearwater (*Puffinus gravis*) may also be seen in winter in the north Atlantic as far north as the Denmark Straits and Norwegian Sea. Manx Shearwaters (*Puffinus puffinus*) breed on Iceland, and may be seen as far north as northern Norway in winter.

It is difficult to imagine a stronger contrast in flight characteristics between the shearwaters and their cousins the storm-petrels. While the former glide above the waves, the storm-petrels flutter just above the water surface pecking at small fish and planktonic crustaceans. In the southern hemisphere, several storm-petrels have polar distributions, with Wilson's Storm-petrel (*Oceanites oceanicus*) breeding at several sites on the Antarctic mainland. In the northern hemisphere, there are no truly Arctic species, but both the European Storm-petrel (*Hydrobates pelagicus*) and Leach's Storm-petrel (*Oceanodroma leucorhoa*) breed on Iceland and winter in the north Atlantic to northern Norway. Leach's Storm-petrel and the Fork-tailed Storm-petrel (*Oceanodroma furcata*) also breed on the Aleutian Islands, from Kamchatka to Alaska, wintering in the central Bering Sea.

## Gannets

Gannets are members of the Sulidae, a family of plunge-diving birds that occur in a broad band across the Earth's oceans, primarily concentrated towards the Equator. The Northern Gannet (*Morus bassanus*) is not a true Arctic dweller, but does breed in northern Fennoscandia, on Iceland and Jan Mayen. Gannets are magnificent birds with white plumage apart from black wing-tips, a darker trailing edge on the wings, and a pale yellow head. Interestingly the birds have no brood patch and so incubate by standing on the single egg, using the blood vessels of the foot webs for warmth.

Landing Gannet, Iceland.

## Cormorants

The cormorants are a successful group of diving, fish-eating birds. They are well-adapted to the aquatic environment: with pelicans and the sulids they are the only birds with webs between all four toes, which gives them a strong paddle stroke. They have a long, flexible neck which aids the capture of prey. The bill is long and hooked at the end, while the tongue is rough, both adaptations to deal with slippery prey: the neck is pouch-like and can distend to allow fish to be swallowed whole (as they are after the bird has juggled them to ensure a head-first descent). Features to assist diving underwater include an eye lens that can be modified to assist underwater vision and dense, buoyancy-limiting bones.

A consequence of the lack of waterproofing is that most cormorants are tropical species, but there are exceptions. One species, the Imperial Shag (*Phalacrocorax atriceps*) breeds on the Antarctic Peninsula, while in the north the Pelagic Cormorant breeds on Wrangel Island, and several other species breed at the Arctic fringe. These species have the additional problem that hanging the wings out to dry results in the body being exposed to chilling temperatures, as the birds have little body fat. Such polar distributions are, therefore, remarkable.

Despite the adaptations of the feet for an underwater life cormorants are much better on land than divers. They can stand upright, and some species even perch and nest in trees. Cormorants are strong fliers, usually skimming low over the water, but because of the need to dry the feathers they are not pelagic, usually fishing close to the shore. They are gregarious, both at fishing and nest sites, the latter usually on cliffs. During the breeding season the adults often acquire a white thigh patch and prominent crests and head plumes, together with coloured throat pouches which are used in mating displays. Nests are masses of seaweed held together with excrement. Up to six eggs are laid, the chicks feeding by reaching into the parental throat for regurgitated fish and aquatic invertebrates. Young cormorants form crèches, returning to the nest site for feeding.

### Waterproofing feathers

Despite all the adaptations cormorants have for a diving lifestyle, one that would seem essential is missing – the cormorant's contour feathers are not completely waterproof, only the inner down layer preventing the skin from wetting. Most birds, and all wildfowl and other water birds, have waterproof feathers, the proofing assisted by oil from the preen gland above the tail. The wetting of the cormorant's outer feathers may help reduce buoyancy, though that does seem a rather drastic solution. But even if that is correct, there is a price to be paid for the lack of waterproofing: the feathers must be dried after dives. This results in the bird's characteristic pose with its wings 'hung out' to dry.

Red-faced and Pelagic cormorants, Unalaska, Aleutian Islands, Alaska.

### Pelagic Cormorant *Phalacrocorax pelagicus*

The smallest of the North Pacific cormorants, and the only true Arctic dweller, breeding on the Chukotka coast and Wrangel Island, as well as across the Aleutian Islands. In winter the birds migrate south as far as Taiwan.

Other cormorants are also seen at the Arctic fringe. The Great Cormorant (*Phalacrocorax carbo*) breeds in southern Greenland, Iceland, northern Norway and the Kola Peninsula, and in North America on Newfoundland and adjacent coasts. Greenlandic birds move to the south-west coast, but Icelandic birds are resident. In North America, the Double-crested Cormorant (*Phalacrocorax auritus*) breeds as far north as Newfoundland to the east, and to southern Alaska and the Aleutians to the west. The Red-faced Cormorant (*Phalacrocorax urile*) breeds in southern Alaska, on the Aleutian and Pribilof islands, and in eastern Kamchatka.

## Wildfowl

For the Arctic traveller, there are few sights that herald the arrival of spring more evocatively than skeins of geese arriving from the south. And there are few that announce the coming of the Arctic winter more than those same geese departing south again. The skein will have the V-formation for which flying wildfowl are renowned, the pattern allowing birds to take advantage

Red-faced Cormorant, St Paul, Pribilofs, Alaska.

of the updraught of air created by the wingbeat of the bird ahead, reduce energy output. The bird at the head of the formation works hardest and is regularly replaced by a bird further back. Geese can fly at astonishing heights, formations having been reported at over 10,000m. At such a height a human could not survive, let alone work hard, because of reduced oxygen levels. However, in general geese fly at much lower altitudes, probably because flying at altitude risks wing-icing, which would be catastrophic.

Wildfowl – swans, geese and ducks – are a successful group with more than 150 species spread across every continent except Antarctica. Wildfowl share some general characteristics. They are broad-bodied, long-(or very long-) necked aquatic birds. They have flattened bills with a horny 'nail' at the end, the edges of the mandibles having comb-like lamellae for straining food from water, and rasp-like tongues for manipulating food. In a few species, the sawbills, the lamellae are replaced by 'teeth'. Vegetarian wildfowl do not have bacteria in their gut to break down cellulose, so they gain nutrients only from cellular juices. The plant structure is broken down in the bird's gizzard, grit being ingested to aid the process. One consequence of this is that the birds eat a great deal, spending virtually the entire day feeding. They take the most nutritious parts of plants, the new growth, which has not had time to build up structural fibres.

Wildfowl have webbed feet (though in a very small number there are either no webs or the feet are partially webbed (semi-palmated). The legs are set far back on the body, making walking difficult. This is especially true of the swans and ducks, whose legs are also short. In geese, the legs are longer and set more centrally on the body. These birds are therefore more mobile on land, in keeping with their more terrestrial habits. Though geese can occasionally look awkward on land, they are surprisingly quick, as a traveller who blunders close to a goose nest will rapidly discover. The cryptic incubating bird can show a remarkable burst of speed which, allied to a very aggressive nature, can lead to a worrying few seconds.

Swans and geese usually mate for life, but ducks do not, taking mates seasonally. These different mating systems account for the plumage differences between swans, geese and ducks. In the swans and geese the sexes are alike. In ducks, most males are brightly coloured, with the female cryptic brown.

Wildfowl moult their flight feathers simultaneously and so undergo a flightless period, which varies in length from about three weeks for small ducks to six weeks for swans and large geese. During this flightless phase the male ducks of most Arctic species moult to an 'eclipse' plumage which closely resembles that of the female. In general eclipse plumage is seen in the autumn, the males then moulting to their breeding plumage. Drake courtship displays are limited to simple movements such as tail wagging and water flicking. For swans and geese the displays are even more limited (as pair bonds are lifelong). There is, however, a triumph ceremony, performed by the pair when a real (but sometimes imaginary) intruder is expelled from the nesting territory. The male usually chases the intruder, or engages in the mock chase of a non-existent intruder to illustrate his aggressive, protective intent for the sake of the pair-bond. He returns to the female and they stand facing each other, extending their necks and calling loudly 'in triumph'. Wildfowl habitually nest close to freshwater – even the sea ducks. Female wildfowl lay large clutches. In swans and geese the female incubates the eggs while the male stands guard. In the ducks the males usually abandon the females once incubation has begun. Goslings and ducklings are downy, and can swim, dive and feed themselves as soon as they are hatched. They are, however, brooded and cared for by their parents, in the case of swans and geese, and the female alone in the case of the ducks. In some duck species, the chicks of many females form crèches under the care of a small number of females. Swans and geese tend to migrate as family units after the young have fledged, and even stay together as a unit during the winter. Young swans and geese breed at three or four years, but depart from the family unit when the breeding site is reached after the spring migration.

## Swans

Swans are the largest and heaviest of the wildfowl. There are seven species, of which three are Arctic birds. The distribution of the seven species is curious: apart from the northern swans there are two in southern South America and one in Australia and New Zealand. The northern species are predominantly white and have very long necks. Male, female and winter plumages are similar. Swans are strong fliers once they have taken off. Take-off follows an energetic (and awesome) race across the water. Landing swans are a great sight, feet splayed out as brakes and landing carriage, with a good touchdown seemingly as much of a surprise to the swan as to the onlooker. Landings on ice are hilarious, though presumably traumatic for the bird, and a great deal more traumatic for birds already on the surface as they are skittled by the on-rushing swan.

Swans are wholly or primarily vegetarian, grazing aquatic vegetation and dabbling in shallow water. They also up-end to feed. Dabbling takes in aquatic invertebrates as well as vegetation, and some swans paddle with their feet to bring larvae to the surface. They are, however, opportunistic and have been seen to swallow sizeable fish. Wintering birds feed in cereal feeds and on waste grain, and will grub for tubers and potatoes, the latter a relatively recent addition to the diet.

All northern species are migratory, some travelling great distances, though these are usually accomplished in relatively short flights between 'refuelling' stops. Swans have been seen at heights of more than 8,000m, but chiefly fly at 2,000–3,000m.

## Whooper Swans and the Vikings

Each year the Norse settlers of Iceland watched as the highly visible, huge white Whooper Swans disappeared south across the sea. The settlers held the birds in awe, believing they were imbued with supernatural powers. They did not know that some Icelandic birds migrated to northern Britain for the winter, and would not have believed that so big a bird could cross such hazardous waters. Myths grew that the swans headed to Valhalla or the Moon, returning each spring.

The Norse view is understandable, as the journey the swans undertake is incredible – more than 800km across hostile seas. Today the tracking of individual birds has shown that the journey can take as little as 13 hours, but if the weather is poor birds may take several days, sometimes resting on the sea to recover. In particularly bad weather, many birds will not make it.

## Swans in legend and history

The grace and beauty of swans has inspired people for thousands of years. The Greek legends of Leda and the Swan, and of Phaeton and Cygnus (the latter story giving its name to the genus – *Cygnus* – and has given us 'cygnet' as the name of a young swan) are early examples, while a later Scandinavian tale was the basis for the ballet Swan Lake. For sheer elegance, a swimming swan is hard to beat. The bird is occasionally mocked, the suggestion being that the above surface elegance is belied by frantically paddling feet below the water. In reality, the paddle stoke is usually leisurely and every bit as elegant.

The grace that excited poets did not always stretch to the more pragmatic human: swans are large birds and make good eating. Northern native peoples have always prized them for food, and for the luxury of its feathers: the preferred bed of the Inupiat of northern Alaska was swan skin.

Whooper Swan family, Dalarne, Sweden.

### Whooper Swan *Cygnus cygnus*

As a diagnostic, Whoopers have the most extensive yellow on the bill of any of the northern swans.

Breeds on Iceland and across Arctic Eurasia to Kamchatka (though uncommonly in Norway). Some wintering birds are resident, but most move to southern rivers and coasts.

### Tundra Swan *Cygnus columbianus*

The status of the tundra swans of North America and Eurasia is contentious, with some authorities believing they are independent species, while others differentiate the two as subspecies – the Eurasian Bewick's Swan (*C. c. bewickii*) and the North American Tundra Swan (*C. c. columbianus*), the latter occasionally called the Whistling Swan to add further confusion. The bill of each is black with minimal yellow at the base.

North American birds breed on the Aleutian islands, western Alaska and across the northern continent to the western Ungava Peninsula, including the southern Canadian Arctic islands. Eurasian birds breed across Arctic Russia from the Kanin Peninsula to Chukotka. Those authorities seeing separate subspecies consider Bewick's Swans breed in western Chukotka, with Whistling Swans breeding in eastern Chukotka, and consider interbreeding may occur.

Whistling Swan, Southampton Island, Canada.

### The Trumpeter Swan (*Cygnus buccinator*)

This swan is not truly Arctic, breeding in south-eastern Alaska, though it is occasionally seen in western and central Alaska where it may overlap the Whistling Swan. In those areas, the all black bill of the Trumpeter is usually diagnostic, though Whistling Swans with all black bills are known. The Trumpeter's scientific name derives from the Latin for a military trumpet. As a family, the swans are a noisy bunch (though the Mute Swan – *Cygnus olor* – while definitely not mute is less noisy), the name 'swan' apparently deriving from the Saxon for noise. The Trumpeter has a long, twisted windpipe and its deep, resonant bugling is the loudest of all wildfowl and ranks with the loudest of all birds.

## Geese

The 'true' geese – birds of the genera *Anser* (the grey geese) and *Branta* (the black geese) – are all found in the northern hemisphere, though the colour groupings are not exact. The sexes of all northern geese are alike. Of the fifteen species, twelve are Arctic dwellers. These are essentially terrestrial, though they do feed on water and may even be seen up-ending in search of food. On land they consume vegetation, while the larger species also grub for roots and tubers. In winter they feed on agricultural land, taking waste grain, but also more substantial foods such as beets and potatoes. Geese migrate considerable distances. They cannot afford the weight of a large digestive system that would maximise the nutrient extraction from their poor diet and therefore take fresh green shoots which are more easily digested and are rich in protein and carbohydrate. Migrations are timed so the geese move north into continuous springs, with green shoots in abundance. They will also occasionally move to higher ground to gain advantage of an 'altitudinal' spring. Though it was long assumed that females laid eggs as soon as they reached their nest sites, it is now known that many geese spend the first few days feeding continuously to replenish bodily reserves lost during the long flight north. Egg-laying cannot be delayed for long, however, as the best strategy for raising young to the point where they can migrate is to have them hatch early. They need to be in good condition for laying, but laying early is a dilemma for the female goose. She therefore grows many ovarian follicles, but chooses how many eggs to produce. In bad years, when poor weather means the migration flight has been poorly timed and food resources are limited, the female may resorb the follicles and lay no eggs. But some females may lay too many eggs and if they cannot then find enough food to continue incubation have two choices, some deserting their eggs, saving themselves to breed the following year, some being known to die of starvation on the nest.

Even if incubation is successful, chick-rearing is stressful and fraught with danger. There are both avian and mammalian predators to whom goslings are an easy and welcome meal. The climate, too, plays a part – a spell of bad weather may kill off the vulnerable chicks, or a poor growing season for plants may not allow some chicks to fatten sufficiently to make it to winter quarters. Life for northern geese is harsh, and that is without the extra problems caused by human hunters.

## Grey Geese

Grey geese differ from their black cousins not only in colour, but in having serrated mandibles and vertical furrowing of the neck feathers. The latter is prominent, particularly when the feathers are vibrated, a sign of aggression.

### (Greater) White-fronted Goose *Anser albifrons*

Breeds in Russia eastwards from the White Sea, including the southern island of Novaya Zemlya, in west Greenland around Disko Bay, Alaska and in isolated areas of northern Canada. All populations migrate south for winter, Greenlandic birds to Iceland, Ireland and Scotland. European Russian birds head for the British Isles and Low Countries.

### Lesser White-fronted Goose *Anser erythropus*

Similar to its cousin, but smaller and with a shorter neck. Breeds in northern Fennoscandia, where it has been re-introduced following severe depletion due to overhunting. Wintering birds are seen across northern Europe and in parts of China.

### Pink-footed Goose *Anser brachyrhynchus*

Breeds in east Greenland, Iceland and Svalbard. In winter, western birds head to northern England and Scotland, while Svalbard birds fly to Norway, Denmark and Holland.

White-fronted Geese in flight over Alaska's North Slope.

The Lesser White-fronted Goose is virtually identical to its larger cousin apart from being abour 25% smaller. It also has a prominent yellow eye ring. Some Greater White-fronts also show such a ring, particularly the so-called Tule Goose (*Anser albifrons elgasi*) which breeds in southern Alaska. Tule Geese are named for tules (bullrushes) which they eat at their Californian wintering grounds. The eye-ring of the Lesser goose is more prominent, and the ranges of it and the Tule do not overlap.

Pink-footed Goose in southern Iceland.

Bean Geese, Orre, Norway.

Snow Geese above Yellowknife, Canada.

Comparison of the heads of a Snow Goose (above) and Ross's Goose (below). The distinctive 'grin' of the former, and caruncles of the latter are diagnostic.

### Bean Goose *Anser fabalis*

Very similar to White-fronted Geese, but lacking the dark striping on the underparts. There are two distinct forms, which have occasionally been considered separate species, the 'Tundra' Bean Goose (*A. f. rossicus*) and the 'Taiga' Bean Goose (*A. f. fabalis*) which is larger and has a shorter, thicker bill which is black, with a yellow band close to the nail.

Taiga Bean Geese breed in Scandinavia east to the Urals. Tundra Bean Geese breed on the northern Russian tundra east of the Urals and on the southern island of Novaya Zemlya. Migrates to eastern Britain and central Europe.

### Snow Goose *Anser caerulescens*

Unmistakable. There are two colour morphs. The white morph is entirely white apart from black primary feathers and pale grey coverts. The 'blue goose', has blue-grey lower neck, breast, belly, flanks and mantle. Intermediate forms are also seen, with most flocks including a range of birds. Greenlandic birds are considered a subspecies, as is the Greater Snow Goose of Baffin, Bylot and Ellesmere Islands.

Breeds across North America and in north-west Greenland. A small population also breeds in north-east Russia, with a more significant population on Wrangel Island. In winter, North American and Greenlandic birds fly south to the southern states and Mexico. Many Russian birds join them, but some head south to China and Japan.

### Ross's Goose *Anser rossii*

Plumage as the Snow Goose, though blue morphs are much rarer. Some authorities believe that blue morphs result from hybridisation between Ross's Goose and blue-morph Snow Goose (perhaps following egg-dumping by a blue goose into a Ross's nest, with the chicks imprinting on its foster parent). The bill of Ross's Goose differs from that of its larger cousin, having no, or minimal, 'grin' and blue-grey protuberances (called caruncles) around the base.

Breeds in just a few places in Arctic Canada – Banks and Southampton Islands and on the mainland near Bathurst Inlet and western shores of Hudson Bay. Wintering birds are seen in Mexico and California.

### Emperor Goose *Anser conagica*

Small and exquisitely marked, the only grey goose to show an exotic plumage. Emperors (which were named for the Russian Tsar) are the most musical of the grey geese, with a trisyllabic call. They are also the most maritime, though rarely seen far from the coast.

They breed on the west Alaskan coast and, in small numbers, on the east coast of Chukotka. Many birds are resident, but some move to the Aleutians and Kodiak Island and, occasionally, further south along the western seaboard of the USA.

Though not a true Arctic species the Greylag Goose (*Anser anser*), the largest grey goose, breeds in Iceland and northern Norway, but otherwise only in more southerly areas of Eurasia. The Greylag is the ancestor of domesticated geese, these being famous for the racket they make when disturbed: the honks of domestic geese are reputed to have saved Rome from a Gaulish horde in 390BC.

## Black Geese

While the black geese are generally darker than the grey, the distinguishing features of the northern species are bold patterning and the lack of prominent mandible serrations.

### Brent (occasionally Brant) Geese *Branta bernicla*

Breeds in northern Greenland, Svalbard, Franz Josef Land, across Arctic Russia east from the Taimyr Peninsula, including Severnaya Zemlya, the New Siberian Islands and Wrangel Island, and across northern North America from Alaska to western Hudson Bay and on all Canada's Arctic islands. The geese of western North America (Black Brent – *B. b. nigricans*) are generally dark.

In winter, birds from Greenland, Svalbard, Franz Josef and western Russia fly to the Low Countries and the British Isles. Those from eastern Russia move to Japan, while American birds winter on both coasts of the USA.

### Canada Goose

The taxonomic status of the Canada geese has recently changed with Canada Goose *Branta canadensis* (with seven subspecies) and Cackling Goose *B. hutchinsii* (with four) now being recognised, a replacement of the long-held belief that there was a size difference between 'Greater' and 'Lesser' Canada geese. However, the variation in size of subspecies means that large 'lesser' (i.e. Cackling) geese may be larger than small 'greater' (i.e. Canada) geese. To add further confusion the Cackling Canada Goose was a name once reserved for the Aleutian/south-west Alaskan subspecies, while most authorities consider Cackling Geese to be the most northerly breeders. A fifth subspecies of the Cackling Goose, *B. h. asiatica*, which bred on the Kuril and Commander islands, is now thought to be extinct as no specimen has been verifiably sighted since 1914.

The two forms of Canada Geese are North American species, though a small population of Cackling Geese has established in west Greenland. The geese are attractive

Brent Geese at Randeberg, Norway.

On tussock tundra, incubating Canada Geese can occasionaly be difficult to spot, but the goose soon leaves the visitor in no doubt that their presence is not welcomed. Churchill, Manitoba, Canada.

## The Aleutian Cackling Goose

The Aleutian Cackling Goose (*B. h. leucopareia*) once bred on all the islands of the Aleutian chain. To increase fur production, Russian settlers introduced Arctic Foxes to all the readily accessible islands. The effect on the geese, and other ground nesting birds, was catastrophic, and it was feared they had become extinct. But they were found breeding on inaccessible islands, and a programme of fox elimination and goose re-introductions has ensured the survival of the subspecies.

birds and Canada Geese have been introduced into, or escaped captivity in, the British Isles, Scandinavia and other areas of northern Europe. The Cackling Goose has not been recorded in the UK.

### Barnacle Goose *Branta leucopsis*

A beautiful small goose. Barnacle Geese have the most restricted range of the Arctic black geese, breeding only in north-east Greenland, Svalbard, the southern island of Novaya Zemlya and nearby parts of Russia's northern coast. In recent years Barnacle Geese may have also bred in Iceland. The Greenland and Svalbard birds winter in Scotland and Ireland, Russian birds flying to Denmark, Germany and the Low Countries.

### Red-breasted Goose *Branta ruficollis*

Perhaps the most attractive of the black geese. Formerly bred only on Russia's Taimyr Peninsula, but in recent years seems to have extended its range both east and west. Despite this good news, the position of the goose remains vulnerable. Wintering birds are concentrated in a small number of places on the Black Sea where

## Geese and barnacles

Each winter little black-and-white geese appeared in northern Europe and the people who observed them were puzzled by their origin – so much so that a belief arose that they hatched from barnacles attached to flotsam wood that occasionally washed ashore. Writing in the early 17th century, the herbalist Gerard noted an extension of the legend – 'There are found in the North parts of Scotland and the Islands adjacent, called Orchades, certain trees whereon do grow certaine shells of a white colour tending to russet, wherein are contained little living creatures: which shells in time of maturitie do open, and out of them grow those little living things, which falling into the water do become fowles, which we call Barnacle Geese … and in Lancashire, tree Geese: but the other that fall upon the land perish and come to nothing'. So the myth had extended, the fact that barnacles washed ashore were invariably attached to wood meaning that there must, somewhere, be a tree with very strange fruit. Eventually, of course, the truth was discovered, but the name (Barnacle Goose rather than Tree Goose) stuck, in part because barnacles, being 'fish' could be eaten on Fridays by Christians (especially Catholics) and the goose, being a 'fish goose' had meat that could also be eaten as fish.

Red-breasted Geese are the most beautiful of the northern geese.

changes to local agricultural methods represent a threat. The birds are also hunted as they migrate.

## Dabbling ducks

Dabbling ducks are the largest duck group with representatives on all continents except Antarctica (though some species breed on Southern Ocean islands). The group is named for their habit of working the surface of the water for food. They feed chiefly in fresh water, though they are not unknown in marine areas. They primarily pluck aquatic vegetation, but also take aquatic invertebrates. They do not take fish. Dabbling involves taking in a volume of water that is then squeezed out through lamellae at the edges of the mandibles, seeds and other particles then being swallowed, a feeding method analogous to that of the large whales.

Ducks have long, broad wings. These allow not only a fast flight but also a short take-off, some ducks being able to rise almost vertically from the water if necessary. The ducks are sexually dimorphic, males (drakes) usually being brightly coloured during the breeding season, and the females (ducks) always remaining cryptic brown as camouflage. After the breeding season the males also adopt a more cryptic 'eclipse' plumage. Drakes have a coloured speculum, which is maintained in eclipse. Females usually have a speculum as well, but it may be much smaller or less clear-cut. The drakes of many species forcibly inseminate available females; in some circumstances several drakes will attempt to forcibly mate with a female (sometimes the female is held underwater and drowns due to the relentless activity of squabbling drakes). In general, insemination is the only contribution of the drake, the males deserting the females after she lays.

## Prey and predator

Since goose chick-rearing is fraught with danger of predation, it is odd that some species deliberately choose nest sites close to those of a predatory bird. This apparently nonsensical arrangement is beneficial to both. If the nest is close to a raptor or owl nest, the bird of prey will see off other potential predators, while the raptor gains since geese and ducks are noisy if a predator is viewed, acting as a useful early-warning system. Most famous of these mutual arrangements are those between Red-breasted Geese and Snowy Owls or Peregrine Falcons. The geese occasionally nest within 5m of the bird of prey and occasionally a colony of a dozen or more goose nests will be placed near the same raptor nest. Falcons rarely hunt close to their nests so the geese are afforded a measure of protection. Snowy Owls are much less particular, and once hatched the goslings must run the gauntlet of the owls, though predation is limited if the owl chicks have yet to hatch. The unusually short fledging time for Red-breasted goslings may be a response to this potential predation. The extent to which the geese benefit from the relationship became apparent when Peregrine numbers fell dramatically due to DDT use in the 1960s and 1970s. The crash in falcon numbers was mirrored by that in goose numbers, with the population of Red-breasted Geese climbing as falcon numbers recovered after the ban on DDT was introduced.

Although the Red-breasted Goose is the most obvious example of this arrangement, it is not the only one. Long-tailed Ducks have been observed nesting closer to Red-breasted Goose colonies (and therefore raptor nests) than would normally be expected. King Eiders occasionally nest close to Long-tailed Skuas or Snowy Owls, Steller's Eiders may nest near Pomarine Skuas, and Brent Goose nests have been seen unusually close to nests of Snowy Owls.

### Mallard *Anas platyrhynchos*

The most widespread and probably most recognisable of the dabbling ducks. Breeds in west and east Greenland, Iceland and across Eurasia (though only to the northern coast in Fennoscandia and then increasingly southerly to the east, though occurs in Kamchatka). In North America, Mallard breed throughout Alaska, to the Mackenzie delta then more southerly across Canada. Greenland and Icelandic birds are resident, but northern

## Greenland's Mallards

Greenlandic birds are considered a separate subspecies, *A. p. conboschas*. They are the most Arctic of all Mallards, breeding on the west coast from northern Disko Bay to the island's southern tip, and around Ammassalik on the east coast. All the island's birds spend the winter in the open water area off the south-east coast, the east coast birds crossing the ice sheet to reach the wintering grounds. The ducks have a precarious existence: although the open water area can be relied upon, its extent varies and some birds need to search for areas of unfrozen ocean.

Drake Northern Pintail, Chruchill, Manitoba, Canada. While females are, as with other species, drab for nesting camouflage, the drake Pintail is among the most attractive of the ducks.

birds in Eurasia and North America move south for the winter.

### Northern Pintail *Anas acuta*

With their elongated tail feathers, drake Pintails are unmistakable. It breeds across northern Eurasia, mostly to the coast (but not on any Arctic islands) and on Iceland, but is absent from Greenland. Breeds across North America as far east as the Ungava Peninsula, to the north coast except on the Boothia Peninsula, and on southern Banks Island. Wintering birds are found in central Africa, central Asia, Japan, the southern United States, Mexico and Central America.

### Teal (Eurasian or Common Teal) *Anas crecca*

The smallest of the northern *Anas* ducks, breeding in Iceland and across Eurasia, but rarely to the north coast.

Drake Green-winged Teal, Great Slave Lake, NWT, Canada. Until recently Nearctic and Plearctic birds were considered a single species, but the differences between them have resulted in them now being classified as separate species.

### Green-winged Teal *Anas carolinensis*

Until recently considered a subspecies of the (Eurasian) Teal but now accorded species status. Breeds across North America, including west and north Alaska and northern Yukon/western North-West Territories. It breeds around southern Hudson Bay and near Ungava Bay and in northern Labrador.

### Eurasian Wigeon *Anas penelope*

Breeds in Iceland and across Eurasia, though not to the northern coast except in Fennoscandia and European Russia. Also breeds on Novaya Zemlya. Birds from northern Iceland migrate to the south of the island, though some move to the east coast of North America. Eurasian birds head south to Japan and central Asia.

Eurasian Wigeon, Orre, Norway.

**The American Wigeon (*Anas Americana*)**

It breeds across northern North America, but is only 'Arctic' in western Alaska, near the Mackenzie delta and southern Hudson Bay.

## Pochards

Pochards, often called Bay Ducks because in winter they are normally found in coastal bays, are members of the genus *Aythya*. They share the same diet as dabbling ducks, but primarily feed with shallow dives, usually to depths of a few metres. Pochards are heavier than dabblers, have longer necks, and feet set further back on the body to act as more efficient paddles: this makes them awkward on land and they are rarely seen away from water. Their heavier weight means they have more difficulty becoming airborne, with take-off following a run across the water, a sharp contrast to the dabblers. In general, pochards are drabber than dabbling ducks, and lack the colourful speculum. However, they share some behavioural habits, e.g., drakes desert females after the latter have begun incubation. They also interbreed, as some of the dabbling species do, but do so much more readily, a fact that can cause the observer problems. Why ducks and geese in general, and pochards in particular, are apt to interbreed is not understood.

**Greater Scaup** *Aythya marila*

Breeds throughout Alaska, though rare in the north, to the Mackenzie delta, then more southerly to Hudson Bay. Absent in eastern Canada, apart from the Ungava Peninsula. Also breeds in Iceland and across Eurasia, though only to the north coast in Fennoscandia and Russia to the Taimyr Peninsula and on Kamchatka. Western Palearctic birds head to the coasts of the British Isles, the Low Countries, France, and the Mediterranean: eastern Palearctic birds reach the Black, Azov and Okhotsk seas. Nearctic birds winter on the southern coasts of the USA.

**Tufted Duck** *Aythya fuligula*

Breeds in Iceland and across Eurasia, though only to the northern coast in Fennoscandia. Also breeds on Kamchatka. Wintering birds are largely resident but some move to the Mediterranean coasts of southern Europe and north Africa.

Greater Scaup drake, Chruchill, Manitoba, Canada. The green head sheen is usually the best way of distinguishing the two Scaup species.

Tufted drake, Stavanger, Norway.

**The Lesser Scaup (*Aythya affinis*),**

This species is distinguished from its Greater cousin by having a purple (rather than green) head gloss (though this is not always a safe method in identification as the head sheen can sometimes vary in different lights). It breeds across North America, though to the north coast only near the Mackenzie delta and also on the south-eastern shore of Hudson Bay, but otherwise it is sub-Arctic. The Canvasback (*Aythya valisineria*) is also sub-Arctic, though it may be seen at the Mackenzie delta and in central Alaska.

## Eiders

The four eiders are the most maritime of all wildfowl, and are pelagic for much of their lives. They are dive-feeders, making the deepest dives of any duck. Though all eiders normally dive to only about 5m, Common Eiders are known to dive to 20m at least and it is believed that the Spectacled Eider, probably the deepest diver, reaches depths of 50m. During dives the birds use their feet as paddles, with occasional wing strokes. Prey comprises molluscs, which are prised from rocks, crustaceans and other benthic animals. Prey is crushed either by the powerful bill or in the grit-free, muscular gizzard. Steller's Eider, the smallest of the group, feeds extensively on aquatic larvae, especially in freshwater. As an adaptation for diving the birds are heavy, take-off requiring a lengthy run across the water. Once airborne they are strong fliers, but not manoeuvrable, a fact that is particularly noticeable when they land: landings have none of the grace of the smaller ducks, involving more of the crash-landing technique of the swans.

Drake eiders are finely patterned, the females having the familiar cryptic brown plumage required of tight-sitting nesters. The drakes parade their plumage in mating displays while calling their musical three-syllable *coo*. They also coo at sea, the call, when heard through an opaque Arctic sea mist, being ethereal and evocative. Female eiders have an additional protective technique when incubating: if forced to flee they will, just before departing, defecate evil-smelling faeces on their eggs.

Common Eider, Kjorholmane, Norway.

This may well deter Arctic Foxes, but is of little benefit against gulls and skuas, which have a limited sense of smell. Eider ducklings frequently form large crèches of up to 100 birds (500+ have been seen) in the care of one or more females.

### Common Eider *Somateria mollissima*

Breeds on both coasts of Greenland (though not to the far north), Iceland, Svalbard, Franz Josef Land, in Fennoscandia and the southern island of Novaya Zemlya. In Arctic Russia breeding is patchy, occurring on the New Siberian Islands and Wrangel Island, with isolated breeding sites on the mainland (mostly in Chukotka). In North America, the birds breed on the Aleutians, west and north Alaskan coasts, the northern Canadian mainland coast (including Hudson Bay), and southern Arctic islands.

## Eider down

Eider females pluck down from their breasts to insulate their eggs. The value of eider down to humans in keeping northern winters at bay has been obvious since at least the 7th century when St Cuthbert, the first Bishop of Lindisfarne, set up a sanctuary for Common Eiders on one of the Farne islands off the Northumberland coast: eiders there are still known as Cuddy Ducks by locals in his memory. Commercial farming was begun by the Vikings two centuries or so later, and until the early 20th century down remained a major export of Iceland. The down was collected from nests after the chicks had departed. Each nest produces about 15g of raw down, the cleaning process reducing this to about 1.5g of usable material. A kilogram of exported down therefore required the input from 700 nests: at the industry's height, Iceland exported over 4 tonnes of raw down annually, representing the output from almost 300,000 nests. Though now reduced in scale due to competition from Chinese goose down producers and other sources, Iceland still exports down worth around $2 million annually.

The scientific names of the bird reflect this usage. *Somateria*, the genus name for the three large eiders, derives from the Greek for 'down body' while *mollissima*, the specific name for the Common Eider, derives from the Latin for 'softest'.

Greenland birds winter in Iceland, where the ducks are resident. The birds of the Eurasian Arctic islands move to the southern Scandinavian coasts or the Bering Sea. Nearctic birds move to the both coasts of the United States.

### King Eider *Somateria spectabilis*

Drakes justify their name with a regal head pattern that includes an orange and black forehead shield which the Inuit sometimes bite off and eat immediately after killing a bird. It breeds on both coasts of Greenland, on Svalbard, Novaya Zemlya and along the north Russian coast from the White Sea to Chukotka, including the New Siberian Islands. North American Kings breed along the north coast from Alaska to Hudson Bay and on all Canada's Arctic islands. Absent from Canada's coast east of Hudson Bay, in winter, east Greenland birds join resident Icelandic birds. From western Greenland the birds make for the open water off the south-west coast. Svalbard and west Russian birds move to the Barents and Kara Seas, with east Russian birds heading to the Bering Sea. American birds move to the Bering Sea and Labrador coast.

Until the mid-1990s the wintering range of Spectacled Eiders was completely unknown. Only with the advent of radio-tracking did it become possible to follow tagged birds, this leading researchers to the central Bering Sea, where photography revealed perhaps thirty flocks comprising a total of at least 150,000 birds. It is now known that the birds winter in large numbers in areas of open water, chiefly near St Lawrence and St Matthew islands. Aerial photography suggests that the birds help keep leads free of ice by their diving and swimming.

Drake Spectacled Eider, Barrow, Alaska.

Drake Common and King Eiders, Sirevaag, Norway.

### Spectacled Eider *Somateria fischeri*

The least-known of the three *Somateria* eiders, males, females and chicks all have spectacular goggles (the spectacles of the name). It breeds on the Asian Russian coast east of the Lena delta, on the New Siberian Islands and Wrangel and also on the western and northern coast of Alaska. The Alaskan population crashed in the 1990s for unknown reasons and has continued to decline ever since.

### Steller's Eider *Polysticta stelleri*

The smallest of the four eiders is named for Georg Wilhelm Steller the German naturalist who accompanied Bering's second journey. It breeds on the Asian Russian coast east from Khatanga Bay, on the New Siberian Islands and may also breed on other, isolated sections of the Russian coast, perhaps as far west as the Kola Peninsula. Also breeds in western and northern Alaska. Birds have been seen off the Fennoscandia coast and in the Baltic Sea in winter, but it also winters in the Bering Sea.

Drake Common (top) and Steller's Eiders, Varangerfjord, Norway.

215

Barrow's Goldeneye, Mývatn, Iceland.

Harlequin Duck pair on the surging waters of the north Pacific at Starichkov Island, off southern Kamchatka's eastern shore.

Long-tailed Ducks, Mackenzie Bay, north-east Greenland.

## Sea ducks

Sea ducks are a group of pelagic waterfowl (which, technically, should include the eiders, though they have been considered separately here as their flamboyant plumage makes them different from the other, essentially black-and-white, ducks). Sea ducks dive much deeper than the diving ducks and occasionally pursue prey, particularly the sawbills, which take fish and swimming invertebrates, while the rest chiefly feed on benthic animals. As with the eiders, shellfish and crustaceans are often swallowed whole and crushed by the muscular gizzard.

As with the dabbling and diving ducks, drakes abandon incubating females. The precocial chicks have good down coverings and some subcutaneous fat, which allows them to dive in icy waters almost from hatching. Northern sea ducks are partially migratory, moving south ahead of the sea ice in winter.

### Barrow's Goldeneye *Bucephala islandica*

Named after Sir John Barrow, the man behind the Royal Navy's expeditions in search of the North-West Passage in the 19th century. Breeds in Iceland (mostly at Lake Mývatn where the population is declining). Otherwise breeding is restricted to the Nearctic (the breeding birds of west Greenland are now thought to be extinct), and covers central and southern Alaska, southern Yukon and British Columbia. Icelandic birds are resident: Nearctic birds move south in winter to inland and coastal sites.

### Harlequin Duck *Histrionicus histrionicus*

The drake is the most colourful of the non-eider sea ducks. They breed in the south-west and the central coast of east Greenland, Iceland, Chukotka, Kamchatka and around the northern shore of the Sea of Okhotsk. Also in coastal southern Alaska, Yukon, North-West Territories and in southern Quebec and Labrador. In winter the ducks are resident or partially migratory.

### Long-tailed Duck *Clangula hyemalis*

The drake is unmistakable, with long, upcurved tail feathers. It breeds throughout the Arctic, including most of the Canadian Arctic islands. On Ellesmere Island the ducks vie with King Eiders for the title of most northerly breeding waterfowl. They overwinter in the Bering Sea, off eastern Canada, southern Greenland, around Iceland, north Norway and in the Baltic and North Seas.

216

## The Oldsquaw

An older name for Long-tailed Duck is Oldsquaw, apparently a reference to the noisy drake of the species, which reminded early travellers of the chatter of elderly Inuit women. In this age of political correctness, the name was considered inappropriate and changed, though it is still sometimes used in North America. The scientific name for the bird translates as 'noisy winter bird'.

### Common Scoter *Melanitta nigra*

Breeds in Iceland and across Eurasia to the Lena delta, but only to the north coast west of the Urals. Wintering birds move to the North and Baltic seas.

The Velvet Scoter (*Melanitta fusca*) of Eurasia and White-winged Scoter (*Melanitta deglandi*) of North America, were formerly considered to be conspecific, but are now considered separate species. The former breeds across Eurasia, but only to the north coast west of the Urals. Largely absent from Chukotka, but breeds on Kamchatka. The latter breeds across North America from Alaska to the western shores of Hudson Bay, but only to the north coast near the Mackenzie delta.

The Bufflehead (*Bucephala albeola*) breeds in central Alaska, but more southerly across Canada, though it is occasionally seen north of the Arctic boundary as defined here. The Common Goldeneye (*Bucephala clangula*) breeds in Fennoscandia and across Russia, but only to the north coast around the White Sea. Also breeds on Kamchatka. North American birds breed in southern Alaska, at the Mackenzie delta, then south towards Hudson Bay's southern shore and in southern Quebec and Labrador. Where the two Goldeneyes overlap, drakes can be distinguished by the head gloss and the patterning of the upperparts. The Black Scoter (*Melanitta americana*) was formerly considered conspecific with the Common Scoter and is very similar to that species. It breeds in Asian Russia east of the Lena delta (but not to the north coast) and on Kamchatka and also breeds in west and south Alaska, southern Quebec and Labrador. The Surf Scoter (*Melanitta perspicillata*) breeds patchily throughout Alaska (though is rare in the north), the Yukon and North-West Territories, and in southern Quebec and Labrador.

## Mergansers

The *Mergus* ducks are commonly known as sawbills because of their elongated, thin bills, which have serrations on the mandibles to grasp fish and hooked tips to aid their capture. There are three Arctic and subarctic breeders, though a fourth species, the Hooded Merganser (*Lophodytes cucullatus*), while essentially a temperate bird, has recently been increasing its range northward.

Mergansers are the only ducks capable of catching fish, but they also take other prey opportunistically, including amphibians, molluscs, crustaceans and even small mammals.

### Red-breasted Merganser *Mergus serrator*

Breeds on west and east Greenland, Iceland and across Eurasia, though rarely to the north coast. Also breeds on Kamchatka and across North America, often to the north coast, including the southern part of Baffin Island. Most birds are resident or partially migratory, though western American birds move along the west US coast.

Evening light illuminates a Red-breasted Merganser pair in southern Hudson Bay, Canada.

Red-breasted Mergansers in flight, Hafrsfjord, Norway.

Red-breasted Merganser pair, Hafrsfjord, Norway.

The mating displays of the Arctic raptors often involves some form of 'sky-dance', in which the male circles the chosen nest site. There may even be mutual dancing, the pair touching or even linking talons. Sky-dancing is less common in falcons, though the 'high circling' flight of males above the nest site is an equivalent. Courtship-feeding is also common. In general, the pair bond is monogamous and may be long-lasting, though in some raptors it appears to be seasonal. The nest sites of raptors are often re-used annually. Traditional cliff sites may be highlighted by the 'whitewash' streaks from years of defecation. The arboreal stick nests of Steller's Sea Eagles, added to annually, sometimes attain such vastness that they overwhelm the tree, causing major branches (or even the entire tree) to break. By contrast, falcons make no nests, females making only a scrape on a ledge or usurping the old nests of other species: Gyrfalcons may use an old Raven or Rough-legged Buzzard nest, or even that of an eagle.

The Goosander (*Mergus merganser*) breeds in Iceland and across Eurasia, though only to the north coast in Fennoscandia. It also breeds on Kamchatka and across North America the continent, but is essentially subarctic, being rare in west and central Alaska. They do, however, breed on the southern shore of Hudson Bay. The Smew (*Mergus albellus*) breeds in southern Fennoscandia and across Russia to southern Chukotka and northern Kamchatka, but is essentially subarctic. Smew drakes are beautiful white birds with black sculpting.

## Raptors

Raptors are superbly equipped for the task of locating and catching live prey. The wings can be broad and long to allow the birds to soar with minimal effort as they seek relatively slow-moving animals or carrion, or tapered to allow fast flight, the better to overhaul speedier prey. The birds have talons for gripping fast-moving or struggling prey and hooked beaks for tearing flesh.

Grey-phase Gyrfalcon and chick, Hood River, Nunavut, Canada.

The larger raptors lay 1–3 eggs, the falcons more. The eggs are incubated by the female, which is fed by the male. The chicks have minimal down and are altricial or semi-altricial. They are fed by the female at first, with food brought by the male. The female also broods. Later, when the chicks have down and can thermoregulate, both birds will hunt and feed the young. The chicks are independent soon after fledging. Raptors take a long time to mature, and often do not breed until they are two or three years old, though falcons may breed at one year. Steller's Sea Eagles do not breed until they are at least four or five years old, and perhaps even as late as eight years.

Apart from the harriers, hawks and some falcons, the sexes are similar in plumage. In all species except the Osprey females are larger than males, a reverse of the norm (though not unique to raptors). It is not entirely clear what the basis for this reversed sexual dimorphism is: the size of prey taken by the two birds (smaller males are more agile and can catch smaller prey) is clearly important since it reduces interspecific competition, but why it should be the females that are larger than the males is not understood.

Breeding and winter plumages are similar in all Arctic raptor species.

## Falcons

Falcons are the only true Arctic raptors, although other species, as we shall, are also found close to our defined Arctic line.

### Gyrfalcon *Falco rusticolus*

This is the world's largest falcon. High Arctic individuals are almost pure white – small wonder that for many travellers the Gyrfalcon is the ultimate tick on the species list. The bird also has other colour phases. While 'white' birds are white, some with black barring, 'grey' birds have steel-blue upperparts and white underparts, again with black barring, while 'dark' birds have dark brown upperparts and white underparts, the latter with heavy dark brown barring. Birds of the different forms will mate, often producing broods with chicks of differing, intermediate forms. Gyrs hunt birds and small mammals. The prey is taken after a low quartering flight, or by steep, fast dives ('stooping').

They breed in Greenland (both west and east coasts), Iceland, across Eurasia, but not on Russia's Arctic islands, also across North America, including Canada's Arctic islands as far north as Ellesmere Island. Resident, but move locally in search of prey.

White-phase Gyrfalcons, Bylot Island, Nunavut, Canada.

Dark-phase Gyrfalcon, Norway.

Grey-white phase Gyrfalcon, northern Iceland.

**Peregrine Falcon** *Falco peregrinus*

One of the world's best-known birds – an avian icon for its beauty and speed. The population went into rapid decline in the 1950s and 1960s due to the widespread use of organochloride pesticides (specifically DDT), but has since recovered.

It breeds on Greenland and across Eurasia, including the southern island of Novaya Zemlya throughout Alaska and Arctic Canada, including the southern Arctic islands. There are numerous subspecies across the Peregrine's vast range, the most significant being the Nearctic and Palearctic 'tundra' Peregrines which are much paler than southern birds. Northern birds are migratory, moving south in winter to join resident birds.

Male Tundra Peregrine Falcon, near the Hood River in the Barren Lands of Nunavut, Canada.

**Merlin** *Falco columbarius*

The smallest Arctic falcon, it breeds in Iceland, Fennoscandia, and then more southerly across Russia. It is absent from Kamchatka but breeds in central and southern Alaska (although rare throughout) then more southerly across Canada to Hudson Bay and in southern Quebec and Labrador. Some Icelandic Merlins stay on the island, but most Arctic breeding birds move south in winter.

## Other Raptors

**White-tailed Eagle** *Haliaeetus albicilla*

Breeds in west Greenland, Iceland and across Eurasia, reaching the north coast in Fennoscandia and

Female Merlin, Mývatn, Iceland.

## The fastest bird?

The Peregrine Falcon is often thought of as the world's fastest bird, with speeds of 300km/hr often quoted for the stoop speed. However, how fast a domesticated falcon can fly and how fast a wild bird does fly are very different things as reaching high speed takes time for acceleration under gravity and for a wild bird the imperative is reaching prey not breaking a speed record. Since Gyrfalcons are heavier than Peregrines it is likely they would travel faster in hunting stoops.

Adult White-tailed Eagle, Sea of Okhotsk, Russia

Chukotka. Some birds are resident, but others migrate south.

### Rough-legged Buzzard (Rough-legged Hawk)
*Buteo lagopus*

Breeds in Fennoscandia and across Russia, to the north coast except on the Taimyr Peninsula: absent from all Russia's Arctic islands. It also breeds across North America and on southern Canadian Arctic islands. Eurasian, Kamchatka and North American birds are considered separate subspecies. Arctic breeding birds move south to southern Europe, Ukraine, southern Russia, and southern US states.

### Northern Harrier *Circus hudsonius*/ Hen Harrier (*Circus cyaneus*)

Very recently (early 2016) it was decided to split the Northern Harrier from the Eurasian Hen Harrier on the basis of DNA differences, male plumage and wing length. Initially the two were considered subspecies. The two species are very similar in habits and prey.

Northern Harriers breed in eastern Alaska, Yukon and western North-West Territories to the north coast, then more southerly to Hudson Bay and into Quebec and Labrador. In winter the birds head to the southern US states. Hen Harriers are Arctic fringe species, breeding in southern Fennoscandia, then more southerly across Russia to Kamchatka.

Several other raptors which are Arctic fringe breeders may be seen at higher latitudes. The Golden Eagle (*Aquila chrysaetos*) is the only northern *Aquila* eagle. It breeds in Fennoscandia, though is absent from the far north, then more southerly across Russia. In 2016 conservationists were alarmed at the decision of the Norwegian government to cull 200 eagles

### The eagle and the child

Despite the numerous legends and tall tales, there is only one documented record of an eagle having picked up a child. On 5 June 1932 a White-tailed Eagle snatched four-year old Svanhild Hansen, said to have been a particularly small child, at Leka in Norway and carried her for more than a kilometre to a ledge close to its nest site, almost 250m up a mountain. The girl was scratched but otherwise unharmed.

Adult Steller's Sea Eagle, Kamchatka, Russia.

to appease Sámi reindeer herders and some sheep farmers who claimed the birds were killing young calves/lambs, despite evidence to suggest they were responsible for no more than 2% of all losses. Golden Eagles also breed in Alaska (though is common only in central Alaska) on the Mackenzie delta, then more southerly to southern Hudson Bay and across southern Quebec and Labrador. The Bald Eagle (*Haliaeetus leucocephalus*), the symbol of the United States, is the Nearctic equivalent of the White-tailed Eagle. Adults have a white head (the origin of the name) and white tail, though these are only acquired when they are five years old. Bald Eagles breed in central and southern Alaska, including the Aleutians, and across North America, but rarely north of the timberline. Steller's Sea Eagle (*Haliaeetus pelagicus*) is the world's largest fish eagle and arguably the largest bird of prey: the Steller's huge bill is without doubt the largest of any raptor. The eagles breed on Kamchatka, the northern Kuril islands, along the coasts of the Sea of Okhotsk and on Sakhalin Island. The Osprey (*Pandion haliaetus*) is another fish hunter. They are seen as far north as the timberline, breeding in southern Fennoscandia, but are more southerly across Russia, in southern Alaska (but are rare) and across Canada, including the southern shores of Hudson Bay. The (Northern) Goshawk (*Accipiter gentilis*) is essentially aboreal and breeds to the treeline in both the Palearctic and Nearctic.

## Grouse

Grouse are a group of northern gamebirds, and a highly successful one, several having wide latitudinal ranges: the (Rock) Ptarmigan resides from about 38°N to 82°N. Grouse are stocky birds, their size a reflection of their diet, particularly the winter diet. Boreal species eat the needles of pine and spruce. Though abundant, the needles have a very low nutritional value so very large quantities are ingested to provide the bird's energy requirements. Large crops and long intestinal tracts are therefore necessary.

In winter, Arctic birds moult into a dense white plumage offering high insulating qualities and excellent camouflage (both avian and mammalian predators share their winter quarters). The birds have strong claws that allow them not only to dig in the snow for

Rough-legged Buzzard, Orre, Norway.

Golden Eagle, Bjerkreim, Norway.

[Above] Adult Bald Eagle at its nest, Great Slave Lake, NWT, Canada.

[Below] Osprey, Sirdalsvannet, Norway.

food, but also to excavate shelters to escape the worst of the weather. In spring males moult to a distinctive breeding plumage, while females acquire a cryptic plumage – a necessity for these ground-nesting birds, which provide a high fraction of the vertebrate biomass of the tundra and boreal fringe and are important prey for many species.

The males of the Arctic species tend to perform solo displays involving tail fanning, dropped wings and exaggerated strutting: the birds perform to intimidate rival males, since the holding of a territory is key for mating success. The males of some southern species take a different strategy, creating 'leks', communal display grounds in which the males strut their stuff. Females visit the leks, observing the displays before making a final choice.

Once mated a female grouse will lay a large clutch of up to a dozen eggs, though 5–8 is more usual. Only the female incubates. Grouse chicks are downy and precocial, well capable of feeding themselves. Unlike their parents, the chicks feed chiefly on invertebrates, changing to a vegetarian diet as they grow. They are fully grown by their first winter, but do not always breed the following spring.

Grouse are resident in many Arctic areas, though they will move if local food sources become too poor or unavailable due to bad weather (e.g. rain followed by frost, which seals food beneath an impenetrable coat of ice). Some grouse species do migrate, occasionally travelling considerable distances. In general, though, grouse do not fly often and they are primarily terrestrial birds. This limitation on movement and the wide geographical range of the birds has led to the evolution of many subspecies; as many as 30 have been recognised for the (Rock) Ptarmigan alone.

### Willow Grouse (Willow Ptarmigan) *Lagopus lagopus*

Breeds across Eurasia to Kamchatka, north to the coast and on the New Siberian Islands also across North America including the southern Arctic islands. Resident. Absent from Greenland, Iceland and Svalbard.

### (Rock) Ptarmigan *Lagopus mutus*

In summer the colour of the male is diagnostic when comparing to the Willow Grouse, though females are extremely difficult to differentiate. The stouter bill of the Willow Grouse is the only distinguishing feature

Willow Grouse, Churchill, Manitoba, Canada. The female (below) is superbly camoflaged for security while incubating, in contrast to the male (above) emerging from his winter white plumage, who is conspicuous, the better to be seen by rivals for his territory. Males also occasionally take prominent positions, trading their own short-distance speed of flight which often evades an attacker against the extra security for their mates drawing the predator's attention.

Male Rock Ptarmigan, north-east Greenland. It is autumn and the bird is moulting from breeding to winter plumage.

Male Rock Ptarmigan in the Barren Lands of Nunavut, Canada. It is spring and the bird is moulting to its breeding plumage.

between the two species in winter, though male Ptarmigan have black lores.

A resident that breeds on Greenland, Iceland and Svalbard, and to the north coast across Eurasia to Kamchatka and may breed on Russia's Arctic islands. It also breeds across North America, including all Canadian Arctic islands to northern Ellesmere Island.

The White-tailed Ptarmigan (*Lagopus leucurus*) is an upland tundra species, so not a true Arctic dweller. They breed in central and southern Alaska, but are not common. Black Grouse (*Tetrao tetrix*) sit, ecologically, between tundra and forest grouse, and so may be seen by Arctic travellers. They breed in southern Fennoscandia, then more southerly across western Russia. Spruce Grouse (*Falcipennis canadensis*) are a boreal species that can be seen to the timberline in central and southern Alaska (but are rare in the west) and across Canada. Ruffed Grouse (*Bonasa umbellus*) are a boreal species that prefers deciduous woodland, and may be seen by travellers enjoying Alaska's Denali National Park.

## Cranes

Cranes are the tallest flying birds, a creature seemingly too delicate to withstand the rigours of the Arctic. Yet two species are true Arctic breeders. Cranes are omnivorous, opportunistic feeders, taking seeds, berries, invertebrates, amphibians and reptiles, and even small birds. Wintering Sandhill Cranes feed in fields, taking waste grain and potatoes, but the Siberian Crane is less adaptable, feeding only in wetlands. This specialism represents the greatest threat to the species, as wetland drainage along its migration routes and at its wintering sites have reduced feeding sites.

In general cranes are monogamous. Though gregarious on migration and at wintering sites, they tend to be solitary while breeding (though the Sandhill Crane is an exception). Cranes are renowned for their 'dancing' displays – wing-stretches, head-tosses and vertical leaps, and calling in unison with extended necks – that all species perform. Most cranes breed at 3–5 years, but are long lived. They make large nests in which 2 (sometimes 3) eggs are laid. Both birds incubate. Crane chicks are precocial, following their parents into feeding areas at an early age, but they are initially fed by both parents. The chicks do not develop flight feathers until they are 3–4

Female of the Svalbard subspecies of Rock Ptarmigan (see also page 179). It is winter and the bird is in full winter plumage.

months old. Siberian Cranes differ from other species in laying just a single egg. The combination of a single chick and its vulnerability during its long fledging period pose a problem for conservationists. Thankfully programmes are now in hand in Russia aimed at stabilising, then raising, the threatened population.

**Sandhill Crane** *Grus canadensis*

Breeds in coastal Chukotka east of the Kolyma delta and south to the Gulf of Anadyr. It breeds across North America from Alaska to western James Bay, but is uncommon in northern Alaska, though breeding to the north coast along much of the Canadian mainland, and on Victoria and Baffin islands. Migratory, with Russian birds joining those from North America in the southern United States.

Dancing Sandhill Crane, Churchill, Manitoba, Canada.

Siberian Crane pair at nest site, northern Yakutia, Russia.

**Siberian (White) Crane** *Grus leucogeranus*

Breeds in two areas, with a small population near the Ob River and larger one in northern Yakutia. The Ob birds migrate to the Iranian Caspian Sea and northern India, where hunting along the route (especially in Pakistan and Afghanistan) adds to the threats the species faces. Yakutian birds move to China.

## Waders

Waders (or shorebirds) represent the largest group of Arctic birds. They work the intertidal region of the Arctic coast, and the lakes and marshes created by the summer thaw, feeding on insects, worms, crustaceans and molluscs, together with some plant material, and occasionally other foods such as small fish. There are several wader groups (particularly the calidrids) in which, although the birds generally have distinct breeding, wintering and immature plumages, these are very similar and confusion between species is high. In the species summaries that follow some pointers for identification are included, but it must be stressed that these are general; the identification of waders requires patience and perseverance.

Waders generally nest on the ground (though Wood Sandpipers use the old nests of other species, often high above the ground). The nest is little more than a scrape in the ground or a depression in vegetation, often with minimal lining. Between 3 and 5 eggs are laid. The eggs are large – as they must be to accommodate a chick which is born downy and precocial – and consequently represent a large energetic investment for the female: for some of the smaller calidrids (e.g. the stints) the clutch can weigh up to 90% of the female's body weight. Care differs between the species. In most it is shared by the parents, but in some it is by the female alone, and there are species in which the male takes responsibility. In species in which both birds incubate the eggs and care for the young, the female often abandons the chicks before they fledge, with the male continuing with care. Young waders leave the nest within a few days of hatching and are, in general, self-feeding. Waders breed in their second or third year.

## Plovers

Stocky, short-billed birds with large eyes, characterised by their feeding method of 'look–run–peck'. The pecking is of insects on land forays, or of crustaceans and molluscs on the shore. Some plovers also use a foot-paddling technique when feeding, the bird standing on one leg and paddling the water surface with the other, the paddling causing prey to move and betray itself.

In the 'tundra' plovers the pair bond is monogamous and probably lifelong: initial pairing takes place after a display that involves the male running at the female (the 'torpedo run'), which looks far from subtle but seems to work. The ringed plovers have similar displays, though the pair bond, while monogamous, is seasonal. Things are different in the Dotterel with females taking the lead, luring males with raised wing and fanned-tail displays. Though some bonds are monogamous, polyandry

## Broken-wing displays

Plovers are famous for drawing potential predators away from their nests and eggs, drooping a wing and calling plaintively as though badly injured as they head away from the nest. Humans seem especially gullible to the performance, following the bird even when they know exactly what is going on. But birders and naturalists are, in general, much less dangerous than other predators, which can rapidly overhaul the faking bird, causing it to take off in a hurry or to become a genuine casualty.

Nesting Dotterel, Jotunheimen, Norway.

is common and males are occasionally polygamous. Usually the female will abandon her eggs soon after laying to seek out another male for a second clutch, leaving the first male to care for eggs and young. Female Dotterel rarely share brood responsibilities.

**(Eurasian) Dotterel** *Charadrius marinellus*

Breeds in Fennoscandia and across Russia to Chukotka, to the northern coast and on Novaya Zemlya's southern island, but absent from Greenland, Iceland and Kamchatka. Has bred in western Alaska. In winter the birds fly to the Middle East and North Africa as far west as Morocco. East Siberian birds may travel more than 10,000km during migration.

**(Common) Ringed Plover** *Charadrius hiaticula*

Common Ringed Plover, Varangerfjord, Norway.

Breeds in Greenland, Iceland, Svalbard, Fennoscandia and along the northern coast of Eurasia including Novaya Zemlya's southern island and the New Siberian Islands. Has also bred on St Lawrence Island, northern Ellesmere and western Baffin Island, where it is now thought to be established. In winter, European birds move to southern Europe and North Africa, Asian birds to the shores of the Caspian Sea and the Middle East.

**Semipalmated Plover** *Charadrius semipalmatus*

Virtually identical to the Ringed Plover in summer and winter, but lacks the prominent white ear patch, having an indistinct crescent (and sometimes no white at all). The name derives from the webs between the inner and middle, and middle and outer toes.

Breeds across North America, to the north coast and on Banks, Victoria, Southampton and southern Baffin islands. Wintering birds move to the southern Pacific

Semi-palmated Plover, Cold Bay, Alaska.

Eurasian Golden Plover, Varangerfjord, Norway.

and Atlantic coasts of the USA, the Caribbean islands and the coasts of Central and South America.

### Lesser Sand Plover (Mongolian Plover)
*Charadrius mongolus*

Breeds in Chukotka, Kamchatka and the Commander Islands. Has also bred in western Alaska, but is not an established species there. In winter the birds move to the Philippines, Indonesia and Australia.

## Tundra plovers

### American Golden Plover *Pluvialis dominica*

Breeds in western North America from Alaska to Hudson Bay, but also on Banks, Victoria, Southampton, and western Baffin islands. Has also bred on Wrangel Island but is not established. Wintering birds occur primarily in the southern USA and Central America, but fly as far as Argentina.

### Eurasian Golden Plover *Pluvialis apricaria*

Very similar to American Golden Plover though slightly larger. Breeds in north-east Greenland (in small numbers), Iceland, Scandinavia and northern Russia as far east as the southern Taimyr Peninsula. Absent from Svalbard (though breeding records do exist) and the

western Arctic islands of Russia. Greenland and Iceland birds winter in Iberia, Scandinavian and Russian birds in Iberia and North Africa.

### Pacific Golden Plover *Pluvialis fulva*

Very similar to Eurasian Golden Plover, but with longer, darker grey legs, a uniform grey-brown underwing, and less white (and less well-defined white areas) on the flanks.

Breeds in Russia east of the Yamal Peninsula and in western Alaska (though uncommon there), and to the north coast, but not on any Arctic islands. In winter the birds move as far south as Australasia.

### Grey Plover (Black-bellied Plover) *Pluvialis squatarola*

Breeds in Greenland (but rare), across Russia as far north as the coast, and on the New Siberian Islands and Wrangel, and discontinuously in North America, being found on the west and south-west coasts of Alaska, but rare in the north, in Yukon and western NWT. Also breeds on Banks, Victoria and Southampton islands and on the nearby mainland, and on western Baffin Island. Eurasian birds winter in south-west Europe, south-east Asia and Australia. North American birds move to the Atlantic and Pacific coasts of southern USA, and into Central and South America.

## *Calidris* sandpipers and related species

Calidris waders differ from the plovers in having bills adapted for probing rather than pecking. The bills vary enormously in size and shape, that of the Spoon-billed Sandpiper being among the most remarkable of any bird. The bills have an array of touch sensors (called Herbst's corpuscles) that allow the birds to 'feel' prey beneath the sand or mud.

Some species can move the bill tip independently of the rest of the bill. This ability, termed rhynchokinesis, means that the bird can open the tip while the rest of the bill stays closed, and so can grab a worm or insect larva that can then be taken by the tongue and consumed without the bill being withdrawn. Together with the amazing touch sensitivity of the bill, this allows the bird to feed quickly and efficiently. This useful characteristic is shared by some snipes, *Tringa* sandpipers, godwits, curlews and dowitchers.

Dunlin in breeding plumage, Varangerfjord, Norway.

In general, *Calidris* sandpipers have long wings and legs, and relatively short tails. The sexes are similar. Most calidrids are northern breeding species and migrate great distances, reaching Tierra del Fuego and Australia. Indeed, Australia is a major wintering site for wintering northern waders, with close to a million birds of more than 50 species occurring in the north-west.

### Dunlin *Calidris alpina*

The black belly patch allows Dunlin in breeding plumage to be easily separated from similar shorebirds.

## Calidrid identification

Differentiating between shorebirds is one of the major problems of Arctic bird identification. Some general rules will help, but there is no substitute for close observation over time. First, gain some impression of the bird's size – not easy if it is on its own, but of value if the there are other species close by. Shape is also useful, but bear in mind that this can vary with activity. Look at the bill to judge its size relative to the bird, and its shape – is it drooping towards the tip? Many calidrids look very similar but have different leg colours – try to be sure that you are seeing the true colour and not the mud from the local habitat. Check the habitat: is it wet or dry, sandy or rocky? This can help, but again is not an absolute. Finally, listen for the call, but be cautious – in North America calidrids are often called 'peeps' because of their voice, and they can sound depressingly similar to first-time observers who have yet to attune themselves to the subtle differences between species.

It breeds in east Greenland, Iceland, Jan Mayen, Svalbard, Fennoscandia and across Arctic Russia to Chukotka, including both islands of Novaya Zemlya, the New Siberian Islands and Wrangel, and in west and north Alaska and the northern Canadian mainland west of Hudson Bay, including Southampton Island. Some Icelandic birds stay on the south coast during the winter, but most Palearctic birds move to western Europe, North Africa and southern Asia. Nearctic birds winter on the cast and west coasts of North America and in Central America.

### Baird's Sandpiper *Calidris bairdii*

Baird's can be distinguished from other calidrids by the wing-tips, which project beyond the tail and describe an oval as the bird walks and pecks. However, the wings of White-rumped Sandpiper also project beyond the tail.

It breeds in Chukotka and on Wrangel, and across North America from northern Alaska to Baffin Island; on the Canadian mainland it breeds only to the west of Hudson Bay. Also breeds in north-west Greenland. Wintering birds fly to South America.

### White-rumped Sandpiper *Calidris fuscicollis*

Very similar to Baird's. Breeds in Canada from the Yukon to Hudson Bay and on the southern Arctic islands, but rare in Alaska and east of Hudson Bay. Wintering birds head for southern South America, flying as far as Tierra del Fuego and the Falkland Islands.

### Semipalmated Sandpiper *Calidris pusilla*

Very similar to Baird's Sandpiper, but smaller and without the distinctive black spotting.

A Semipalmated Sandpiper, Seward Peninsula, Alaska.

In its breeding plumage the Curlew Sandpiper is one of the more distinctive Calidrids, eastern Chukotka, Russia.

Red Knots at Randaberg, Norway.

Breeds across northern North America, including the southern Arctic islands. In winter seen on Caribbean islands and in South America as far south as Uruguay.

### Western Sandpiper *Calidris mauri*

Very similar to Semipalmated Sandpiper but head and upperparts are distinctly chestnut and there is darker, heavier spotting on the underparts.

Breeds in Chukotka and western Alaska. In winter, Russian birds cross the Bering Sea to join Alaskan birds on flights to California and the Caribbean islands.

### Curlew Sandpiper *Calidris ferruginea*

Male Curlew Sandpipers stay on the breeding grounds for just a couple of weeks, spending their entire time there attempting to copulate with as many females as they can before departing.

Breeds from the Taimyr Peninsula to Chukotka, and probably also on the New Siberian Islands. In winter the birds move to Africa, southern Asia, Indonesia and Australia.

### Red Knot *Calidris canutus*

Similar in pattern and colour to the Curlew Sandpiper, the main diagnostic being the short, straight bill which is very different to the longer, downcurved bill of the Curlew Sandpiper. The Red Knot's legs and feet are also dark olive, rather than black.

Breeds on Russia's Taimyr Peninsula, the New Siberian Islands and Wrangel, and inland Chukotka, northern Alaska (but not consistently), Canada's Melville Peninsula and many Arctic islands. Also breeds in north Greenland. However, this distribution is patchy, the birds often being very local. In winter, Greenland and some east Canadian birds move through Iceland to western Europe. North American birds move to the coasts of the southern USA, Central America and as far south as Tierra del Fuego. Central Russian birds move to Africa, while east Russian birds fly as far as Australia.

The Red Knot was once one of the most abundant of all North American waders, but relentless hunting during its annual migration severely reduced numbers: thankfully hunting has now diminished. Red Knot have exceptionally long migration flights, with estimates of individual birds flying up to 30,000km. Some birds cross the Atlantic in one flight, accumulating subcutaneous fat before the flight that can total as much as 80% of the bird's body weight.

**Pectoral Sandpiper** *Calidris melanotos*

Adult males are larger than females and tend to have darker breasts. Males have a curious hooting call amplified by the expansion of an air-filled sac that causes the neck/upper breast to puff out. This strange *wow–wow* call carries over large distances and can be eerie on misty days.

Breeds from the Taimyr Peninsula to Chukotka, to the north coast but absent from all islands except Wrangel. In North America it breeds from north-western Alaska along the mainland coast to Hudson Bay and on southern Arctic islands as far north as southern Ellesmere Island. In winter both Russian and North American birds move to South America, with small numbers in Japan and Korea.

**Sharp-tailed Sandpiper** *Calidris acuminata*

Very similar to the Pectoral Sandpiper, but the Sharp-tailed in breeding plumage has brighter upperparts and much more extensive spotting on the underparts so there is no clear pectoral band and it has a distinct white eye-ring. In winter the birds are duller, and virtually indistinguishable from Pectoral Sandpipers.

Purple Sandpiper, Badlanddalen, north-east Greenland.

Breeds in northern Yakutia between the Lena and Kolyma rivers. Wintering birds fly to Australasia.

**Purple Sandpiper** *Calidris maritima*

Breeding adults are the darkest of the calidrids and tend to eat more vegetation than other calidrids, particularly algae, which is picked up as the bird searches among seaweed for other morsels. More likely to be seen swimming than other sandpipers, this ability probably relates to their usual habitat – very close to the waterline – which means they are occasionally collected by tides or waves.

Breeds on Iceland, Jan Mayen, northern Fennoscandia, Svalbard, Franz Josef Land, Novaya Zemlya, the Taimyr Peninsula and the southern island of Severnaya Zemlya. In North America the bird breeds on Baffin and Southampton Islands and southern Ellesmere Island. Also breeds in southern Greenland. In winter, many Icelandic birds are resident. Greenland birds move to Iceland and the British Isles, while birds from Svalbard and Russia to the Norwegian coast or western Europe. Canadian birds move to coasts of north-east USA.

**Sanderling** *Calidris alba*

Wintering birds are much paler than other calidrids. In summer confusions are possible, but Sanderlings are more active than other species and tend to be found on drier ground. It feeds at the waterline, but never seems to be overwhelmed by the waves, timing its runs forwards and backwards with perfection.

Breeds in northern Greenland, Svalbard, the Taimyr Peninsula, Severnaya Zemlya and the New Siberian Islands, and on the eastern Canadian Arctic islands north to Ellesmere Island. In winter, Greenlandic birds

## The rodent run

Many calidrids have their own version of the distraction display of plovers. In this case, rather than persuading the potential predator that they are injured, the birds make a crouched run away from the nest while emitting high-pitched squeaks. The effect is of a rodent running across the tundra and most predators will immediately give chase, only to discover that, perplexingly, the 'rodent' can fly.

Sanderling, Orre, Norway.

move to west Africa, Asian birds go to Australia and southern Africa, and Canadian birds fly to both coasts of the USA, to Central, and South America as far as south as Tierra del Fuego.

**Least Sandpiper** *Calidris minutilla*

The smallest calidrid. Breeding adult very similar to Semipalmated Sandpiper but has yellow or yellow-green, rather than black, legs.

Breeds from western Alaska to Labrador, but not on the Canadian Arctic islands. In winter the birds fly to southern USA, Caribbean islands, Central and northern South America.

**Little Stint** *Calidris minuta*

The smallest Palearctic calidrid that breeds in northern Fennoscandia and in Russia from Cheshskaya Guba to Chukotka, including the southern island of Novaya Zemlya and the New Siberian Islands. Wintering birds move to North Africa, the Middle East and southern Asia.

**Temminck's Stint** *Calidris temminckii*

Similar to Little Stint, but more subdued, less active and with yellow-green, rather than black, legs and feet.

Breeds from Fennoscandia to Chukotka, but only on the New Siberian Islands. In winter, European birds move to Mediterranean coasts, while Asian birds head for southern Asia and Japan.

**Red-necked Stint** *Calidris ruficollis*

Breeds in Russia, patchily from the Taimyr Peninsula to Chukotka. Also breeds sporadically in western Alaska. Migrating birds fly to southern China, Indonesia, Australia and New Zealand.

Other calidrids which breed at the Arctic fringe are Great Knot (*Calidris tenuirostris*), the largest calidrid, which is rare, with a very limited range, breeding in north-east Siberia (though the range is imprecise). The Rock Sandpiper (*Calidris ptilocnemis*) is very similar to the Purple Sandpiper (indeed, so similar that Aleutian birds are almost identical and some authorities consider the two taxa conspecific). Breeds in eastern Chukotka and the Commander and Kuril islands, in western Alaska, the Pribilof and Aleutian islands. The Long-toed Stint (*Calidris subminuta*) is very similar to both Little Stint and Least Sandpiper (though the ranges do not overlap). The toe of the name is the middle one (though the hind toe is also longer than in other calidrids), but the value of this difference in the field is somewhat limited. Found on upland tundra, but rare. Breeds in isolated pockets in central Siberia east of the Ob River, particularly near Magadan, in southern Chukotka, northern Kamchatka and on the Commander Islands.

## Other sandpipers

**Buff-breasted Sandpiper** *Tryngites subruficollis*

Attractive birds which were once numerous, but overhunting in the 19th century caused a dramatic decline from which the species has not recovered: it remains rare. It differs from many other shorebirds in being found almost invariably in dry areas, and in pecking rather than probing for food.

Breeds on Wrangel Island and north-eastern Chukotka, in northern Alaska and the Yukon, and on southern Canadian Arctic islands. In winter, all populations fly south to Argentina and Paraguay.

**Ruff** *Philomachus pugnax*

Exhibits extreme sexual dimorphism, males (*ruffs*) being much larger than females (*reeves*) and having exotic plumage. In winter, male birds lose this, becoming rather anonymous grey-brown birds. Breeding male Ruffs have elaborate lek display behaviour, the leks being traditional sites that may have been used over hundreds of years. The displaying Ruff flutters his wings and jumps in the air, with the ruff expanded to its full circle and the ear-coverts raised. Ruffs will mate with any interested reeve, so polygamy is common. Display and copulation are the ruff's only contribution to the next generation, the reeve incubating the eggs and caring for the chicks.

Lekking male Ruffs, Varangerfjord, Norway.

As if to finally confirm female prejudices regarding the character of males, the testes of the ruff do indeed weigh more than his brain. It is, though, worth noting that reeves frequently mate with several ruffs in the arena, studies suggesting that more than half of the broods contain chicks with different fathers.

Breeds in Fennoscandia and across Russia to the border of Chukotka, to the north coast in places, but absent from all of Russia's Arctic islands. Wintering birds occur in western Europe, along the coast of the Mediterranean, and as far south as the coasts of India and southern Africa.

**Spoon-billed Sandpiper** *Eurynorhynchus pygmaeus*

Perhaps the most remarkable of all Arctic breeding species, but also one of the rarest, with a small population that is declining due to habitat loss at the breeding sites, on migration and at wintering quarters. It is hoped that the recent initiative by the WWT to captive breed birds in Britain for reintroduction is successful in increasing the stock. The bill is broad at the base, then tapers before flattening into a diamond shape. The bill is so unusual and is diagnostic if it can be clearly seen, but with less-than-perfect views the bird can be confused with Red-necked Stint.

Breeds only in far eastern Chukotka and northern Kamchatka. Migrating birds head to south-east Asia.

**Stilt Sandpiper** *Micropalama himantopus*

Breeds from northern Alaska (where it is rare) to Bathurst Inlet and on southern Victoria Island. Also breeds at the south-west corner of Hudson Bay, in northern Fennoscandia and in isolated areas of Asian Russia as far east as the Kolyma river. In winter the birds move to central South America.

The Broad-billed Sandpiper (*Limicola falcinellus*) is a rare and elusive Arctic fringe breeder with a bill which is broad at the base (when viewed from above) and tapers towards the downcurved tip.

## Snipes and dowitchers

These shorebirds are all generally stocky birds with long bills that share feeding characteristics. Snipes have identical summer and winter plumages.

**Common Snipe** *Gallinago gallinago*

Snipes are among the few birds that deliberately make a mechanical (i.e. non-vocal) sound. During display flights the birds dive at an angle of c.45° while fanning the tail. The two outer feathers of the tail are stiff and have asymmetric vanes, the leading edges being narrow. The bird holds these feathers at right-angles to the body: at speeds of about 60km/hour or higher they vibrate. The air through the vibrating feathers is modulated, creating a loud drumming noise (also called winnowing), which can be heard over considerable distances. It is often heard before the observer has seen the bird. Although it is male snipe that are chiefly responsible for the drumming, females also make the sound, particularly early in the breeding season.

Common Snipe, Varangerfjord, Norway.

Breeds in Iceland, northern Fennoscandia, and across Russia, though rarely to the north coast and not on any Arctic islands. Icelandic birds are resident in winter, but European birds move to Mediterranean coasts and the sub-Saharan belt, while Asian birds fly to India and south-east Asia.

### Long-billed Dowitcher *Limnodromus scolopaceus*

When feeding, the birds probe with their bills very quickly, an action which is frequently referred to as being 'sewing machine-like'. They also often feed in shallow water, so for the photographer, getting a shot with the long bill exposed rather than half-buried is tricky.

Breeds in Russia east from the Yana River and south to Anadyr, on Wrangel and patchily in North America eastwards to the western and southern shores of Hudson Bay. Also breeds on southern Victoria Island. Both Russian and American birds winter in the south-west United States and in Mexico.

## Curlews, Godwits and *Tringa* Sandpipers

The Tringinae shorebird subfamily varies considerably, from the elegant godwits and the curlews with their distinctive bills to the exotic tattlers, encompassing rather more nondescript birds along the way. Only one is an Arctic breeder, and even that species is only found at the edge of the area as defined in this book.

### Hudsonian Godwit *Limosa haemastica*

Breeds patchily in western North America, at sites on the southern coast of Hudson Bay, near the Mackenzie delta and, rarely, in western and southern Alaska. In winter the birds fly to southern Argentina.

Other species are Arctic fringe breeders. Bar-tailed Godwits (*Limosa lapponica*) breed in northern Fennoscandia (chiefly Finland), and patchily across western Russia, though more consistently east of the Taimyr Peninsula to western Chukotka and near Anadyr; it also breeds in western Alaska. Bar-tailed Godwits are believed to fly non-stop from their breeding grounds on the western coast of Alaska to New Zealand. If this is the case then the journey – of about 12,000km – is the longest single flight of any bird. Recently, telemetry has confirmed that one bird flew non-stop from New Zealand to North Korea, a distance of over 10,000km,

Long-billed Dowitcher, Barrow, Alaska.

Hudsonian Godwit, Churchill, Manitoba, Canada.

at an average speed of 55km/h. Black-tailed Godwits (*Limosa limosa*) breed on Iceland and patchily on the Norwegian coast, but are otherwise more southerly. Whimbrels (*Numenius phaeopus*) breed on Iceland, in Fennoscandia and western Russia, then more patchily across Siberia. They also breed in western and central Alaska, but more rarely in the north, on the Mackenzie delta and around the south-western and southern shores of Hudson Bay. The Bristle-thighed Curlew (*Numenius tahitiensis*) breeds in a few places in western Alaska. The name derives from the stiff leg feathering. This is diagnostic against the Whimbrel, which is very similar and which may overlap in Alaska, though it is of little value in the field. Bristle-thighed Curlew breed in a few places in western Alaska (the Yukon delta and upland Seward Peninsula). In winter the birds move to Polynesia. The birds are rare, probably numbering less than 5,000. This is strange because in Alaska the available habitat could accommodate many more. What limits the population is the size of the winter habitat, this being confined to a handful of islands in the south Pacific. To add to the pressure of limited wintering grounds, feral cats and dogs on those islands hunt the birds when they are flightless during moulting. The birds have a strange winter diet, taking invertebrates and rodents, but also feeding on gull eggs, occasionally stealing eggs from beneath incubating birds. The stolen eggs are pierced with the bill or broken by being dropped. A bird will sometimes drop a rock on to an egg – one of the few observations of avian tool-use.

Common Redshanks (*Tringa tetanus*) are essentially sub-Arctic but breed on Iceland and western Fennoscandia. Spotted Redshanks (*Tringa erythropus*) breed in northern Fennoscandia, on the eastern shores of the White Sea, and from the Ob east to Chukotka, north towards the coast, except on the Taimyr Peninsula. In North America the Lesser Yellowlegs (*Tringa flavipes*) breeds in central and southern Alaska (but is rare on the west and north coasts), around the Mackenzie delta, then more southerly to the southern shores of Hudson Bay. Wood Sandpipers (*Tringa glareola*) are a boreal species, found at the timberline but rarely on the tundra. They breed in Fennoscandia, and across Russia to central Chukotka and throughout Kamchatka.

### Is the Eskimo Curlew (Numenius borealis) extinct?

Overhunting and loss of habitat caused a drastic depletion of this once-numerous species. Migrating birds fed primarily on the Rocky Mountain Grasshopper (*Melanoplus spretus*) that became extinct when the prairies were claimed for agricultural land and fires on the remaining areas were suppressed. The present position of this species is unclear. There has been no confirmed sighting since the early 1980s and most authorities consider the species to be extinct, pointing to the fact that a bird with a migration path that took it across the southern USA and on to southern South America can hardly have escaped detection for 30 years. Even the most optimistic consider the population to be less than 50–100 birds and the long-term survival of the bird to be bleak.

## Turnstones

Turnstones have bills similar to those of the plovers, being short for pecking at prey rather than probing for it. The prey is often found beneath stones and other debris, the name giving the search method away – unlike the plovers' 'look–run–peck' foraging, turnstones use powerful neck muscles to overturn objects that look as though they would be good hiding places. Occasionally several birds will team up to overturn a larger object if something that can be shared, such as a dead fish, lies underneath. Stone-turning can reveal all sorts of things, alive or dead, the birds consequently having the most varied diet of any shorebirds, as they will eat just about anything that is revealed.

### Ruddy Turnstone *Arenaria interpres*

Breeds in northern Greenland, Iceland, Svalbard, in Fennoscandia and across Russia, including Novaya Zemlya's southern island, the New Siberian Islands, and Wrangel. In North America, it breeds in west and north-west Alaska, along the northern coast of western Canada, and on southern Arctic islands. Icelandic birds and (probably) some from Greenland remain on Iceland's south-west coast in winter. Eurasian birds move to the coasts of western Europe, Africa, the Middle East, India, south-east Asia and Australasia, American birds

Ruddy Turnstone, Helglandskysten, Norway.

## Role reversal

In all three phalaropes the females are larger than the males, and have much brighter breeding plumage. The biggest size difference is in the non-Arctic breeding Wilson's Phalarope (*Phalaropus tricolor*), in which females are about 35% larger. In the Grey Phalarope the difference is about 20%, and in the Red-necked about 10%. The females court males with aerial and terrestrial displays. After mating and laying a clutch of eggs, the female will abandon the male, leaving him to incubate the eggs and care for the young. The female seeks another male if one is available, laying a second clutch before abandoning that as well. The female stays on the breeding grounds, so if either clutch is lost she can mate and lay again. Such polyandry is uncommon, though not unique, among birds.

move to the coasts of southern USA, Central America, the Caribbean islands and South America to northern Chile and Argentina.

The Black Turnstone (*Arenaria melanocephala*) is subarctic, breeding only in west and southern Alaska.

## Phalaropes

Two of the three species of phalaropes (all of which are northern birds) are true Arctic breeders and make long migration flights: they are clearly tough little birds, belying their fragile looks. The two Arctic species share many traits, including polyandry. They also share a feeding habit, swimming in tight circles ('spinning') to stir up the water and bring prey to the surface. As well as spinning, phalaropes feed at the shoreline, taking insects found among seaweed and tidal debris. Wintering birds are chiefly pelagic, taking small fish and free-swimming crustaceans, and even picking parasites from the skins of whales.

### Red-necked Phalarope *Phalaropus lobatus*

Breeds in southern Greenland (but rare in the north), Iceland, Svalbard, in Fennoscandia, and across northern Russia, but not on any Russian Arctic islands. It also breeds across North America and on southern Arctic islands. Full details of the species' wintering sites are not known. Asian birds are seen off western South America, in the Arabian Sea and north of Indonesia, but though it is known that the birds congregate in the Bay of Fundy, as many as two million birds being reported, no other Atlantic wintering sites have yet been confirmed.

### Grey Phalarope (Red Phalarope) *Phalaropus fulicarius*

The name difference reflects the fact that Europe sees only much duller wintering birds. Breeds in north-west Greenland (and in the north-east, though much more rarely), Iceland (where it is among the rarest of breeding birds), Svalbard (where it also very rare), on the north coast of Asian Russia east of the Yenisey, the New Siberian Islands and Wrangel. It also breeds in northern Alaska, around the Mackenzie delta, on Canada's Boothia Peninsula and the north-west shore of Hudson Bay, and on Canada's Arctic islands north to southern

Red-necked Phalaropes, the more colourful female is closer to the camera, Varangerfjord, Norway.

Female Grey Phalarope, Igloolik, Canada.

Ellesmere Island. The population on Southampton Island has declined drastically in the last few years, for unknown reasons. In winter, European birds fly to the Atlantic Ocean off central and southern Africa, while Asian and American birds move to the Pacific Ocean off southern South America.

## Skuas

Of the seven members of the Stercorariidae or skuas, four are true Arctic dwellers. Skuas have webbed feet and sharp claws that, combined with a powerful hooked bill, make them formidable predators. They take a variety of foods, not only the fish that might be expected but also birds and eggs, small mammals, and even berries and insects. Both Pomarine and Long-tailed Skuas tend to specialise in hunting rodents and are frequently seen far from the coast and in upland areas: in Norway; the Long-tailed Skua is actually known as the Mountain Skua.

The northern skuas are also kleptoparasites, obtaining food by chasing gulls and terns and forcing them to drop or disgorge their recent catches, then retrieving the food, usually in mid-air. This behaviour is characteristic of the Arctic Skua, which obtains most of its food by piracy. This species even tends to migrate with Arctic Terns, so the terns can be pirated along the way. Pomarine and Long-tailed Skuas are piratical in winter, but in summer they tend to hunt their own food. Great Skuas are also less frequently piratical.

The northern skuas are sexually dimorphic, females being larger than the males by 10–15%. The three smaller northern species have both light and dark morphs (though dark-morph birds are very rare in Long-tailed Skuas).

Skua mating displays are limited, amounting to little more than the male adopting an upright posture in front of the chosen mate, though the bond is reinforced by courtship feeding, this also occurring in long-paired birds. The pair bond is monogamous and life-long or, at least, long-lived. Two eggs are laid in a rudimentary nest. These are incubated by both birds (though mainly the female in Great Skuas). The chicks are semi-altricial and are fed by regurgitation. Skuas are long-lived and mature slowly, Great Skuas not breeding until they are seven or eight years old. The parent birds are highly aggressive in defence of their eggs and young, dive-bombing intruders to scare them off. Such attacks are frequently not bluffs: I have been knocked down by a Great Skua and had blood drawn by a Long-tailed Skua. Interestingly, the smaller skuas will also adopt the plover technique of luring a predator away by feigning injury, though this approach is used much less often than a direct attack.

Skuas take long migration flights, these often being overland, hungry birds then incurring the wrath of farmers by attacking chicken or duck flocks. The southern species also migrate over long distances, and vagrants can drift vast distances off course: one South Polar Skua *Catharacta maccormicki*, a species that breeds on the Antarctic continent, was, sadly, shot near Nuuk, Greenland.

### Great Skua *Stercorarius skua*

Great Skuas feed mostly on fish (primarily sandeels - *Ammodytes* spp.), but they are opportunistic and will take birds (species as large as geese), mammals (hares having occasionally been seen taken) and even chicks from adjacent skua nests; some Hebridean birds feed nocturnally on returning storm-petrels.

The Great Skua expanded its breeding range in the last half of the 20th century, reaching Svalbard in the mid-1970s. In places the species has ousted the Arctic Skua from breeding sites it once held, with the bigger bird killing both adults and chicks.

Breeds on Iceland, Jan Mayen, Svalbard and Bear Island, and on the northern coast of Norway. In winter the birds are pelagic in the North Atlantic from Ireland

## Skua or jaeger?

In Britain and some other countries, the northern members of the Stercorariidae are known as skuas. The origins of the word are unclear but it is likely to derive from the Shetland name *skooi* for the Arctic Skua, though the derivation of *skooi* is obscure. One delightful suggestion is that the word has its roots in *skoot*, the Norse word for what would be best termed 'excrement' in polite company, due to the widely-held belief that Arctic Skuas consumed the excrement of other seabirds, with the skuas frightening gulls then cleaning up afterwards. In reality, of course, the skuas were frightening the birds into dropping their fish catch, well deserving their American name of Parasitic Jaeger. The word jaeger almost certainly derives from the German for hunter, which also fits the birds rather well. It is interesting to note that the Shetland islanders gave the Great Skua its own name, *bonxie*, which probably comes from the Norse *bunski*, an untidy mess, a word usually applied, in a very definitely non-PC way, to untidy, wizened old women.

to central Africa, with some Icelandic birds flying to waters off Newfoundland.

### Arctic Skua (Parasitic Jaeger) *Stercorarius parasiticus*

The most piratical of the skuas. Breeding adults have two colour morphs, pale and dark. Intermediate forms are also seen.

Breeds on the west and north-east coasts of Greenland, Iceland, Svalbard, Fennoscandia, and across northern Russia to Kamchatka, including Franz Josef Land and Novaya Zemlya. Breeds across North America from west Alaska, including Canada's Arctic islands north to southern Ellesmere. Wintering birds move to the coasts of South America, western and southern coasts of Africa, the Arabian Sea and the coasts of Australasia.

### Long-tailed Skua (Long-tailed Jaeger) *Stercorarius longicaudus*

The smallest northern skua. Dark morphs are very rare and perhaps do not occur at all, reports being fanciful, though juvenile birds do show the colour variation, with dark-morph juveniles moulting into the 'standard' adult plumage.

Great Skua, southern Iceland.

Light and dark morph Arctic Skuas, Runde Island, Norway.

Long-tailed Skua, Fosheim Peninsula, Ellesmere Island.

When Per Michelsen and Richard Sale were on Ellesmere Island they camped close to a nest of Long-tailed Skuas. The birds rapidly learned the meaning of breakfast, and were also partial to using the two men (particularly Per Michelsen as he is taller) as look-out posts, even maintaining their position as the two strode about the island since the perches were more energy-efficient than flying when searching for lemmings. But between the camp site and the nest was an invisible line which the men dared not cross. If they did, the friendly birds became instantly hostile, defending their territory with vigorous dive-bombing that soon had either man retreating quickly. Once the men had recrossed the boundary line to their own side the birds returned to friendlier ways.

It breeds in west and north Greenland, Jan Mayen, Svalbard, in mountainous and northern Fennoscandia, and across Russia to Kamchatka, but not on the Taimyr Peninsula, Novaya Zemlya's southern island and Wrangel, across Alaska and the northern Canadian mainland to the western shores of Hudson Bay. Absent from eastern Canada apart from a colony on the east-central shore of Hudson Bay. Also breeds on all Canada's Arctic islands to northern Ellesmere. Wintering birds are pelagic off both coasts of southern South America and the west coast of southern Africa.

### Pomarine Skua (Pomarine Jaeger) *Stercorarius pomarinus*

Two morphs, though dark morphs represent a smaller percentage of the population than in Arctic Skuas.

It has been suggested that Pomarine Skuas breed on west Greenland, but the lack of lemmings there lead many to consider this doubtful. It definitely breeds on Svalbard, and on Russia's northern coast from the White Sea to the Bering Sea. In North America it breeds on the west and north coasts of Alaska, on the Canadian mainland around Bathurst Inlet, on north-west and north-east Hudson Bay, on Banks and Victoria islands, and southern Baffin Island. May also breed at remote sites on other islands. In winter the birds fly to the waters of the Caribbean, the Atlantic off northern Africa, the Arabian Sea and the Pacific Ocean off eastern Australia, northern New Zealand, Hawaii and north-western South America.

## Gulls

Gulls represent not only one of the most readily identifiable groups of birds for the casual observer (even if the different species are often more difficult to distinguish), but one of the more successful bird families, in part due to gulls being supremely opportunistic. People in temperate regions may have seen Black-headed Gulls following ploughs to feed on earthworms. The gulls will also readily visit fields where Northern Lapwings (*Vanellus vanellus*) are feeding: in the absence of a plough the gulls have little success at finding worms, but Lapwings are excellent at the task, the gulls waiting their opportunity and thieving a meal. Gulls are often seen on rubbish dumps, while in winter they feed on waste grain in cereal fields. Many gulls are also efficient predators, with the larger species feeding on smaller birds and chicks. Some gulls will cannibalise chicks in their own colonies: studies have suggested that in Herring Gull colonies up to 25% of the chicks will be consumed by adult birds each year. Most extraordinary of all, Ivory Gulls feed on the faeces of Polar Bears and cetaceans.

Gulls have large crops that aid courtship feeding and the feeding of chicks, each of which involves regurgitation. The birds fill their crops during a day's voracious feeding and then digest the food at leisure during the nightly roost. In some species adults have a red gonys spot on the lower mandible – the chicks peck at this to stimulate regurgitation. Gulls are also good fliers, excellent at soaring and gliding, and with the high manoeuvrability essential for the cliff nesting that most favour (though some species are ground nesters). Gulls are chiefly birds of colder waters, and are particularly numerous in the northern hemisphere. The majority are primarily marine, though coastal during the breeding season of course. In winter, they are no longer tied to land and some species, such as the kittiwakes, become pelagic.

The northern gulls are divided into two major groups, the 'white-headed' (e.g. Herring Gulls) and the 'hooded' (e.g. Black-headed Gulls). In the latter, the hood is entirely, or mostly, moulted during the winter. Sabine's Gull is hooded, but is not considered a true member of the hooded group. It is, though, one of the trio of much-sought after Arctic species, the others being Ross's Gull and the Ivory Gull. Though head colour may vary, all northern gulls (indeed, almost all gulls) are essentially white. It is assumed that this is to camouflage the birds when they are fishing, fish having more problems seeing the birds against the normally cloudy sky. Unlike auks, gulls are not, generally, countershaded, presumably because being large they suffer little avian predation.

The speciation of gulls is a fertile subject for biogeographers. Herring Gulls and Lesser Black-backed Gulls are sympatric where they overlap, but do occasionally interbreed successfully. It is suggested that these species arose when two populations of an ancestral gull were separated during one of the Pleistocene ice ages. one ancestral form is thought to have been yellow-legged and confined to central Asia, evolved into the Lesser Black-backed Gull and spread west towards Europe. In the 1920s it reached Iceland and has now bred in southern Greenland, though it is not yet fully established there. The second population was isolated in north-eastern Asia, evolved pink legs and spread east, crossing North America. However, this simple story is complicated by the curious distribution of the species in the Palaearctic. Speciation of the two gulls is further enlivened by the existence of subspecies of each, which some authorities classify as full species. In the far east of Russia a smaller, paler form of the Herring Gull is occasionally considered a separate species (the Vega Gull *Larus vegae*), while throughout North America the American Herring Gull (in which adults are indistinguishable from the Eurasian bird, though juveniles are darker) is frequently awarded specific status as *L. smithsonianus*. There are also several Lesser Black-backed Gull subspecies, one of which, a larger bird with paler upperwings (but black wing-tips) that breeds from the White Sea to the Taimyr Peninsula, is considered by many a full species – Heuglin's Gull *L. heuglini*. To add yet more complication, some authorities consider that not only is Heuglin's Gull a full species, but that it also has subspecies!

Biogeographers also argue about the status of Iceland and Thayer's gulls. Again it is assumed that an ancestral form was isolated by an ice sheet which separated eastern and western populations which evolved into Iceland and Thayer's gulls respectively. However, some authorities consider the two taxa as sibling subspecies. An extra complication is afforded by Kumlien's Gull *L. kumlieni*, considered a full species by some and a subspecies of Iceland Gull by others. Add to this mix of species and sub–species the occasional hybrids and the result is a recipe for argument and discussions that will last for years to come.

The displays of gulls are usually territorial, the holding of a territory by males attracting females. Herring Gulls pluck grass from near the nest site as a way of establishing proprietorial rights. Males also have a 'long call', in which the head is thrown back while the bird emits the familiar trumpet-like 'seagull' call. The position of the head before and during the long call and the exact note are diagnostic of the species. Mated birds engage in head flagging, a sideways flick of the head (which is also used as a threat gesture to intruders, particularly by the hooded gulls, since the movement exposes the full hood), and in courtship feeding. Pair bonds are usually monogamous and long-lasting. Nests are made of seaweed or local vegetation, sometimes minimal, occasionally non-existent. In general gulls lay two or three eggs, these being incubated by both birds. The chicks are downy and semi-precocial or precocial. The clutch does not hatch simultaneously, with a third chick often being the victim of a cannibalistic (or other predatory) attack, or starving to death if there is insufficient food.

## Gull plumages

Gulls are long-lived and take time to mature, the smaller gulls sometimes breeding in their second year, the larger ones not until they are four or five years old. Juveniles do not acquire their adult plumage for several years, and as the differences in juvenile plumages can be slight, identification of juvenile forms can be tricky. Species are classified as two-, three- or four-year gulls, the time being the number of winters before a juvenile acquires adult plumage, i.e. a four-year gull shows its adult plumage in the fourth winter of its life and will show full adult breeding plumage in its fourth spring. The sexes are essentially similar in all species.

Some authorities now consider the American Herring Gull to be a separate species from the European Herring Gull rather than a subspecies, though the plumage differences are slight, Great Slave Lake, NWT, Canada.

### Herring Gull *Larus argentatus*

This is four-year gull that breeds on Iceland and northern Fennoscandia, and patchily across Russia to Chukotka. In North America it breeds in central Alaska and across Canada, but is rare on the north coast, though breeding on Southampton Island and southern Baffin Island. As noted above, some authorities consider American birds a separate species. In winter birds move south to wherever food is available.

### Lesser Black-backed Gull *Larus fuscus*

A four-year gull that breeds in southern Greenland, Iceland, northern Fennoscandia and Russia east to the Taimyr Peninsula. Increasing numbers are now seen in North America, but it is not yet an established breeding species. In winter, Greenlandic and Icelandic birds move to Iberia and north-west Africa, while Scandinavian and European Russian birds head for the eastern Mediterranean, the Arabian Sea and east Africa.

Common Gull, Ytre Ryfylke, Norway.

### Common Gull (Mew Gull) *Larus canus*

A three-year gull that breeds on Iceland, in northern Fennoscandia and across Russia, but always sub-Arctic. It also breeds throughout Kamchatka and in North America breeds throughout Alaska (but rare in the north), in Yukon and North-West Territory (again, rare in the north).

North American birds are considered by some authorities to be a full species. The plumage is similar, though the wing-tips tend to be whiter, but mtDNA studies show marked differences.

Greater Black-backed Gull, Kjorholmane, Norway.

**Greater Black-backed Gull** *Larus marinus*

Another four-year gull and the largest. Breeds on Canada's Labrador coast, the central west coast of Greenland, Iceland, Jan Mayen and Svalbard (where it became established only in the 1930s), northern Fennoscandia, Novaya Zemlya's southern island and the nearby mainland coast of Russia. Partially migratory, it is rarely seen beyond the limit of the continental shelf.

**Glaucous Gull** *Larus hyperboreus*

A four-year gull and a true Arctic breeder and formidable predator throughout the region.

## Watching the flight of competitors

Predatory gulls that hold territories on cliffs otherwise occupied by the species whose eggs and chicks, and perhaps adults, they take, will watch conspecifics (and other predatory species) closely if they come too close. The taking of eggs and chicks requires a gull to make a slow pass of the cliff, and this, in turn, requires the bird to fly upwind. So if the watching gull sees a competitor moving downwind it will ignore it, knowing the intruder lacks the flight control needed to seize prey. But those moving upwind will need to be seen off.

Glaucous Gull and chicks, Hvalrossbukta, Bear Island.

It breeds on Greenland, Iceland, Jan Mayen, Svalbard, northern Fennoscandia and across Russia, including all Arctic islands and across the northern mainland of North America and on all Canadian Arctic islands to northern Ellesmere. Wintering birds move south, but only as required by the sea ice, and are both coastal and pelagic.

### Iceland Gull *Larus glaucoides*

Despite the name, the bird is only seen in Iceland during the winter. This is a four-year gull that breeds in west and southern Greenland, on northern Baffin Island and south-west Ellesmere. Kumlien's Gull (often considered a race but regarded as a full species by some authorities) has darker grey wing-tips. It breeds on southern Baffin Island, western Southampton Island and the extreme northern tip of Quebec. Wintering birds are seen across the North Atlantic from Newfoundland to Iceland, but rarely further east, and are both coastal and pelagic.

### Thayer's Gull *Larus thayeri*

A four-year gull that breeds on the western shore of Hudson Bay and the Canadian Arctic islands from Banks to Baffin, and north to Ellesmere, though apparently absent from the Parry Islands. Winters on the west coast of the USA.

An Icelandic Gull coming in to land, west Greenland.

### Black-legged Kittiwake *Rissa tridactyla*

Juveniles of this three-year gull have distinctive black barring on the wings in the form of an M (or W, depending on the direction of flight!). The bird's curious name derives from its call, a trisyllabic *kitt–ee–wake*.

Breeds in Greenland (apart from north-west), Iceland, Jan Mayen, Svalbard, northern Fennoscandia, at a limited number of sites on Russia's northern mainland (though common in Chukotka and Kamchatka), but on all Russia's Arctic islands. It also breeds on all Bering Sea islands, western Alaska (but rare on the north coast) and eastern Canada, including eastern Baffin Island. Is is pelagic, moving ahead of the sea ice in winter to feed in the North Atlantic and North Pacific.

Black-legged Kittiwakes, Ekkeroy, Varangerfjord, Norway.

Sabine's Gulls, Storfjorden, Svalbard.

### Sabine's Gull *Xema sabini*

This two-year gull breeds in northern Greenland, Svalbard, northern Chukotka and on Wrangel. Also breeds in west and north Alaska, northern Yukon, and on Banks, Victoria, Southampton and western Baffin islands. In winter the birds migrate – the only truly migratory Arctic gull – and are pelagic, feeding in the Benguela Current off south-western Africa, and the Humboldt Current off north-western South America.

### Ivory Gull *Pagophila eburnea*

One of the Arctic specialities that all visitors will want to see: the gulls are often viewed above sea ice, occasionally appearing ghost-like from out of the mist, a magical, ethereal sight. A two-year gull; the only pure white gull (all others white gulls will be albinos).

Breeds in north and east Greenland, Jan Mayen, Svalbard, and Russia's Arctic islands (apart from Novay Zemlya's southern island), also breeds on all Canada's northern Arctic islands. Moves south with the advancing ice edge.

An Ivory Gull, Svalbard.

### Ross's Gull *Rhodostethia rosea*

If the Ivory Gull is the most magical of Arctic gulls, Ross's must qualify as the most beautiful. The two are equally elusive, much more so than Sabine's Gull, the third of the Arctic triumvirate of must-see gulls. It is a two-year gull.

Nesting Ross's Gull, north-east Russia.

The breeding range is poorly studied. Known to breed at sites on the southern Taimyr Peninsula, on the Lena and Kolyma deltas and in Chukotka it has bred in north-east Greenland and on Svalbard but has not become established at either. Remarkably the gull has also bred sporadically at Churchill on the southern shore of Hudson Bay. It is possible the gull breeds at remote sites on Canada's High Arctic islands. In winter the birds are seen in the Bering Sea. Flocks were once regularly seen passing Barrow during the autumn, but they are very much less common now.

Two other gulls breed at the Arctic fringe. Bonaparte's Gull (*Larus philadelphia*), named after the French zoologist Charles Lucien Bonaparte (a nephew of Napoleon Bonaparte) breeds in southern Alaska (rarely on the west and north coasts), northern Yukon, then southerly to southern shores of Hudson and James Bays. It nests in trees, which is a remarkable sight. The Slaty–backed Gull (*Larus schistisagus*) breeds in far eastern Chukotka, on Kamchatka and the northern coast of Sea of Okhotsk. Occasionally seen in Alaska (e.g. Pribilof Islands), its breeding is not proven and certainly not established.

## Terns

Terns are found on all the continents, though they are mostly tropical birds. The three far north breeders belong to the *Sterna* group of black-capped terns. These are slender birds with tapering wings and deeply forked tails. They are fish eaters, catching their prey by plunge diving: they do not swim underwater, the fish being taken just below the surface. Terns often seek shoals that have been driven close to the surface by the attentions of predatory fish or aquatic mammals.

Tern displays usually involve a 'high flight' in which the male flies fast and high, usually carrying a fish, with an interested female chasing him. Courtship feeding is an important part of pair formation and reinforcement. Monogamous, the pair bond long-lived or life-long. The birds are gregarious, with colonial breeding leading to nests so close together that a pair must defend its small territory. This occasionally means the female stays on the nest site, defending it while the male fishes and feeds her. Two to three eggs are laid, these incubated by both birds. Chicks are downy and semi-precocial, and are fed by both birds. Terns are long-lived and take up to five years before first breeding.

### Two summers each year

Most terns are migratory, with the Arctic Tern making the longest migration flight of any bird, flying 17,500km each way to Antarctica to enjoy two summers each year. Recovered birds suggest that although most birds travel over water, which would aid feeding, some travel overland, flying across central Russia at very high altitudes. It has been calculated that including feeding flights at both poles, some birds may fly up to 50,000km annually. They also see more hours of daylight each year than any other animal. In 2016 a bird satellite tagged on the UK's Farne Islands clocked a total round trip flight of 96,000km. With a probable lifetime of 30 years, a tern might then fly more than 3 million km (about the same as Earth to the Moon and back four times). And with limited ability to soar and glide, the tern would flap most of that distance.

Nesting Aleutian Tern, Seward Peninsula, Alaska.

### Aleutian Tern *Sterna aleutica*

Breeds in southern Chukotka, on Kamchatka and the Commander Islands, in western Alaska and on the Aleutians, but it is uncommon throughout the range. Pelagic outside the breeding season.

### Arctic Tern *Sterna paradisaea*

Extremely aggressive to intruders when nesting and will attack, and strike, humans who roam too close. Breeds in Greenland, Iceland, Jan Mayen, Svalbard, Fennoscandia and across Russia to the Bering Sea, including Russia's Arctic islands and across North America from Alaska to Labrador, including the Canadian Arctic islands. As noted, spends the winter in Antarctica.

The Common Tern (*Sterna hirundo*) is non-Arctic but breeds at a small number of sites in northern Norway, on the southern shore of the White Sea, and on Kamchatka.

## Auks

Auks are often referred to as the northern equivalent of the penguins, though the anatomical adaptations of auks are much closer to those of the diving-petrels of the Southern Ocean. However, auks do share some penguin-like characteristics, adaptations that have resulted from the two bird families facing similar evolutionary pressures. The legs are set well back on the body, the feet webbed to act as efficient paddles for swimming. Larger auks adopt the same upright stance as the penguins, though some of the smaller auks lie on their bellies on land. The wings are short and used for underwater propulsion: penguins have lost the power of flight, but the auks have had to retain this ability as northern regions have terrestrial predators. In flight the short wings beat furiously. The two families share an extraordinary diving ability. Only the larger penguins dive deeper than the larger guillemots: one confirmed dive of a Brünnich's Guillemot reached 210m. Most dives are to shallower depths, but the larger auks regularly dive to 60m and can stay submerged for up to three minutes. Auks can swallow food underwater, an adaptation that allows a longer dive time.

The auks are a northern group, but the distribution of species is highly asymmetric, there being many more in the Pacific than in the Atlantic. Although the reasons for this are still debated, it is likely that in part it derives from auks having evolved in the Pacific: the only Atlantic species without a Pacific equivalent is the Razorbill (though there was also no Pacific equivalent of the flightless Great Auk). Auks are birds of the continental shelf, only the puffins being found in deeper waters, particularly in winter, when they are truly pelagic.

Auks show marked differences in bill shape, these related to the choice of prey. Fish-eaters tend to have dagger-like bills like other piscivorous birds, while plankton-feeders have shorter, wider bills. The Parakeet Auklet, which feeds on jellyfish (as well as crustaceans), has a curious, rounded bill, rather like a scoop. The bills of the Razorbill and the puffins are laterally compressed and seem, particularly for the puffins, to be important in mating displays. Puffin bills are encased in nine plates that are shed during the autumnal moult, the 'new' bill being much more subdued. Puffins also shed their flight feathers simultaneously, something they share with the other larger auks. This means that the birds are flightless for a period, but the wing loadings of these birds mean that the successive loss of flight feathers would leave them essentially flightless, and therefore more vulnerable to predation, for a longer period.

Some auks have distinctive head plumage that acts in a similar way to the puffin's bill, the plumes being displayed to rivals or potential mates. The Tufted Puffin has both crests and a splendid bill. Apart from the bills and head plumage, auks have other mating displays: these usually involve head-shaking and bowing and, after the pair has formed, bill-nibbling or clacking as bond reinforcement. The murrelets, which find walking particularly difficult, tend to have sea-based displays, often involving parallel swimming. Auk pair bonds are monogamous and long-lived or life-long.

Most auks are colonial breeders, though some colonies are rather loose, and several of the smaller Pacific

## The Great Auk

There has been a long association between auks and people, northern dwellers probably collecting both eggs and birds from auk colonies. Collecting auks for food continues to this day: the people of Iceland's Westmann Islands have an annual hunt, with Atlantic Puffin regularly seen on the menus of island restaurants. The Great Auk, a flightless bird and the largest of the auks at up to 80cm tall, nested on remote Atlantic islands from Newfoundland to Scotland (and further south in prehistory). Probably never common, it was ruthlessly hunted for its meat and feathers by fishermen and, when it had become rare, by egg and skin collectors. Ironically, the last known specimens – a breeding pair – were collected on the island of Eldey off Iceland's southern shore by a party financed by an egg and skin dealer. The two birds were killed in 1844. There are credible, but unauthenticated, sightings of Great Auks after that date, but it is now generally agreed that the species was extinct by the late 1850s at the latest.

auks are solitary nesters. Nests vary considerably within the family: some auks nest on cliffs, laying eggs directly onto a ledge with little or no nesting material. Some nest in burrows, usually excavating these themselves. Kittlitz's Murrelet occasionally breeds far inland, a curiosity for a bird that finds walking so troublesome: it makes a scrape in bare tundra, sometimes at altitudes to 600m and often amid snowfields. The Marbled Murrelet (not an Arctic breeder) is exceptional in nesting in trees: its nest sites are extraordinary, the egg being placed on an old, wide branch or where epiphytes create platforms. Even more remarkably, the birds are usually nocturnal at the nest sites, flying through the woodland in darkness.

In most species, a single egg is laid (two is more common in a few species). The eggs of the larger guillemots are pyriform i.e. pointed at one end, an adaptation to minimise the chances of the egg rolling off the narrow nesting ledge. The eggs are of highly variable colour, particularly for the Common Guillemot, which has the most variable egg of any bird: the base colour varies from white to a beautiful turquoise, with a patterning of dark scribbles and splotches. The egg or eggs are incubated by both birds.

The developmental strategies of auk chicks are as varied as the nesting strategies. The chicks of most species are semi-precocial and are fed by the parent birds until they are at, or close to, adult weight, at which point they become independent. The Little Auk differs in one respect, with fledglings joining the adult male on the sea for a short time. The larger guillemots and the Razorbill have a markedly different strategy. The chick is fed until it is about 25–35% of adult weight. At that stage the still-flightless chick is encouraged to leave the nesting ledge by adult calls, and it glides to join its parents at sea. It is assumed that this strategy has developed because the adult birds, which have high wing loadings, can no longer carry the quantities of food required to allow continued development of the chick at the nest site: the chick is therefore fed and continues to grow at sea. The disadvantage of the strategy is the requirement of the chick to reach safety at sea by gliding. If the nesting ledge is on a sea cliff the glide is straightforward, but if it is inland then the glide angle must be low and some chicks land short: they must then survive both the landing impact and a hurried scramble to the sea,

Crested Auklet, St Paul, Pribilofs, Alaska.

guided by the calls of anxious parents. At such inland sites the ground below the cliff is patrolled by Arctic Foxes, which eagerly snatch up chicks, while skuas and gulls patrol the skies to pick off the gliders. A third strategy has evolved among some smaller auks. Here the chicks are precocial, with adult-sized feet, so they can swim. They leave the nest soon after hatching (usually within two or three days) and go to sea, where they are fed for several weeks before being abandoned. This strategy has presumably developed to save the adults the energetic cost of flying to the nest, though together with nocturnality, which is a feature of some smaller auks, may also be a defence against avian predators.

Auks are long-lived birds, their chicks taking time to mature, and not breeding until they are two or three years old, this applying even to the smaller auks. The plumages of the sexes are similar.

### Razorbill *Alca torda*

Breeds on south-west Baffin Island, northern Quebec and Labrador, west-central and south-west Greenland, Iceland, Jan Mayen, Bear Island, Svalbard (in small numbers), and northern Fennoscandia. There are also

Razorbills, Kjorholmane, Norway.

## Bridled Guillemots

In the two guillemots and the Razorbill there is a narrow channel in the feathers leading away from the eye. This seems to aid the flow of water past the eye when these large auks swim rapidly underwater. Some birds in the North Atlantic Common Guillemot population have this channel picked out in white: these birds also have a white orbital ring. Birds of this form are known as 'bridled' guillemots. The percentage of bridled birds increases as the observer travels north. There are no bridled birds in the south of the species' range (e.g. Iberia), but around 50% of birds are bridled on Iceland, Svalbard and Novaya Zemlya.

breeding colonies on the western White Sea coast, but the bird is absent from the remaining Russian Arctic. Partially migratory, it moves to waters off north-eastern USA and Canada, the western Atlantic, North Sea and western Mediterranean.

### Common Guillemot (Common Murre) *Uria aalge*

Breeds in south-west Greenland, Iceland, Jan Mayen, Svalbard, northern Fennoscandia, Novaya Zemlya's southern island, but absent from the Russian coast except for eastern Chukotka and Kamchatka. It also breeds on the Aleutians and islands of the Bering Sea, and Alaska's west coast, but not on the northern Neartic mainland coast (apart from northern Labrador) or the Canadian Arctic islands. Dispersive rather than migratory, with some birds being resident and others moving to nearby but more southerly waters.

### Brünnich's Guillemot (Thick–billed Murre) *Uria lomvia*

More northerly than its Common cousin it breeds on west and north Greenland, Iceland, Jan Mayen, Bear Island, Svalbard, northern Fennoscandia, and all Russia's Arctic islands. There are also isolated colonies on the northern Russian mainland and on eastern Chukotka and Kamchatka. It also breeds on the Aleutians and

Brünnich's Guillemots, St Paul, Pribilofs, Alaska.

Little Auks, Forkastningsdalen, Spitsbergen, Svalbard.

Bering Sea islands, west and north-west Alaska, on Baffin and Ellesmere islands, and in northern Quebec and Labrador. It is dispersive in winter, to the North Atlantic and North Pacific.

### Little Auk (Dovekie) *Alle alle*

Related to the guillemots and Razorbill despite its similarity to the smaller auks, it breeds in west-central and northern Greenland, on Iceland, Jan Mayen, Svalbard, Franz Josef Land, Novaya Zemlya and Severnaya Zemlya. There is also a small colony on western Baffin Island. Remarkably, in recent years small but expanding colonies have been found on the Diomede Islands and St Lawrence Island in the northern Bering Sea. Partially migratory, it moves south to cold currents of the North Atlantic and North Pacific.

### Black Guillemot *Cepphus grylle*

Breeds in west and north Greenland, Iceland, Jan Mayen, Svalbard, Franz Josef Land, Novaya Zemlya and Severnaya Zemlya, northern Fennoscandia, and the Taimyr Peninsula, the New Siberian Islands and Wrangel and the adjacent Russian coast to eastern Chukotka. It also breeds in north-west and north Alaska, on Canada's northern coast from the Yukon to Franklin Bay, northern Hudson Bay coasts and adjacent islands, western Baffin and southern Ellesmere islands. Resident or dispersive, it stays as far north as conditions allow.

Black Guillemot, Helgelandskysten, Norway.

### Kittlitz's Murrelet *Brachyramphus brevirostris*

A rare, small auk with legs set far back so that standing and walking is difficult; the bird rests on its belly on land. Breeds on Wrangel, the coasts of eastern Chukotka and north-western Kamchatka, on the Aleutian islands, and in isolated colonies on the west and south Alaskan coasts. The 1989 *Exxon Valdez* spillage is thought to have killed 10% of the world population. Northern birds fly south in winter to join resident southern birds.

### (Atlantic) Puffin *Fratercula arctica*

The three puffins are unmistakable – the clown princes of the bird world.

Breeds in Labrador (and, recently, in northern Quebec and south-western Baffin Island where, hopefully, they will become established), west Greenland, Iceland, Jan Mayen, Bear Island, Svalbard and Novaya Zemlya, and northern Fennoscandia. In winter the birds move to a broad band of the North Atlantic from Newfoundland

Atlantic Puffin, Kjorholmane, Norway.

## Strutting puffins

Having acquired its burrow – either by excavation or by taking ownership of a shearwater burrow (or a rabbit burrow in southern parts of the range) – a male Puffin will display ownership with a comical upright walk in which the bill is lowered against the chest and the feet are brought high on each step. The walk has, not surprisingly, been compared to that of a guardsman.

Horned Puffins gathering nesting material, St Paul Island, Pribilofs, Alaska.

and southern Greenland to the British Isles, and to the North Sea and the coasts of southern Iberia and North Africa.

### Horned Puffin *Fratercula corniculata*

Breeds on Wrangel, eastern Chukotka, Sea of Okhotsk shores, Sakhalin, the Commander and Kuril Islands, the Aleutians, the Alaska Peninsula, Bering Sea islands and the west coast of Alaska. In winter the birds are found in a broad band of the North Pacific from northern Japan to the central USA.

### Tufted Puffin *Fratercula cirrhata*

Breeds in eastern Chukotka, Kamchatka, at limited sites in the Sea of Okhotsk, Sakhalin, Commander, Aleutian and Bering Sea islands, western Alaska, the Alaska Peninsula, and south/south-east Alaska. In winter the birds are seen in a broad band across the North Pacific from northern Japan to the western USA.

In addition to the above, the Pacific has several auks which breed at the Arctic fringe. The Pigeon Guillemot (*Cepphus Columba*) breeds on eastern Chukotka, eastern Kamchatka, the Kuril, Commander and Aleutian Islands, and in west, south and south-east Alaska. Crested Auklets (*Aethia cristatella*) breed on the Aleutians and Bering Sea islands, on Chukotka near Providenya, northern Kamchatka, northern Sea of Okhotsk coast, Sakhalin and Commander islands. Least Auklets (*Aethia pusilla*) breed on islands off the southern Alaskan Peninsula, the Aleutian, Bering Sea and Commander islands, Chukotka near Providenya, and islands near Kamchatka and in the Sea of Okhotsk. Least Auklets are the smallest of the auks and appear far too fragile for their marine environment with its tempestuous weather. On occasions this view is borne out when a bird is seen flying very hard into the

Tufted Puffin, St Paul Island, Pribilofs, Alaska.

251

wind, but travelling slowly backwards as its efforts are overwhelmed. Parakeet Auklets (*Aethia psittacula*) breed on islands in Sea of Okhotsk, the Kuril, Commander, Aleutian and Bering Sea islands, eastern and southern Chukotka.

## Owls

One of the most recognisable of bird families though, paradoxically, as they are mostly nocturnal few people have ever seen one. The smaller owls take insects, but the larger ones can take mammals to the size of hares and even small deer.

Owls are densely feathered, the feathers having soft fringes so the bird's flight is quiet, the normal hunting technique being a pounce onto the prey from a slow or hovering flight. Tundra species, which hunt by day, have long wings that allow an efficient quartering flight as they visually search for prey. The hooked bill is small and often almost completely hidden in the feathers. The talons are sharp and the outer toe can be reversed, improving both capture area and grip. Owls usually carry

Looking rather more like a Russian doll than a bird, this Short-eared Owl, photographed at Orre, Norway, is showing the flexibility of its neck by turning its head almost completely round.

## Sight and hearing

Owls are squat birds with wide heads and huge eyes. Head size and shape is an adaptation to help improve the acuity of both sight and hearing. In cross section, the owl's eyes are cylindrical (as opposed to spherical as in mammals such as humans). This allows a larger pupil and lens, and so improved light capture. The eyes are also mounted frontally to improve binocular vision. The disadvantage of these adaptations is that they reduce the field of view (to about 110° compared to 180° in humans). To compensate, owls have highly flexible necks that allow the head to be turned almost 270°. An owl will also bob its head sometimes, allowing it to get a better fix on its target. Owl vision is only about 2–3 times better than that of humans, but this improvement is highly significant in low light.

The wide head aids hearing. A notable feature of owls is the facial disc, a dish of stiff feathers (much more developed in nocturnal species than in diurnal owls; the day-flying Snowy Owl has a much less pronounced disc than most other owls). The feathers channel sound to the ears, much as parabolic dishes aid sound reception for human observers attempting to catch bird sounds

(or for more clandestine reasons). The ears are far apart, allowing the bird to detect small differences in the arrival time of a sound across the head. This gives information on the location of the noise in the horizontal plane. Some owls also have asymmetric ears (usually the right ear set high, the left lower), providing information in the vertical plane. Those owls that do not have such asymmetric ears can move ear-flaps that alter the size and shape of the ear to obtain the same information. So good are owls at locating sound that in experiments they have been able to catch prey in complete darkness, hunting by sound alone. Indeed, Snowy Owls (and others, such as the Great Grey Owl – *Strix nebulosa*) can hunt rodents through several centimetres of snow with no visual clues. One further adaptation is that the ears are particularly sensitive to high frequency sounds, such as the rustling of a rodent making its way through dead leaves.

Despite the name of some owls, (such as the Short-eared Owl), the prominent 'ear' tufts of these species have nothing to do with hearing or, indeed, ears. The tufts are for communication between conspecifics.

their prey in their bills, though occasionally larger prey may be carried in the feet (in the manner of raptors). The prey is ingested whole, with the indigestible portions – fur, bones etc. – being periodically disgorged as pellets. The examination of pellets allows a study of the owl's diet. In times of abundant prey owls may cache food. Prey abundance has a dramatic effect on owl numbers, with periodic fluctuations following the rhythm of prey population. In 'lemming years' the numbers of Snowy Owls increase substantially, though many birds will starve when lemming numbers crash. The owls may also migrate in search of prey, being seen far south of their normal range.

Not surprisingly for essentially nocturnal creatures, plumage colour is generally not important, with many forest-dwelling owls having a cryptic plumage that allows the bird to rest undisturbed during the day. Some species that inhabit a latitudinal range encompassing both deciduous and coniferous forests change their plumage base colour from brown in the south to grey in the north. In general plumages are similar for the sexes and for both summer and winter.

Because sight is less important for territorial and mating purposes, sound has replaced it as the primary means of communication in owls. Pair formation is therefore based on the male hooting his territorial claim. For the two true Arctic breeders, visual displays are more appropriate, male Snowy Owls having a flight with the wings raised in a V, and wing-raising ground display. Male Short-eared Owls clap their wings. In general pairs are monogamous and seasonal, though males may be polygamous in good prey years. Owls breed early in the season so that young rodents are plentiful when their chicks are learning to hunt, and when moulting adults hunt less efficiently. Boreal owls nest in tree holes, but the two Arctic owls are ground nesters. The number of eggs is highly dependent on prey numbers and can be as low as 1 or 2 or as high as 10–12. The female incubates while the male feeds her. The owlets are nidicolous/altricial, and are dependent on their parents for a relatively long period.

### Snowy Owl *Bubo scandiacus*

Magnificent birds that provide one of the great sights of the Arctic they breed in north Greenland, northern Fennoscandia, northern Russia (including Novaya

Snowy Owl, Hudson Land, north-east Greenland.

Zemlya, Severnaya Zemlya and Wrangel), west and north Alaska and northern Canada including the southern Arctic islands. However, the distribution is dependent on prey numbers: Snowy Owls have bred on Iceland and Jan Mayen, probably on Svalbard and the other Russian Arctic islands, and on Canada's Arctic islands north to Ellesmere. The owls are essentially resident, but will move south if prey numbers decline.

### Short-eared Owl *Asio flammeus*

Breeds on Iceland, northern Fennoscandia and across Russia to Kamchatka, though rarely to the north coast. Breeds in Alaska, Yukon and NWT, around Hudson Bay, in Labrador and on southern Baffin Island. Ranges are dependent on prey density, but northern owls do move south to north-west Europe, central Asia and southern USA.

Northern Hawk Owl and chicks, Sirdal, Norway.

The (Northern) Hawk Owl (*Surnia ulula*) is a boreal species, but can be seen at the treeline. It breeds in Fennoscandia, across Russia to Kamchatka, in central Alaska (but uncommon in west and south) and across Canada to the Atlantic coast.

## Larks

Larks are ground–dwelling passerines that walk (rarely running and almost never hopping) in search of seeds and insects. Because of their habitat and lifestyle, they are, in general, cryptic brown, and have relatively long legs and a long hind claw to aid standing. Only one species breeds in the Arctic. The Shore (or Horned) Lark is typical of the family. Flocks form for migration and at wintering sites. The male holds a territory and entices a female by calling. The pair bond is monogamous and seasonal. The usual clutch is 3–4 eggs, which are incubated by the female. Chicks are, however, fed by both parents.

### Shore Lark (Horned Lark) *Eremophilia alpestris*

Shore Larks have a complex taxonomy, with more than 40 subspecies having been described. In the Palearctic the bird is a northern dweller only, with southern habitats being occupied by the Eurasian Skylark *Alauda arvensis*. Shore larks are the only Nearctic lark, the species breeding in both northern and southern habitats.

They breed in Fennoscandia, on the southern island of Novaya Zemlya and across Russia to the Kolyma delta, north to the coast, and throughout Alaska and northern Canada, including Arctic islands north to Devon Island. In winter, Palearctic birds are seen in south-eastern Europe and southern Russia, while Nearctic birds fly south to join resident southern birds.

## Swallows

Among the most popular of birds as their arrival in northern Europe heralds the arrival of summer. The birds would be even more popular with the Arctic traveller if only they occurred there in greater numbers, since they take insects on the wing and so reduce – but sadly only by a minimal amount – the number of mosquitoes.

These birds are superbly adapted for the task of chasing down insects on the wing, with long, narrow, pointed wings and deeply forked tails. The high aspect-ratio wings

Incubating Shore Lark, Varangerfjord, Norway.

make the birds fast but with reduced manoeuvrability, the forked tail restoring that so the birds are quick and agile. The specialised diet of winged insects means that northern species are migratory, heading far south where insects are found during the northern winter.

The birds have limited displays. The pair-bond is monogamous and seasonal, but males will attempt to mate other females. Nests are often made of mud pellets, but they may be more conventional if a suitable crevice is found. 4–5 eggs are incubated by the female only, and she also takes responsibility for brooding the chicks, though both parents feed them.

Tree Swallow, Potter Marsh, Alaska.

### Tree Swallow *Tachycineta bicolor*

Breeds throughout Alaska (but is uncommon in the north), in central Yukon and NWT, around the southern shores of Hudson Bay, in southern Quebec and in eastern Labrador. In winter the birds fly to the southern United States, Mexico, and the east coast of Central America.

## Pipits and wagtails

The members of the Motacillidae are seed- and insect-eating passerines, which can be usefully divided into two groups, the wagtails, which are generally boldly coloured and the more cryptic pipits. Wagtails have long tails, the name deriving from the endearing habit of tail-wagging, which the birds do almost continuously while on the ground. Pipits are renowned for the display songs of the males, the birds flying high and delivering territorial claims while apparently hanging motionless (by flying into the wind). Some species fly so high that they are out of sight, reaching over 100m. The flights can also be long-lasting, occasionally taking several hours, though they are often much shorter, and end with the bird 'parachuting' back to earth, this involving fluttering wings and a near-vertical descent.

A male pipit's song defines its territory and encourages females. Wagtails also have a territorial song, this usually starting from a perch. Male wagtails point their bills upwards to expose their breast patterns, and may also fan their tails. The pair-bond is monogamous and seasonal. The nest is a cup of vegetation well-hidden on the ground in which 2–5 eggs are laid. These are generally incubated by both birds in the wagtails, the female only in the pipits, but there are exceptions. The chicks are nidicolous/altricial and are cared for and fed by both parents.

### Buff-breasted Pipit (American Pipit) *Anthus rubescens*

Breeds in Alaska, across mainland Canada to the north coast and on southern Arctic islands from Banks to Baffin. Also breeds in north-west and west-central Greenland, in Chukotka and Kamchatka. In winter, Nearctic and Greenlandic birds fly to the southern USA and Central America, while Asian birds head for southern Asia.

### Meadow Pipit *Anthus pratensis*

Breeds in south-east Greenland, on Iceland and Jan Mayen, in northern Scandinavia and Russia east to the shores of the White Sea. Wintering birds are seen in southern Europe, north Africa, the Middle East and central Asia.

### Pechora Pipit *Anthus gustavi*

Breeds from the Pechora River eastwards to Chukotka, but to the north coast only in Chukotka. Also breeds throughout Kamchatka and on the Commander Islands. Wintering birds fly to Indonesia and the Philippines.

### Red-throated Pipit *Anthus cervinus*

Breeds in northern Fennoscandia and across Russia to Chukotka and Kamchatka, to the north coast except on the Taimyr Peninsula. Also breeds in small numbers in western Alaska. In winter the birds are seen in central Africa and south-east Asia.

Red-throated Pipit, Varangerfjord, Norway.

### White Wagtail *Motacilla alba*

Breeds in south-east Greenland, Iceland, Jan Mayen, northern Fennoscandia, and across Russia to Kamchatka, north to the coast except on the Taimyr Peninsula, on Novaya Zemlya's southern island and Wrangel. Also breeds irregularly in north-west Alaska. Wintering birds are seen in north and central Africa, the Middle East, and south/south-east Asia.

### Yellow Wagtail *Motacilla flava*

Breeds in northern Fennoscandia and across Russia to Kamchatka, north to the coast except on the Taimyr Peninsula. Also breeds in small numbers on the western and northern coasts of Alaska. In winter the birds head for central and southern Africa, India and south-east Asia.

## Thrushes and chats

The Turdidae are a large and widespread family of birds which includes some of the best-known species to northern peoples – the Eurasian Robin *Erithacus rubecula*, American Robin, Eurasian Blackbird *Turdus merula* and the bluebirds. The family is, however, diverse, the many species having few characteristics in common. The birds are principally invertebrate feeders, supplementing this with fruit. Many species are excellent singers (the Nightingale *Luscinia megarhynchos* being famous for its voice), males calling to establish territories and for pair formation. Pair-bonds are usually monogamous and seasonal. There are usually 4–6 eggs, these laid in a well-built nest and normally incubated by the female only, though both birds usually contribute to the care of the chicks.

### Northern Wheatear *Oenanthe oenanthe*

Breeds on Greenland and Iceland, irregularly on Svalbard, and from northern Fennoscandia to Chukotka, north to the coast in most areas, and Novaya Zemlya's southern island, western Alaska, but rare in the north, and the eastern Canadian High Arctic (Baffin and Bylot Islands, southern Ellesmere Island and northern Quebec and Labrador).

Given the circumpolar distribution of the Northern Wheatear it would be thought that in winter the birds would be found equally well-spread around the globe. But this is not the case, the entire population flying to central Africa. This curiously asymmetric winter distribution indicates the original breeding range of the bird – it was a northern European species that has

Bluethroat, Varangerfjord, Norway.

257

Northern Wheatear, Skaftafell National Park, Iceland.

American Robin, Great Slave Lake, North West Territories, Canada.

spread east and west. This accounts for the Nearctic distribution, where the bird is absent from a huge central section of the continent.

**American Robin** *Turdus migratorius*

The largest Nearctic thrush. Breeds throughout North America, including Alaska (though uncommon in the north), the Yukon and much of NWT. Its range is more southerly in eastern Canada, but includes the southern shore of Hudson Bay, southern Quebec and Labrador. Wintering birds are seen in southern USA and Central America.

Other thrushes are subarctic breeders. The Fieldfare (*Turdus pilaris*) breeds in northern Fennoscandia, and western and central Russia, though not to the north coast. It has bred in Iceland, but is not established. Fieldfares nest colonially, a very unusual habit among thrushes. The reason is a highly coordinated defensive system against predators. When a potential predator is noticed, the birds fly towards it. Each in turn then dives at the intruder, letting loose a bomb of excrement. The bird's aim is usually good, with the intruder rapidly becoming peppered with spots of excrement. On a predatory bird such spotting can be dangerous as it affects the aerodynamics of the feathers, and on terrestrial predators it can require a lot of grooming to remove. Not surprisingly the strategy is extremely effective in persuading predators to retreat, and is so effective that other birds choose to nest close to Fieldfares to benefit. Redwings (*Turdus iliacus*) breed on Iceland, northern Fennoscandia and across Russia (though not in Chukotka or Kamchatka). Some Icelandic birds are resident, but in general all populations move to southern Europe for the winter. Naumann's Thrush (*Turdus naumanni*) breeds in Russia from the Yenisey to central Chukotka and throughout Kamchatka. Varied Thrushes (*Ixoreus naevius*) are extremely attractive birds that breed throughout Alaska (apart from the extreme north) and in Yukon and western NWT.

## Warblers

Only two species of warblers are northern breeders in the Palearctic, and only one of these is a true Arctic dweller. The Willow Warbler (*Phylloscopus trochilus*) is a boreal species, but may be seen in forest-tundra. They breed in northern Scandinavia and across Russia to Chukotka.

Arctic Warbler, Varngerfjord, Norway.

Willow Tit, Varangerfjord, Norway.

### Arctic Warbler *Phylloscopus borealis*

Also essentially boreal, but more likely to be seen in forest-tundra, it breeds in northern Scandinavia and across Russia to Chukotka, though infrequently to the north coast. It also breeds on Kamchatka and the Commander islands, and has bred on Wrangel and in western Alaska. In winter the birds are seen in the forests of south-east Asia.

## Tits and chickadees

Tits are small passerines with pale heads beneath dark or coloured crowns. The bill is short and pointed, ideal for catching the birds' main prey of insects, though as any European garden-owner will know the birds also feed on nuts and seeds, particularly in winter when insects are harder to find. Tit chicks are fed exclusively on insects and insect larvae (requiring upwards of 10,000 individual items before fledging), taking seeds only when they accompany their parents to local gardens. Nearctic chickadees look very similar to Palearctic tits and have a similar diet, but are not exclusively tree hole or box nesters. All the birds mentioned here are boreal and, consequently, Arctic-fringe species.

The Great Tit (*Parus major*) breeds surprisingly far north, reaching the northern coast of Fennoscandia, where it is resident in areas well north of the Arctic Circle. The Willow Tit (*Poecile montanus*) breeds in northern Fennoscandia and across Russia to Kamchatka, but only to the treeline. The birds of Kamchatka are a subspecies: they are very pale, appearing all white apart from the black crown and bib. The Siberian Tit (Grey–headed Chickadee) – *Poecile cincta* – breeds in northern Fennoscandia and across Russia to Chukotka, but not to the north coast, and in small numbers in a belt across central Alaska to the Mackenzie delta. Some birds are resident surviving winter temperatures as low as -50°C. At temperatures of about -40°C the birds take shelter in a tree hole, tuck in their heads and feet and fluff out their feathers. The feathers have large numbers of barbs and barbules (up to 100/cm³), allowing an insulating layer of still air to form. As the temperature falls the bird drops its body temperature by up to 10°C, entering a state of torpor. To maintain normal body temperature and to build the resources necessary to survive periods of torpor, the bird must consume around 7g of food daily (55–65% of its body weight), a quantity that must be foraged in a day as short as 4 hours.

## Crows

Corvids are among the most familiar of birds: large, noisy and opportunistic, they are well-known to people throughout the world. They are also renowned for their intelligence, being good puzzle-solvers and able to use simple tools to obtain food. The pair-bond is monogamous (though male Common Magpies are promiscuous) and life-long or long-lived. The (usually 3–6) eggs are incubated by the female only (but both birds in the case of the Spotted Nutcracker – *Nucifraga caryocatactes*), with the chicks cared for and fed by both parents. An old saying maintains that 'a crow lives three times longer than a man, and a raven lives three times longer than a crow', but most crows do not make it past 10 years of age. Yet despite this, sexual maturity is not reached until the bird is 6 or 7 in some species, though an age of 2 or 3 is more usual. The plumages of male and female, and in summer and winter, are similar in the species described here. Only one crow is an Arctic breeder, though the Grey Jay (*Perisoreus Canadensis*) breeds in central and western Alaska, central Yukon and NWT, then southerly to the southern shores of Hudson Bay, and in southern Quebec and Labrador, and the Siberian Jay (*Perisoreus infaustus*) breeds in northern Scandinavia and across Russia to the border of Chukotka.

### Raven *Corvus corax*

The Inuit claim that the call of the Raven is *kak*. Their story is that the first Raven was a man. Before setting out on a hunting trip with friends, the man, nervous that the blankets (*kak* in Inuktituk) would be forgotten, repeatedly told his companions to look out for them. So fed up did the others become that they deliberately forgot the blankets and then sent him back to fetch them. In his panic the man ran, then flew and then become a Raven, still calling *kak*.

Breeds throughout Greenland, Iceland, northern Fennoscandia and across Russia to Chukotka, north to the coast except on the Taimyr Peninsula, throughout Alaska and across Canada, north to the coast and on the Arctic islands north to southern Ellesmere. Resident or dispersive. May also breed on Wrangel.

## Wood (or New World) Warblers

The wood warblers of the Nearctic share features with Palearctic warblers, to which they are only distantly related, such as having straight, short, pointed bills ideal for taking insects (though most also take fruit in winter), but they are flamboyant in plumage rather than drab. Indeed, they are the most colourful of the passerines of the Arctic fringe and Low Arctic, and a true delight. Male wood warblers sing to establish a territory and attract a mate, although there are usually some visual displays

## Crows in history

Crows, particularly the Raven, the largest and most powerful of the corvids (and the largest passerine) have been incorporated into myths and stories from ancient times. They appear in the epic of Gilgamesh, in the Bible, and in the legends of Romans, Greeks and Celts. One Celtic myth is the likely source of the tale that if the Ravens leave the Tower of London, Britain will fall. In Japanese mythology crows were created to save mankind, seeing off a monster who threatened to swallow the sun. Crows are also common in Chinese myths and stories.

The Vikings associated the Raven with Odin, the main god of their pantheon, who was known as the Lord of Ravens. The Raven could be friendly: in *Flokki's Saga* a released Raven leads Norse settlers to Iceland. But the bird was also associated with death, an association that probably arose from the birds feeding on battlefield dead. Viking chiefs used the standard of a Raven with outstretched wings as a battle emblem, symbolising the future of the enemy as forage for crows.

In medieval Europe, the birds also fed on the gibbeted remains of the executed, reinforcing the association with death and encouraging the belief that crows – already sinister because of their black plumage – were birds of ill-omen: in Scandinavia a Raven croaking outside a house foretold a death in the family. This association has endured: in Edgar Allen Poe's famous poem *The Raven* the bird again signifies foreboding.

For the native peoples of the Arctic, crows were invariably seen in a positive light. Ravens are at the heart of the creation myths of the Inuit, Chukchi and Koryak peoples, suggesting a tale inherited from a common ancestor. Shamans often took Ravens as their familiars and the birds are frequently seen on the totem poles of the Tlingit and Haida peoples of southern Alaska. The Inuit also believe that the Raven helps them with their hunt: if a raven flies overhead an Inuk will ask it if it has seen Caribou or other prey, and if it has the Raven will dip its wing to point the direction.

Raven in flight above the Barren Lands of Nunavut, Canada.

as well. The pair-bond is monogamous and seasonal, but some males with exceptional territories may be polygamous. The 3–6 eggs are incubated by the female. Chicks are brooded by the female, but fed by both parents.

**Yellow Warbler** *Dendroica petechia*

Breeds throughout Alaska (but rare in the north), Yukon and NWT, but rare in Nunavut, the southern shores of Hudson Bay and in southern Quebec and Labrador. In winter, northern birds join resident birds on Caribbean islands and in northern South America.

While the Yellow Warbler has a claim to be a true Arctic breeder, other species almost match its northern credentials. The Yellow-rumped Warbler (*Dendroica coronate*) breeds in west and central Alaska, central Yukon and NWT, around the southern shores of Hudson Bay and in southern Quebec and Labrador. The Blackpoll Warbler (*Dendroica striata*) also breeds throughout Alaska (but is rare in the north), across Yukon and NWT (though not to the north coast except at the Mackenzie delta), south-western and southern Hudson Bay, and in southern Quebec and Labrador. Wilson's Warbler (*Wilsonia pusilla*) breeds in southern Alaska (but is uncommon in the west), Yukon, north-west NWT, around the southern shores of Hudson Bay and in southern Quebec and Labrador. In winter, all these species fly to Central America and northern South America.

Male Yellow Warbler, Churchill, Manitoba, Canada. The bird is singing its territorial claim.

Yellow-rumped Warbler on its way to feeding this year's chicks, Great Slave Lake, Canada.

## New World Sparrows and Buntings

'Bunting' is an old English word and would have been taken across the Atlantic by early settlers. This resulted in some species being labelled inappropriately as buntings, and a few of these names have stuck. The situation was further confused by the settlers referring to some species that looked similar to birds with which they were familiar by their Old World name. The classic example is the American Robin, named after its similarity to the Old World, red-breasted Robin *Erithacus rubecula*, but many birds were also named 'sparrows'. In fact, the New World Sparrows are members of the Emberizidae family which also includes the true buntings.

Emberizids evolved in the Nearctic, where most species still reside. They spread to Asia by way of the Bering Straits, and then continued as far as Europe. The birds have the stout bills characteristic of seed eaters. Males sing to establish territories, but usually have an additional repertoire of visual displays. The pair-bond is generally monogamous and seasonal. The 3–5 eggs are incubated by the female. Chicks are usually brooded only by the female, though are fed by both birds.

### Lapland Bunting (Lapland Longspur) *Calcarius lapponicus*

Breeds throughout Greenland (though rare in the north-east), in northern Fennoscandia and across Russia, including the Arctic islands (apart from the northern island of Novaya Zemlya, and unconfirmed to date on Severnaya Zemlya). It also breeds throughout Alaska

Female Snow Bunting, south-west Iceland.

and on the Bering Sea islands, and across Canada, including Arctic islands north to Ellesmere. In winter the birds are seen in a band across eastern and central Europe and southern Russia, and in the southern USA.

### Snow Bunting *Plectrophenax nivalis*

Breeds on Greenland, Iceland, Jan Mayen, Bear Island and Svalbard, in northern Fennoscandia and across Russia to Kamchatka, including all the Arctic islands, also breeds on the Commander, Bering Sea and Aleutian Islands, throughout Alaska and northern Canada including all the Arctic islands. In winter Icelandic and some Bering Sea birds are resident, but most populations fly south to occupy a broad band across central Europe and southern Russia, and a broad band across the northern USA and southern Canada.

### McKay's Bunting *Plectrophenax hyperboreus*

Rare and local. McKay's Bunting breeds on Hall and St. Matthew Islands and, rarely, on the Pribilof Islands. In winter the birds are seen on the west coast of Alaska, from Nome to Cold Bay.

### Fox Sparrow *Passerella iliaca*

A complex species, the Red Fox Sparrow, is believed by some authorities to be a species rather than a subspecies, and breeds in central and western Alaska (but more rarely in the north), the Yukon and NWT, on the southern shores of Hudson Bay and in central Quebec and Labrador. Another subspecies, the Sooty

Male Lapwing Bunting, Varangerfjord, Norway.

## Northern passerines

The Snow Bunting is the most northerly breeding passerine on Earth, breeding in the far north of Greenland and on northern Ellesmere Island. A Snow Bunting was seen at the North Pole, from the deck of a surfaced US Navy submarine.

MacKay's Bunting was once thought to be a pale subspecies of Snow Bunting, but is now considered a species by most authorities, though the fact that hybrids of the two have been observed leads some to still maintain it is a subspecies. It is very rare, with a population of only a few thousand and is much sought after by birdwatchers, though the breeding grounds are small, uninhabited islands that are difficult to reach. The bird was named after Charles McKay, a US Signal Corps observer and bird collector who drowned in suspicious circumstances in 1883.

White-crowned Sparrow, Seward Peninsula, Alaska.

Fox Sparrow, is also believed by some to be a species: it breeds in the Aleutians and southern Alaska. In winter the birds fly to both coasts of southern USA.

**White-crowned Sparrow** *Zonotrichia leucophrys*

Breeds throughout Alaska (though uncommon in the north and south) and across mainland Canada to the north coast, though absent from the Arctic islands. Wintering birds fly to the southern USA and Mexico.

**Savannah Sparrow** *Passerculus sandwichensis*

Breeds throughout Alaska and northern Canada, and on southern Canadian Arctic islands. Wintering birds are seen in the southern USA, Central America and on Caribbean islands.

Savannah Sparrow, Unalaska, Aleutian Islands, Alaska.

American Tree Sparrow, Barren Lands, NWT, Canada.

### American Tree Sparrow *Spizella arborea*

Breeds in western and central Alaska (though rare in the north), throughout the Yukon and much of mainland NWT, on the shores of Hudson Bay and in northern Quebec and Labrador. In winter the birds are seen in the north and central USA.

The Golden-crowned Sparrow (*Zonotrichia atricapilla*) is an Arctic fringe species, breeding in the Aleutians (to Unimak Island) and west Alaska, but uncommon in north and central Alaska. It also breeds in southern Yukon and British Columbia. In winter the birds are seen along the USA's west coast. It has the most distinctive song of all Nearctic sparrows, a song which, once heard, is never forgotten and allows the observer to find the elusive singer. The song is a descending three syllables, often written as 'Oh Dear Me', but which could as easily be rendered as 'Three Blind Mice'.

## Finches

Finches evolved in the Palearctic, the similarities of the finches to the buntings, which evolved in the Nearctic, arising in response to the evolutionary pressures of seed-eating. As some buntings have crossed to the Palearctic, so some finches have crossed the other way, with several Arctic breeders being circumpolar. Finches have an array of bill shapes, ranging from the massive conical bill of the Hawfinch (*Coccothraustes coccothraustes*), evolved for cracking large, hard seeds, to the more delicate bill of the Siskin (*Carduelis spinus*), evolved for taking grass seeds. There are also the curious bills of the Pine Grosbeak (*Pinicola enucleator*), designed for extracting conifer seeds, and the even more extreme conifer-seed extractor of the crossbills.

Finches sing to establish territories and have visual displays. The pair-bond is monogamous and seasonal. The 3–6 eggs are incubated by the female only, but the chicks are usually brooded and fed by both birds.

### Arctic Redpoll (Hoary Redpoll) *Carduelis hornemanni*

Breeds on west and north-east Greenland (but is scarce everywhere), northern Fennoscandia and across Russia, but to north coast only at the White Sea and in Chukotka (though also breeds on Novaya Zemlya's southern island and on Wrangel). Breeds in northern Alaska and across northern Canada to Hudson Bay, around Ungava Bay and on the eastern Arctic islands (Baffin and Ellesmere). Resident or partial migrant, moving short distances only, to southern Scandinavia, sub-Arctic Russia and southern Canada. Resident birds endure winter temperatures down to -50°C. To aid survival the birds have a storage area, called the oesophageal diverticulum, about halfway down the throat, where extra food can be stored. This allows the bird to take in more food than it can digest immediately, storing the rest for later consumption. During the long night hours the birds fluff their feathers to increase insulation, reduce energy output by resting, and transfer the food from storage to the main digestive tract.

### Common Redpoll *Carduelis flammea*

Breeds on Greenland (rare in the north-east), Iceland, northern Fennoscandia and across Russia to Kamchatka, but to the north coast only at the White Sea and in Chukotka. Also breeds throughout Alaska (uncommon in the north) and across the north Canadian mainland, but only on western Baffin Island. Some Greenlandic and most Icelandic birds are resident. Other Greenlandic birds move to Iceland or join European birds in northern and central Europe. Asian birds move to Japan and south-east China, North American birds to southern Canada and the northern USA.

Arctic Redpoll, Varangerfjord, Norway.

Two other finches breed at the Arctic fringe. Until the early 1980s they were considered subspecies of a single species, the Arctic Rosy-finch. Then the Russian and American birds were split into two species, the Grey-crowned Rosy-finch (*Leucosticte tephrocotis*), which breeds in western and central Alaska, throughout the Yukon, and on the Aleutian and Pribilof Islands, and the Asian Rosy-finch (*Leucosticte arctoa*), which breeds throughout Kamchatka and on the Commander and Kuril Islands. But the taxonomy remains controversial. Are the birds of the Commander Islands a subspecies of the Grey-crowned Rosy-finch or of the Asian Rosy-finch? And if they are a subspecies of the latter, where does that leave the Kamchatka birds? Is there, as before, only one species?

Musk Oxen herd, Kuuijua River, Victoria Island, Canada.

# 16 Arctic mammals

In this chapter we explore the mammals, both terrestrial and marine, which make their living in, or close to, the Arctic

## Shrews

Shrews are the only members of the Insectivora that breed in the Arctic. As with most birds, shrews have a cloaca, a single opening for the urinary, digestive and reproductive tracts. They also practice refection, the re-ingestion of excreted food. Shrews have remarkably high metabolic rates and must feed frequently: on average shrews consume their own body weight in food each day, and species will die of hunger if they are unable to feed every one or two hours.

Apart from the Water Shrew (which breeds at the Arctic fringe) Arctic shrews all belong to the genus *Sorex*, the red-toothed shrews, so called because the tips of their teeth are red, a feature caused by iron compounds in the enamel. These teeth are a necessary adaptation as in general these species have the highest metabolic rates. A *Sorex* shrew's first teeth are resorbed before birth, the animal being born with its final set. As shrews do not replace their teeth, the harder tooth-tips of *Sorex* shrews reflect their high food intake. If the shrew lives long enough it will wear its teeth down and die of starvation.

Shrews are solitary animals, usually only coming together for mating. Gestation is about 20 days, the young being weaned after a further 20 days, and even in the Arctic the female may have several litters annually if resources allow. The rigours of life mean males and females rarely survive to their first birthday so that each year, during the late summer, the entire population is likely to comprise that year's young. Winter also takes a toll of the youngsters: shrews cannot hibernate, since their metabolic rate cannot be lowered sufficiently to allow survival on fat stores. The Arctic species shrink both their skeletons and internal organs to reduce food requirements but even so, many die of starvation during the winter.

The short life-cycle of shrews means evolution is rapid: new species are still being discovered regularly, and most species are polytypic across the range. Several very similar shrew species may also be found within a given area: as the animals are omnivorous it is possible for species to co-exist sympatrically by taking different fractions of available prey. Both these factors mean that differentiating shrew species can occasionally be very difficult for experts, and is almost impossible for the non-specialist. Shrews are also difficult to find – in some 30 Arctic field trips I have seen one rarely – so the Arctic traveller who spots one has seen a very rare sight.

### Tundra Shrew *Sorex tundrensis*

Breeds in Siberia east of the Pechora River, north to the northern coast except on the Taimyr Peninsula. Absent from Kamchatka. Also breeds on the Aleutian Islands, in west, central and northern Alaska and, to a limited extent, in northern Yukon. Found on dry tundra and shrub tundra.

Tundra Shrew, east Siberia.

American Masked Shrew, central Alaska.

### Barren-ground Shrew *Sorex ugyunak*

The American Masked (or Cinereus) Shrew (*Sorex cinereus*) which breeds as far north as the Brooks Range in Alaska and the Mackenzie delta, is believed to have been the ancestor of several shrews found in North America/eastern Russia when populations were separated by ice sheet tongues during Quaternary ice ages. Related species are found on St Lawrence Island, the Pribilofs, in Chukotka and Kamchatka. The most northerly of these species is the Barren-ground Shrew which breeds in northern Alaska and across northern Canada, including the Boothia Peninsula, to the western shore of Hudson Bay but it is absent from the southern shore of Hudson Bay and north-east Canada. The species is found in damp and shrub tundra. The Latin name *ugyunak* is the Inuit word for shrew, the animal being the only shrew species the Inuit were likely to have seen.

## Carnivores

Carnivores are a large and diverse group of mammals. The fissipeds (terrestrial carnivores) are found from the Arctic to deserts, and include the familiar terrestrial groups, such as dogs, cats and bears. The difference in weight between the largest and the smallest terrestrial carnivores, both of which are Arctic species, the Polar Bear and the (Least) Weasel, is astonishing – up to 800kg compared to just 25g, a factor of 32,000.

So different are the forms of carnivores that only a few general comments on physiology can be made. Compared to a human, there are two obvious skeletal differences: there is no large clavicle, and a penis bone is present in most species. The function of the clavicle in humans and other primates is to allow sideways movement of the arms. For carnivores, whose main hunting strategy is to run down prey, such a movement would be disadvantageous: they have a smaller clavicle, free at each end, which allows only a forward and backward movement, ideal for fast pursuit. The baculum or *os penis* bone of carnivores prolongs copulation and is assumed to have evolved because in general copulation stimulates ovulation.

Carnivores have a unique and diagnostic dentition. In all carnivores (even in the only species that has become secondarily herbivorous, the Giant Panda *Ailuropoda melanoleuca*), the last upper premolar and first lower molar on each side of the jaw have high cusps and sharp tips. These teeth, called carnassials, allow flesh to be sheared.

In addition to the physical attributes for hunting such as powerful limbs and sharp teeth, terrestrial carnivores also have well-developed senses, sight, hearing and smell all being important. These senses are also used in communication, carnivores marking their presence by urinating and defecating at strategic points throughout their territories.

## Canids

Dogs are noted for pack formation, the pack normally comprising a monogamous 'alpha' pair which breeds, and family members that help with hunting and pup-rearing. However, although this is the normal society for Wolves, Red and Arctic Foxes are more solitary animals. Family groups have been noted in these foxes, but solitary pairs are more likely to be seen in both species. The number of pups produced is dependent on prey density: Arctic Fox numbers increase rapidly in 'lemming years', but the animals may not breed if rodent numbers are low. Many foxes will starve during their first year if the lemming population crashes.

In general canids kill their prey by grabbing the neck or nose, and shaking their heads violently to break or dislocate the neck or spinal cord.

### Wolf *Canis lupus*

The largest wild dog and the first animal domesticated by humans, perhaps as many as 50,000 years ago, and certainly by 14,000BP. However, despite this ancient association, the view of people has undergone a

White Wolf pair, Fosheim Peninsula, Ellesmere Island, Canada. Behind the wolves is frozen Slidre Fjord.

Grey Wolf, Karelia, in the 'No-Man's-Land' between Finland and Russia.

significant change over time. The Wolf was often the familiar of shamans and was revered by early northern dwellers. The Vikings also revered the animal for its speed, stamina and skills as a hunter, Viking chieftains often taking the name Ulf to indicate their prowess in battle: Ulf is still a common name in Scandinavia. But by the early medieval period things had changed, the Wolf becoming an evil presence, the change reflected in a host of folkloric tales in which the animal is represented as deceitful, greedy, ferocious and, most significantly of all, dangerous. Barry Lopez, in his marvellous book *Of Wolves and Men,* conjectures that in portraying the wolf as evil, man was '(externalising) his bestial nature, finding a scapegoat upon which he could heap his sins and whose sacrificial death would be his atonement. He has put his sins of greed, lust and deception on the wolf and put the wolf to death, in literature, in folklore and in real life.' In many European countries, a bounty was placed on Wolves: they were exterminated from England by the early 16th century, and from Scotland about 150 years later. In Scandinavia eradication took longer, but the last Wolf was killed in Sweden in 1966, and in Norway in 1973. They were almost exterminated in Finland too, but recolonisation from Russia prevented total extinction. As attitudes changed, the recolonisation of Sweden and Norway from Finland was officially welcomed, though the animals still face illegal killing by farmers convinced that the Wolf represents a serious threat to livestock, and by people who, fearing attacks on children and the vulnerable, feel that such large carnivores should not be allowed to co-exist in a civilised society.

Unfortunately, Europeans settlers to North America took their prejudices with them, and Wolves were persecuted mercilessly. Only relatively recently has the change in public attitude allowed Wolf re-introduction to

Male Wolf on the Barren Lands of NWT, Canada in winter.

Arctic Fox at sunrise, Dovrefjell, Norway.

Female Arctic Fox, Bellsund, Svalbard.

## Policemen Wolves

The Inuit were familiar with the hunting techniques of the Wolf, the seeking out of the sick or injured animal so that the kill was less dangerous for the hunter. When policemen were first introduced into Inuit societies by the Canadian government they rapidly acquired the name *amakro* (Wolf) because they stood in the background observing, watching for mistakes.

Arctic Fox feeding on the remains of a Polar Bear Ringed Seal catch. In winter the foxes often follow bears to pick up scraps, but need to be careful as they may finish up as prey. The photograph was taken on the sea ice of Kongsfjorden, Spitsbergen, Svalbard.

the Yellowstone National Park. However, in many places outside National Parks the Wolf is still under threat.

The howl of the Wolf is one of the most evocative sounds of the wilderness. Howling helps the pack to reform after a dispersive hunt. In the Arctic, the prey is often Reindeer, though animals as large as Elk (Moose) may be taken. In the High Arctic, Wolves take rodents and hares and, if the pack is large enough or the prey sufficiently weak, they may take Musk Ox. Adult wolves are usually grey, but this can vary (particularly in North American animals) from almost black to white. White Wolves are found in northern Canada and the Canadian Arctic islands, in Greenland and areas of Arctic Russia.

Wolves breed throughout the Arctic, but they are absent from Iceland and Svalbard. Breeding on Russia's Arctic islands is conjectural as the species has not been well-studied in those areas.

### Arctic Fox *Alopex lagopus*

Together with the Polar Bear, the Arctic Fox is the most specialised terrestrial Arctic mammal. Though usually seen on land hunting rodents, the foxes may also be seen on sea ice in winter, where they account for about 25% of the annual kill of seal pups, and follow Polar Bears to feed on the remnants of bear kills (and on bear faeces *in extremis*). The foxes also follow Wolves on land for the same purpose, and regularly patrol bird nesting cliffs to collect eggs and chicks from poorly sited nests, or to take young birds whose first flights (or glides in the case of some auks) end in disaster.

Adults have two colour forms. In one the summer pelage is grey-brown above, paler below and, usually, on the tail. In the winter this form turns white. The second form, the so-called blue fox, is dark chocolate brown or dark blue-grey in summer, with a paler tail, and pale blue-grey in winter. In general white-morph foxes are continental, blue foxes being found on islands, but interbreeding of the forms occurs and litters may comprise both colour forms.

It breeds throughout the Arctic, but has been hunted to extinction on Jan Mayen. It is possible that the island might be recolonised if sea ice allows animals to reach it again. Icelandic animals have been persecuted since human colonisation of the island, with indigenous animals now reinforced by fur-farm escapes.

### Winter fox fur

The luxurious winter coat of the Arctic Fox – so thick it makes the animal look much bigger than it is – is such a good insulator that the animal's body temperature of 40°C can be maintained without the need for shivering, down to ambient temperatures of about -60°C. Unfortunately, the coat was so prized that the animals were relentlessly hunted both in Russia and Canada. Although staggering numbers of foxes were killed (hunting continues, but at a lower level) the species has, thankfully, survived.

### Red Fox *Vulpus vulpus*

The most widespread and abundant of all carnivores, it breeds across the Arctic, but is confined to southerly latitudes, being absent from Greenland and high-latitude Arctic islands (i.e. Svalbard, Russian islands apart from the southern island of Novaya Zemlya, and Canada's northern islands). Also absent from Iceland.

## Unfriendly foxes

In areas where both Red and Arctic Foxes occur, the larger Red will drive the Arctic out or predate it. As the Red cannot thrive in areas of sparse prey, far-northern latitudes favour the Arctic Fox. The northern limit of the Arctic Fox is therefore defined by the availability of food (and of den sites), while the southern limit is defined by the northern limit of the Red Fox.

# Bears

The three northern species of bears are at least double the weight of the largest of their southern cousins, size and a low metabolic rate being adaptations for surviving the northern winter. The northern bears are large and powerful creatures, the Brown and Polar Bears being the largest terrestrial carnivores on Earth.

Both Brown and Black Bears exhibit periods of winter torpor that require the laying down of fat reserves. Pregnant Polar Bear females undergo winter torpor in the maternity den, but males do not, though they will dig a den in which to rest during periods of bad weather. Exactly when the split between Brown and Polar bears occurred is still debated. It is assumed that a population of Brown Bears was isolated from conspecifics during an ice age and evolved to survive the icy habitat they found themselves in. The split would have occurred during the last million years, and very probably within 250,000 years, but perhaps as recently as 100,000 years ago: there are known examples of the two species interbreeding in the wild. The northern bears share a very similar dentition. Brown and Black Bears are omnivorous, and have molars with broad, rounded cusps suitable for grinding plant material, and a diastema – a gap between the canines and the molars created by absent (or vestigial) premolars, which is used for stripping bark. Each of these adaptations for omnivory is a disadvantage for the wholly carnivorous Polar Bear.

Bears do not form long-lived bonds, mating occurring when a male bear encounters a female in mating condition. Male Polar Bears, which do not hold territories, attempt to keep a female away from other males while she is receptive. Male Brown and Black Bears do hold territories, and these usually overlap the territories of several females, all of which the male will mate with. However, females will also mate with other males they encounter while receptive. Ovulation in bears is stimulated by copulation, the fertilized egg developing to a blastocyst whose implantation is delayed. Delayed implantation evolved in Brown and

Black Bears to allow the females to focus on laying down fat reserves in late summer and autumn, rather than cub-rearing. Courtship and mating take place in late spring, the blastocyst being implanted during autumn. Birth occurs in the maternity den during the female's winter torpor. Mother bears lose up to 40% of their body weight during torpor (and up to 50% in some female Polar Bears). At birth the altricial cubs are very small.

### Polar Bear *Ursus maritimus*

The symbol of the Arctic, not only for the traveller but also for the Inuit, who fear and admire the bear in equal measures. Polar Bears are cream or pale yellow, rather than white (though when seen against a dark background they certainly do look white). The fur comprises a thick underfur up to 5cm long with guard hairs that grow to 15cm. A layer of blubber 5–15cm thick lies beneath the skin. So good is the bear's insulation that they are more in danger of overheating than of hypothermia, particularly if they are running. In short bursts they can achieve 40km/h, but such activity rapidly causes overheating. Hot bears lie on their backs and expose the soles of the feet to cool down, an endearing posture. In warm weather the bears will excavate dens in the snow to keep cool. In water the bears travel at about 2–3km/h and can cover great distances, though tales of bears up to 100km from shore almost certainly refer to individuals carried on an ice floe that subsequently melted.

An adult Polar Bear has a characteristic 'roman' nose and lacks the shoulder muscle hump of a Brown Bear. The ears are small, the feet large to aid walking on snow. The forepaws are also used as paddles by the swimming bear: the back legs are not used in water, trailing behind the animal.

Polar Bears are found throughout the Arctic and have been seen as far south as Iceland, mainland Fennoscandia and even Japan's northern island, Hokkaido. Female bears regularly den in James Bay (at the same latitude as London). These bears, forced to live on land when the sea ice retreats, may eat berries and plants while waiting for the sea ice to reform, though many fast for several months. The sea ice of western Hudson Bay forms first at Cape Churchill, explaining why the bears (and bear watchers) congregate there. When the sea ice forms, the bears disappear.

Female Polar Bear and cub, Inglefieldbreen, Svalbard.

Female Polar Bear and twin cubs, Storfjorden, Spistbergen, Svalbard.

Polar Bear hunting on the sea-ice off Spitsbergen's east coast.

An indication of the depth of snow cover on sea ice is indicated by the fact that this bear was around 3m tall on his hind legs.

## Polar Bear hunting techniques

Polar Bears primarily hunt seals, chiefly the Ringed Seal. Female seals give birth to their pups in dens above the sea ice, the pups protected by a layer of compacted snow into which the female digs. When a bear detects a den it rises on its hind legs and crashes down through the den roof. If the bear is lucky it seizes the pup or blocks the escape hole into the ocean. If the bear is unlucky it fails to break through or misses the pup, which escapes. Bears that fail to break through immediately will often try again, but the chances of catching a pup are much reduced if more than one attempt is necessary. On average, a bear succeeds in catching a pup in about one in three attempts: if multiple attempts are needed to break through the roof this percentage falls rapidly. The technique is also used for hunting resting adult seals.

Bears also wait at breathing holes catching seals as they surface. They will also attack Belugas stranded in leads in the sea ice, clawing at a whale each time it surfaces until the animal becomes exhausted by blood loss and anxiety and can be hauled on to the ice. Polar Bears will also search among hauled-out walruses for pups. In the sea a walrus is a match for a bear, and may even inflict fatal damage with its tusks on land, but large walruses are not very mobile and the nimble bear can occasionally steal a new-born pup from an irate, but helpless, mother.

There are many stories of the cunning of bears: these tell of bears that excavate the breathing holes of seals to make capture easier, then shield the hole with their heads so the seal does not observe a change in light level, and of bears hunting with a paw over their noses (the only black mark on them). Though apparently easy to dismiss, there is historical evidence to support claims of such crafty prey-capture techniques. Clements Markham, a midshipman on the British ship *Assistance*, during a Franklin search expedition of 1850–51, reported in a letter to his father that he watched a Polar Bear 'swimming across a lane of water [and] pushing a large piece of ice before him. Landing on a floe he advanced stealthily toward a couple of seals basking in the sun at some little distance, still holding the ice in front of him to hide his black muzzle.' And there is a credible story of a bear stalking a seal in water by swimming towards it when the seal surfaced, then floating, motionless when the seal submerged. Seal eyes are at their best underwater so this strategy allowed the seal to believe the bear was a piece of floating ice. The strategy was apparently successful.

Brown Bear, Karelia, in the 'No-Man's-Land' between Finland and Russia.

### Brown Bear *Ursus arctos*

Arguably the largest bear, though the largest male Polar Bears may be larger. Weights over 1,000kg have been claimed for Kodiak Brown Bears (which exceeds the largest-known Polar Bear weight). Similar weights may also be achieved by the bears of Kamchatka, which, like those of Kodiak, feed on spawning salmon and therefore put on huge fat reserves before winter torpor.

Brown Bears are uniformly dark brown, though some individuals are paler, while others may be almost black. Some bears have paler tips to the guard hairs, giving them a 'grizzled' appearance (hence the name Grizzly Bear). The muscle-hump at the shoulders is prominent in many individuals. In general Brown Bears have a concave head profile i.e. a 'dish' face. The bears are omnivorous, feeding chiefly on plant material but taking mammals opportunistically (they will sometimes excavate rodent burrows) and will occasionally chase

'Grizzly' Bear, Denali National Park.

## Wintering Brown Bears

In winter both Brown and Black Bears enter a state of torpor, a strategy forced on them by a lack of winter food. In each case the heart rate reduces (from 40–70 beats/minute to c.10 beats/minute in the case of the Brown Bear), though body temperature is maintained just a few degrees below normal: this means that the bears can swiftly rouse themselves if danger threatens, this being particularly important for females with cubs. The bears can survive for up to six months not only without eating or drinking, but also without urinating or defecating, an ability not shared by any other mammal.

Black Bear, northern Ontario, Canada.

down sick animals as large as Elk, or young animals, while some feed extensively on spawning salmon.

Breeds throughout the Arctic, but absent from Greenland and all Arctic islands (with the possible exception of Banks Island). Most of the population is in Russia.

The Black Bear (*Ursus americanus*) is an Arctic fringe dweller, breeding in the forests of Alaska and Canada, and rarely being seen above the timberline.

## Cats

The Felidae or cats are the most carnivorous of the Carnivora. They are superbly adapted for the role, having exceptional eyesight, hearing and smell, phenomenal agility and (in all bar one) sheathed claws, the sheath maintaining the claws in good condition by avoiding wear. Although primarily warm-climate predators, some cats do live in cold conditions. The Siberian Tiger (*Panthera tigris altaica*) inhabits southern Siberia, as does the critically endangered Amur Leopard (*Panthera pardus amurensis*), while the Snow Leopard (*Panthera uncia*) lives high among the snows of the Himalayas. However, no cats are truly Arctic. The Old and New World lynxes are primarily boreal hunters, but they may be seen beyond the timberline. Lynxes take rodents, hares, young deer and birds. The Eurasian Lynx (*Lynx lynx*) breeds in central Scandinavia (where it is rare) and across Russia to Kamchatka, though only to the north coast near the White Sea and the Lena delta. The Canadian Lynx (*Lynx Canadensis*) breeds in Alaska (but is rare in the west and north), the Yukon, North-West Territories, Quebec and Labrador, but not to the northern coast.

## Mustelids

The Mustelidae are perhaps the most diverse of the carnivore families, varying in size from the (Least) Weasel to the Sea Otter (the latter being more than 2,000 times heavier than the former). However, all share common features. The head is flattened and wedge-shaped, an adaptation for hunting in burrows, while the body is long and slender, the legs short, again an ideal shape ideal for pursuing prey in burrows (and seen even in mustelids that do not adopt this hunting technique – the Wolverine, for example, which is too large to do so). The body shape has the disadvantage of a large surface

area-to-volume ratio, which is not optimal for reducing heat loss. As a consequence, the metabolic rate of these mammals is high and they must hunt frequently, despite the high energy value of their diet. Some of the smaller female mustelids, when nursing young, must consume over 60% of their body weight daily.

Most mustelids are sexually dimorphic. This reaches an extreme in the Weasel, where the male is sometimes twice the weight of the female. This may be a response to prey availability, the two sizes meaning that males take larger prey and so do not compete with potential mates – but while this might be true of southern mustelids, the prey of most northern species is limited to the same range of rodents. An alternative theory suggests that females are smaller so that their energy needs are reduced, requiring them to catch fewer prey when they are lactating. Males could also be larger to increase the likelihood of their holding territories and competing for females.

In general mustelids kill their prey with a bite to the base of the skull, crushing the brain, or to the neck, severing the spinal cord.

### (Least) Weasel *Mustela nivalis*

Though commonly known as the Weasel in the UK, the northern Eurasian animal is the nominate (*M.n.nivalis*) of a species which includes a larger subspecies (*M.n.vulgaris*) found in southern Europe. Northern weasels moult to a white winter pelage.

Breeds throughout the Arctic, but absent from Greenland, Iceland and the Arctic islands of Canada and Russia.

### Stoat (Short-tailed Weasel) *Mustela erminea*

The North American name refers to the length of the tail relative to the Long-tailed Weasel (*Mustela frenata*) of the southern Canada and the USA. Northern stoats moult to a white winter pelage (known as *ermine* an animal highly-prized for its fur: ermine skins were used to create the upper sections of cloaks worn by members of the House of Lords in Britain, the black tail-tips (which Weasels do not have) providing the spots in the otherwise milk-white attire).

Breeds throughout the Arctic, wherever rodents are found (includes Greenland, but not Iceland or Svalbard) on Canada's Arctic islands as far north as Ellesmere. Not observed to date on Russia's Arctic islands, but probably

Stoat (Short-tailed Weasel) on Victoria Island, Canada. Despite their reputation for ferocity, stoats are prey to large birds of prey and other carnivores.

American Mink, Randaberg, Norway. The species has become established after escaping fur farms and is now found wild throughout Scandinavia.

Wolverine, Karelia, Finland.

breeds on islands with rodents, though in general Palearctic Stoats are boreal. Nearctic Stoats are the most northerly mustelids.

**American Mink** *Mustela vison*

Breeds throughout the Arctic, but absent from Greenland and all Arctic islands (apart from Iceland, where a population has established from escapes). Originally found in the Nearctic only, Palearctic animals are localised but spreading, and now occur at sites throughout Scandinavia and Russia.

**Wolverine** *Gulo gulo*

The largest Palearctic mustelid (in the Nearctic the Sea Otter is, in general, larger). In Europe, the animal is often called the glutton (this being the basis of the species' scientific name) because of its reputation for greed. It is known to take bait from traps set by fur trappers, and to eat trapped animals, these habits hardly endearing the species to people. It also has a reputation for being a cruel killer. The Sámi claim that if a satiated Wolverine catches a Reindeer it will gouge out its eyes so that the deer dares not move. The Wolverine will then, it is said, return when it is hungry to kill and feed on the still-fresh animal. In the absence of significant competition from Wolves this strategy might just work, but historically Wolves and Wolverines have shared ranges. More importantly, Wolverines invariably dismember large prey and cache sections in well-separated places. The Sámi's tale seems to be black propaganda against an occasional Reindeer predator.

Breeds throughout the Arctic, but it is absent from Greenland, Iceland, Svalbard, Russia's Arctic islands and some Canadian Arctic islands.

**American River Otter (Northern River Otter)** *Lontra canadensis*

More Arctic in range than its Palearctic cousin, breeding throughout Alaska (though rare in the far north and absent from the Aleutian and Bering Sea islands), the Yukon, North-West Territories, around Hudson Bay and throughout Quebec and Labrador. However, the species is rarely found as far north as the coast. The European River Otter (*Lutra lutra*) is, in general, a more southerly species, though it is found near the north coast in Fennoscandia and at the White Sea.

Eurasian River Otter, winter in Kajaanijoki, Finland.

## Ungulates

The large terrestrial herbivores of the Arctic are artiodactyls – even-toed ungulates, hoofed animals of the families Cervidae and Bovidae. (The other ungulate group, the perissodactyls or odd-toed ungulates, has no Arctic representatives.) The Cervidae or deer are distinguished by the antlers that are grown by males (and females in Reindeer). In general, antlers are grown and shed annually. They are horn-like extensions that grow out from the cranial frontal bone. The antlers grow within a sensitive, blood vessel-rich covering called velvet. Ultimately the velvet is shed, often hanging in shreds from the antlers for several days until it is finally rubbed free. In southern deer species the antlers are rubbed against trees, staining them (and damaging the tree), while northern species have to be content with rubbing against the ground. Antlers are used during the rut, the annual mating ritual when males compete to establish and maintain harems of females, mating each time a female becomes receptive. Although many competitions involve merely the showing of the antlers to a rival, accompanied by a bellowing that indicates the fitness of the male (the lower-pitched the bellow, the bigger and more worthy the male), occasionally males will lock antlers and engage in battles. Battles are usually trials of strength, the male forced backwards accepting defeat and retreating: antlers are fearsome weapons and could inflict considerable damage, so the males are keen to avoid real

### Reindeer antlers

Uniquely among deer, both male and female Reindeer grow antlers. The reason lies in the need to excavate holes in the snow to reach winter browse. The larger males might allow females to dig holes, then evict them to acquire the food for themselves. Antlers allow the female to prevent such piracy. To support this, females maintain their antlers for longer than the males, usually throughout the winter. Furthermore, females of woodland-dwelling Reindeer, which feed on lichens growing on tree trunks, often do not grow antlers – there is always another tree to browse. As with other deer, the energetic cost of producing antlers is considerable, and Reindeer often eat their shed antlers to regain the minerals within.

Since female Reindeer not only possess antlers but maintain them through the winter, Santa's Reindeer is therefore much more likely to be Rachel than Rudolf.

conflict. That antlers are a good guide to mating fitness has been confirmed by studies that show that females whose partners had impressive sets produced young that were stronger and matured faster. At the end of the rut bone-dissolving cells invade the base of the antlers, and they are then shed.

### Reindeer (Caribou) *Rangifer tarandus*

Reindeer have the widest feet of any deer and have well-developed dew claws, both features being adaptations for walking on snow. The hooves are also sharp-edged to aid walking on ice. As the Reindeer walks, a tendon stretches across a bony nodule, causing a characteristic click. This noise is occasionally suggested as the origin of the North American name for the species, but it actually derives from *xalibu* (pawer), the Mi'kmaq (a Nova Scotian native American group) name, a reference to the animal's pawing search for winter browse.

There are several Reindeer subspecies. Tundra animals are the best known, but in both North America and Eurasia there are Reindeer that forage in forests – the Woodland Caribou and Forest Reindeer. Interestingly, very recently sociologists interviewing the native Dene people of Canada's NWT found that they differentiated three types of Caribou, adding a third – Mountain Caribou. Genetic analysis then proved the Dene view correct.

Three subspecies also breed on islands, one on Svalbard, one on the southern island of Novaya Zemlya, and the third on some northern Canadian islands. In the case of the last two, interbreeding with mainland animals has occurred. This has meant that the pure-bred Novaya Zemlya subspecies has been lost, while the pure-bred Canadian island form (Peary Caribou) may now exist only on the Queen Elizabeth Islands. Ironically, the loss of sea ice coverage, which allowed interbreeding with mainland animals, may now be creating a new problem in Arctic Canada where the lack of ice has eliminated the 'corridors' which allowed the Caribou to transfer between islands and so enhance the genetic diversity of the species, in-breeding becoming a threat to species survival. The Svalbard Reindeer remains pure-bred. All three island forms are smaller than mainland animals,

Wild Reindeer, Forollhogna, Norway.

Svalbard Reindeer, Forlandsundet, Spitsbergen,

Male Musk Ox, Kongeborgen, King Oscar Fjord, north-east Greenland.

## The annual migration

Reindeer use traditional calving grounds, some herds of Caribou travelling more than 5,000km annually between these and winter feeding grounds. The migrations are in part a search for new browse, but also an escape from the predators that feed on the big herds, Wolves and Brown Bears rarely following the Caribou onto the tundra – there are tundra bears and tundra wolves, but there are far fewer of them.

On the northern migration the females forge ahead of the males, which feed on the richer southern tundra to put on weight for the rut. The females' progress is so urgent – the animals may travel more than 100km each day (though 25–65km is more usual, maintaining a steady pace of about 7km/h) – that if they give birth early they may abandon a calf that cannot keep up. Reindeer are good swimmers and they will cross rivers: the Bathurst Caribou herd crosses Bathurst Inlet twice each year, a swim of 10km.

with shorter legs, and are less gregarious, forming small herds or, often, being seen alone. These Reindeer are almost white in winter.

Reindeer were domesticated in Eurasia and truly wild animals are now found only in Russia (including the southern island of Novaya Zemlya and Wrangel), and southern Norway, though there are feral herds in other parts of Fennoscandia. There are also feral Reindeer on Iceland. Caribou were never domesticated in North America and huge herds still cross the tundra on annual migrations. Caribou breed throughout Alaska, mainland Canada and the Arctic islands.

### Musk Ox *Ovibos moschatus*

Although they belong to the same family (Bovidae) as cattle, Musk Oxen are not true oxen, being more closely related to goats and sheep. Adults are unmistakeable, with long straggling hair (individual hairs can be 70cm long), often long enough to cover the short legs. The fine underfur is claimed by many to provide the finest

Caribou on spring migration, Courageous Lake, Canada's Barren Lands.

wool in the world. Known by the Inuit word *qiviut*, it is highly prized for the manufacture of (very expensive) small items of clothing. The upturned horns arise from each end of a central boss across the top of the head. The structure is formidable and is a significant fraction of the animal's total weight. The hooves are broad, with a sharp outer rim and a softer inner one, giving the Musk Ox surprising agility on rock.

Musk Oxen breed in western and northern Alaska (though these are introduced animals originally released near Fairbanks, then on Nunivak Island in 1935 as the native stock was eliminated by hunting – to provide meat for whalers – in the 1860s), on the northern rim of mainland Canada west of Hudson Bay, and on western and northern Canadian Arctic islands. Although absent from Baffin Island, a small herd breeds near Ungava Bay in Quebec. It also breeds in northern Greenland, and has been re–introduced to western Greenland but is absent from Iceland. It was introduced to Svalbard, where it almost certainly never occurred naturally, but is now extinct and has been introduced to Scandinavia and (reintroduced) to Russia, but populations are small.

Other ungulates breed at the Arctic fringe. The Elk (Moose) – *Alces alces* – is the world's largest deer, and apart from extreme bear specimens, the largest terrestrial 'Arctic' mammal. They are rarely seen beyond the timberline, though occasionally stray on to the tundra. The different names of this animal cause much more confusion than the Reindeer/Caribou pairing. The North American name, Moose, derives from an Algonquin name meaning 'twigeater', while the European name is from the Germanic *elch*. The confusion arises because in North America the Elk is another animal – *Cervus canadensis*, also known

## A defence against Wolves

When attacked by Wolves, Musk Oxen form a defensive circle with calves at its centre. Each animal faces outwards, this presenting the pack with a ring of massive horns and head bosses, a blow from which would inflict significant, perhaps fatal, injury to an individual Wolf. As long as the oxen maintain the circle they are almost invulnerable. Stand-offs lasting several hours have been observed, though usually the Wolves give up and leave if attempts to break the circle fail. The Wolves make occasional forays forward, seeking to make the herd break up and run. Even if the Musk Oxen do make a run for it, they can foil the Wolves as long as they remain a cohesive unit. But sometimes the Wolves run into the herd, biting at flanks and seeking out a young or weak animal. Once an individual has been isolated the pack concentrates on it: the outcome is then a foregone conclusion.

The circle is a good adaptation to Wolf predation, but useless against a man with a rifle, and the Musk Oxen of north-east Greenland were almost hunted to extinction in the 1930s by Norwegian fur-trappers, who shot them to feed themselves and their dogs. A similar slaughter in the Canadian Arctic led to the government protecting the species in 1917, and was the primary reason for the creation of the Thelon Game Sanctuary in 1927.

as the Wapiti, which is similar to the European Red Deer *Cervus elaphus*. Dall's Sheep (*Ovis dalli*) also occasionally known as White or Thin-horn Sheep, breeds in upland areas of Alaska and Yukon, but rarely to the north coast. The Snow Sheep (*Ovis nivicola*) breeds in upland areas of eastern Siberia, including Chukotka and Kamchatka.

Female Moose with twin youngsters, Canada's Yukon Territory.

Bull Elk photographed in the light of early morning, Vanern, Sweden.

Snow sheep, Kamchatka, Russia.

Dall's Ram, Denali National Park, Alaska.

## Rodents

With a species list that comprises more than 40% of all mammals, the rodents are one of the most successful and widespread of all mammalian orders. They are also one of the most numerous, a consequence of their prodigious ability to breed. Female rodents can often breed when only six weeks old and have a gestation time of only *c.*20 days: studies on mice having shown that a single breeding pair can yield a population of over 500 within six months.

Rodents have an essential ecological role. Several mycorrhizal fungi (the most famous of which are the truffles) are spread only by being consumed and then excreted by rodents. As the fungi are necessary for the

## Rodere – to gnaw

Rodents vary greatly in size (the smallest weigh just a few grams, while the Capybara *Hydrochoreus hydrochaeris* can weigh 70kg) and form, but all have features in common, one being a pair of incisors in both the upper and lower jaws that grow continuously. These teeth require the animal to gnaw to maintain incisor length – the word rodent derives from the Latin *rodere*, to gnaw. If the animal were to stop gnawing, the teeth of the lower jaw would eventually grow towards the brain, while those of the upper jaw would grow through the lower lip: in either case death through starvation would occur. Feeding generally provides all the gnawing needed, but rodents will sometimes gnaw other materials to help keep their incisors a constant length (as people sometimes discover to their cost, if rodents gnaw through electricity cables).

growth of plants, rodents therefore play an important role in the spread and establishment of plants. Rodents are also pivotal in supporting many populations of predatory mammals and birds.

## Squirrels

Squirrels are arboreal rodents and, as such, cannot be classified as true Arctic dwellers. Nevertheless, several species are found as far north as the treeline and may therefore be seen by Arctic travellers. The Eurasian Red Squirrel (*Sciuris vulgaris*) breeds in Fennoscandia and across Russia to Kamchatka, while the American Red (or Spruce) Squirrel (*Tamiasciuris hudsonicus*) breeds across North America.

## Ground Squirrels and Marmots

### Arctic Ground Squirrel (*Citellus parryi*)

Also known as the Long-tailed Souslik or Parka Squirrel

Despite the alternative name the tail is short, Long-tailed Souslik deriving from the fact that it is longer than in some other ground squirrels and sousliks. Though primarily vegetarian, the animals have been seen feeding on a Caribou carcass and males will kill and eat young they have not fathered. The squirrels form colonies of up to 50 individuals living in underground burrows, with

numerous entrances to offer some protection against predation. The squirrels hibernate in the burrows in a nest of vegetation, laying down a store of seeds prior to sleeping so that they can feed as soon as they wake.

They breed in Chukotka and Kamchatka, in northern Alaska and northern Canada to the eastern shores of Hudson Bay. Absent from Canada's Arctic islands.

### Alaska Marmot *Marmota broweri*

Also known as the Brooks Range Marmot, this is the most northerly marmot, breeding in Alaska's Brooks Range, particularly on the northern slopes. There are unconfirmed reports of breeding to the northern coast.

Other marmots are seen at the Arctic fringe. The Black-capped Marmot (*Marmota camtschatica*), the only northern Palearctic marmot, has a curiously patchy distribution that includes the mountains and tundra of Baikalia, the Upper Yana and Kolyma rivers, northern Kamchatka and parts of central and northern Chukotka.

Arctic Ground Squirrel, Canada's Barren Lands.

Hoary Marmot emerging from its den amoung boulders of the Denali National Park, Alaska.

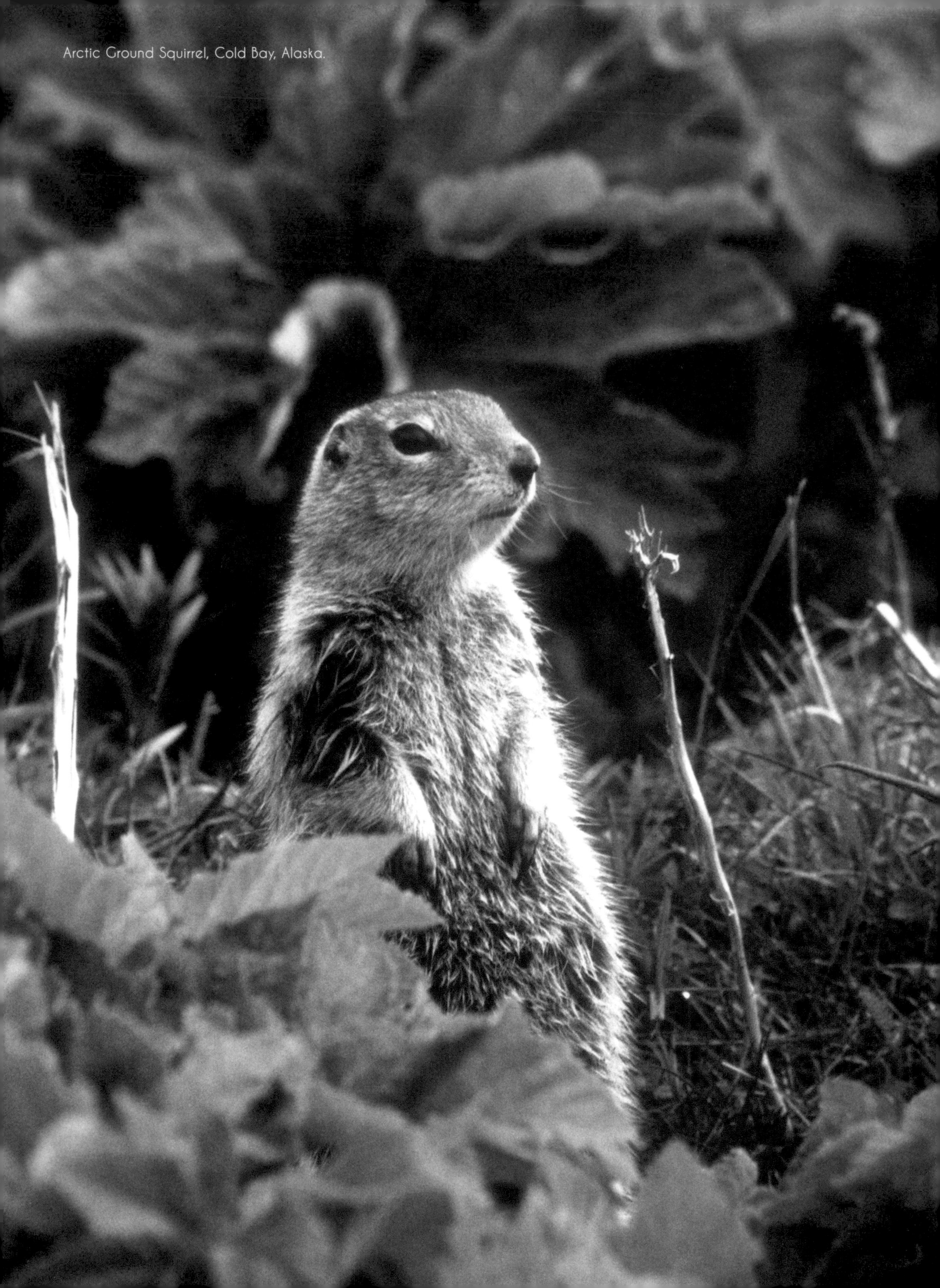

Arctic Ground Squirrel, Cold Bay, Alaska.

## True hibernators

Marmots are the largest of all true hibernators. For Alaska Marmots hibernation begins in September, with the animals remaining in their burrows until early June, having spent almost nine months underground. Black-capped Marmots may spend even longer in their burrows: indeed, so long is the underground period of this species (altitude adds to latitude to create an environment that is harsh almost throughout the year) that they breed and may even given birth before they emerge for the brief summer. For both species, the summer is so short that the young do not mature sufficiently to breed at one year and begin breeding only when they are two. For the Black-capped Marmot the short summer does not allow sufficient resources for females to breed annually, breeding every two years being the norm.

Though not a true Arctic species it is certainly a cold-weather animal, being found at altitudes to 1,900m. The marmots excavate extensive burrow systems that may be up to 100m long and penetrate the permafrost. The Hoary Marmot (*Marmota caligata*) breeds in the Denali National Park.

## Beavers

Although the beaver is a sub-Arctic animal Arctic travellers close to the northern limits of the forest, for instance near the Mackenzie delta, and in the Denali National Park, may see the American Beaver (*Castor canadensi*), or even the Eurasian Beaver (*Castor fiber*) in Scandinavia and Russia. Decimation of the latter in the early years of the 20th century meant the entire population was reduced to around 1,000 animals.

Reintroduction has helped, but the present distribution is very patchy. Interestingly, it was the American Beaver that was reintroduced to lakes in Finland.

## Microtines

Lemmings and voles make up the largest fraction of the Microtinae rodent subfamily, and they include the most northerly of all rodents. Most microtines, and all *Microtus* species, have, in addition to the incisors that characterise rodents, continuously growing molars, which aid the grinding of tough vegetation. Too small to be able to hibernate without starving, microtines are active throughout the winter, using a series of burrows beneath the snow to forage. The snow blanket offers excellent insulation against the Arctic winter, and some protection against predation. However, some northern owls can detect rodents aurally beneath a significant layer of snow, mustelids can follow them in their burrows, and Arctic Foxes also pounce through the snow to reach them: the life of a microtine is one of constant threat from starvation and predation. Many Arctic predator populations are also 'locked' into microtine populations, so that the regular sharp increases in rodent populations are mirrored by increases in predator numbers. For the Arctic traveller, this can be a benefit: microtines are hard to spot, but become much more visible during 'lemming years', while the predators that are often the most sought-after species become more abundant.

There are many northern microtines, all with the same basic characteristics of being small, stocky and brown. Differentiating between them can therefore be difficult, particularly if a view of them consists only of a glimpse of a disappearing animal. In general, where a small rodent is glimpsed, location rather than its colour, is more useful in deciding what it was that you almost just saw.

Eurasian Beaver, Sirdal, Norway.

# Lemmings

### Norway Lemming *Lemmus lemmus*

A beautiful animal, exquisitely coloured that breeds in northern Fennoscandia.

### Siberian Brown Lemming *Lemmus sibiricus*

Breeds on the northern coast of Russia from the White Sea to the Kolyma (including the Taimyr Peninsula), on Novaya Zemlya, the New Siberian Islands and on Wrangel Island.

Norway Lemming, Roskrepp Setesdal, Norway.

## Suicidal lemmings

In one of his modern fables, James Thurber has a man meeting a talking lemming. At the end of their conversation the man asks if he can ask one personal question. The lemming says 'yes' as he also wants to ask one. The man's question is 'why do lemmings occasionally commit mass suicide?' which surprises the lemming as his question is why it is that humans do not.

But despite the oft-heard myth, lemmings do not commit mass suicide. The basis of the myth is the Norway Lemming, because although similar population surges occur in other microtines, mass migrations in these are much less spectacular. The causes of population surges are not well understood, though most authorities favour either climate (an early spring with an abundance of food resulting in early mating and more litters) or a lack of predation. Whatever the cause, the population increases rapidly. The trapping of individuals during lemming years has shown that most migrating animals are immature. Norway Lemmings are very intolerant of their fellows and it is assumed that aggression from older animals forces younger ones out. The topography of Norway, with its high ridges separating narrow valleys, concentrates migrating animals. Conflicts result and the population becomes increasingly panicky, something that is a feature of the mass movements. When a stream, river or lake is reached, pressure from following animals causes the leaders to start swimming. If the lake is large, or if it is the sea, the lemmings, with no concept of lake size or of an ocean, swim regardless, behaviour seen as suicidal by human observers who know that the animals will drown before reaching the far side.

Siberian Brown Lemming, Kolyma Delta, Russia.

Northern Collared Lemming, Victoria Island, Canada.

**American Brown Lemming** *Lemmus trimucronatus*

Breeds in western and northern Alaska and across northern Canada to the western shore of Hudson Bay, on southern Canadian Arctic islands from Banks to Baffin, and on islands in the Bering Sea.

## Collared Lemmings

While the lemmings mentioned above could be described as Arctic fringe species, collared lemmings are true Arctic animals. They belong to the genus *Dicrostonyx*, the name derived from the Greek for 'forked claw', a reference to the growth of a double claw in winter to aid digging in the snow: the pad between the third and fourth toes hardens and enlarges, fusing with the claws. Most species have a 'collar'. If the collar is paler than the upper body it may be visible as a complete band, but often it merges with the colour of the upper body and so is visible only as a chest-band.

Northern Red-backed Vole, St Lawrence Island, Alaska.

**Arctic Collared Lemming** *Dicrostonyx torquatus*

In winter these red-brown animals with a dark dorsal stripe from the head to the tail and a pale grey or grey-buff collar turn white.

Breeds from the eastern shores of the White Sea to Chukotka and eastern Kamchatka, on the southern island of Novaya Zemlya and throughout Severnaya Zemlya and the New Siberian Islands.

**Wrangel Island (or Vinogradov's) Lemming**
*Dicrostonyx vinogradovi*

Breeds only on Wrangel Island, where they are rare.

**Northern Collared Lemming (Greenland Collared Lemming)** *Dicrostonyx groenlandicus*

Breeds in north Greenland, the Aleutian and Bering Sea islands, western and northern Alaska and on the northern Canadian mainland east to Hudson Bay. Breeds on all Canada's Arctic islands to northern Ellesmere Island.

**Ungava Lemming (Hudson Bay lemming, Labrador Lemming)** *Dicrostonyx hudsonicus*

Breeds on the Ungava Peninsula of northern Quebec and Labrador, and on islands of eastern Hudson Bay (e.g. Belcher and King George).

Tundra Vole, northern Alaska.

287

## Voles

As with other rodents, differentiating voles is a job for experts. Voles differ from lemmings in having longer tails, which are invariably bi-coloured.

### Grey-sided Vole *Clethrionomys rufocanus*

Breeds in northern Scandinavia and across Russia to Kamchatka, but only to the north coast on the Kola Peninsula and east to Baydaratskaya Bay.

### Northern Red-backed Vole *Clethrionomys rutilus*

Breeds in northern Scandinavia, across Russia, throughout Alaska and northern Canada east to Hudson Bay. The Canadian range extends to the Boothia Peninsula, but does not include Arctic islands.

### Singing Vole *Microtus miuris*

Adult Singing Voles occasionally sit in exposed places and 'sing' a high-pitched trilling 'song'. This may be an alarm call for young voles, as the singing usually takes place when litters have been weaned. Before winter arrives, Singing Voles make hay balls. These can be huge – up to 30 litres. They are placed above ground and provide winter sustenance.

It breeds in northern and southern Alaska, but is absent from most of central Alaska, from the Alaska Peninsula and the Aleutian Islands; also breeds in central Yukon.

### Narrow-headed Vole *Microtus gregalis*

Has a patchy distribution across Russia, breeding close to the coast of the White Sea and east to the Ob delta, from Khatanga to the Kolyma delta. In the mountains of Kazakhstan and Kyrgizia the species breeds to 3,500m.

### Tundra Vole (Root Vole) *Microtus oeconomus*

One of the larger northern voles. The Latin name means 'house keeper' (the Latin word is also the root of 'economic') and refers to the species' storage of seeds and rhizomes, which supplement its winter diet. In North America these stores were sought after by the Inuit who gathered them for their own use. For the poor voles, this confiscation probably led to winter starvation.

Breeds in northern Scandinavia and across Russia to Kamchatka (to the north coast except on the Taimyr Peninsula), and to the north coast throughout Alaska and the Yukon.

## A new species in the Arctic?

The Sibling Vole (*Microtus levis*) breeds in southern Scandinavia and south-west Russia and is non-Arctic. However, colonies have been found on Svalbard. It is assumed that these have been accidentally introduced from ships. Despite being a southern species the voles appear to have become established, surviving the Arctic winter better than other introduced rodents. It seems that they counteract high winter mortality by increased fecundity in summer, having up to four litters with the young breeding as early as one month after birth.

Muskrat, Great Slave Lake, NWT, Canada.

The Muskrat (*Ondatra zibethicus*) is not an Arctic species, but may be seen from western Alaska to Labrador, but not to the north coast. Because the animals were valued for their fur (called *musquash*), Muskrats were farmed in Europe and escapees have formed breeding populations in Scandinavia and across Russia.

## Rats and Mice

Of the three primary rodent pests to mankind, the House Mouse *Mus musculus*, Brown Rat *Rattus norvegicus* and Black Rat *R. rattus*, the first two are found in association with people even in the Arctic. However, neither has ever become established independently from human habitation, with feral populations invariably dying out during their first Arctic winter.

Arctic Hare, Churchill, Manitoba, Canada. It is spring and the hare is moulting from winter white to summer grey-brown.

## Lagomorphs

Lagomorphs – rabbits, hares and pikas – differ from rodents in having a second pair of incisors known as peg teeth. While picas look similar to rodents, hares and rabbits differ in body shape, having long back legs that allow them to run fast. Though the adaptation is the same in each group, the strategy for avoiding predators differs. While hares simply seek to outrun a predator, rabbits make short runs to their burrows or to available cover. These different escape strategies are reflected in the breeding biology of the two groups. Young rabbits are born underground and are altricial. By contrast, leverets are born above ground and are precocial. To avoid drawing unnecessary attention to the leverets,

the female hare visits them only once each day. At that time the leverets, which spend the rest of the day in concealed places, congregate at their birth place and are suckled for about five minutes. Because of the limited duration of nursing, hare milk has a very high fat and protein content and the leverets are weaned at three or four weeks.

As with shrews, lagomorphs practice refection, the redigestion of 'first-pass' undigested material. The 'second-pass' faecal material is in the familiar form of hard, black 'currants'.

### Arctic Hare *Lepus arcticus*

In summer adults in the south of the hare's range are grey or grey-brown. In winter they are white apart from black ear-tips and inner ear. In northern parts of the range, they are white, the poor forage being inadequate to allow energy to be wasted in fur changes. In the short summer the hares are then highly conspicuous, but nature compensates by making wolves, their main predator, white for the same reason. In winter the hares occasionally burrow into the snow, excavating short tunnels to dens that offer shelter from the wind and intense cold. The hares can reach speeds of over 60km/h and swim well.

They breed on the central and northern coasts of both east and west Greenland, but are rare in the south, also on Canada's northern mainland, and the Arctic islands to northern Ellesmere. The mainland population breeds as far south as Hudson Bay's southern shore. Absent from Alaska.

### Alaskan (or Tundra) Hare *Lepus othus*

The Alaskan equivalent of the Arctic Hare, it breeds in south-western, western and northern Alaska. As with the Arctic Hare, this is a hare of the tundra rather than the forest.

### Mountain Hare *Lepus timidus*

In summer the adult is dark brown or grey-brown, paler below and with grey or white tail. The ears are black-tipped. The winter pelage is white, but not white to the body (as it is for the Arctic Hare), the underfur being slate-grey or blue-grey. This is occasionally visible, creating grey-blue patches on the flanks. Animals in northern Siberia are white.

It breeds in northern Scandinavia and across Russia to Kamchatka, to the northern coast, but is absent from all Russia's Arctic islands.

Three other species may also be encountered. The Snowshoe Hare (*Lepus americanus*) breeds across North America, but is rarely seen above the timberline. Northern Pikas (*Ochotona hyperborean*) breed in Russia from the Yenisey River to Chukotka and Kamchatka, north to the coast (and south into Mongolia). Collared Pikas (*Ochotona collaris*) are much less Arctic than their Palearctic cousins, breeding in southern and central Alaska (including the Denali National Park), and southern Yukon.

Arctic Hare in winter pelage, Barren Lands, NWT, Canada.

Snowshoe Hare in summer pelage, Churchill, Manitoba, Canada.

## Marine mammals

Within the Arctic the summer sea temperature is ≤ 10°C, while in winter it is ≤ 5°C. The sea temperature below sea ice is < 0°C. As the solubility of gases in water increases with decreasing temperature, Arctic waters are oxygen-rich. They are also nutrient-rich, due to the huge inflow of the rivers of North America and Asian Russia. The combination makes the Arctic seas highly productive in summer, attracting not only seabirds but marine mammals. While the pinnipeds that exploit this summer bounty remain in the Arctic year-long, moving with the ice edge, most cetaceans travel to the area to feed in summer, then migrate to more productive waters during the winter.

The marine environment offers advantages to mammals in addition to a copious food supply. Freed from the major constraint imposed by gravity, marine mammals have fewer limits to size. Seals are restricted in size by the need to come ashore to give birth, but whales have overcome this restriction, and become massive: Blue Whales are probably the largest animal ever to have lived on Earth. Giving birth in water would seem illogical for an air-breathing animal, but it has its advantages: female whales do not have to support the weight of their growing foetuses in the way that terrestrial mammals do, and so can carry larger young (within the obvious limitations of the birthing process) and these well-developed offspring are better able to seek air immediately, to contend with the problems of suckling underwater, and to survive the occasionally hostile environment into which they have been born.

The buoyancy of water means that the skeletons of marine mammals do not have to overcome gravity and so can be much less bulky. Honeycombed whale bones are remarkably light in comparison to those of heavy terrestrial mammals such as elephants. The bones have a hard outer shell covering a sponge-like inner layer with numerous blood vessels and a marrow rich in oil: when whales were hunted, about 30% of the oil obtained from a carcass came from the bones. Though strong enough to act as anchors for the whale's huge muscles, some whale bones are so light that they float in water. Similarly, the bones of the larger seals are, despite the time the animal spends out of water, flimsy in comparison to those of terrestrial mammals.

But the marine environment also has disadvantages for mammals. Heat loss in water is significantly higher than in air so mammals must have excellent body insulation, a high metabolic rate, or both. As the volume (and therefore mass) of spheres increases with the cube of diameter, but surface area only with the square, large animals (which can be considered as essentially spherical) have a proportionately smaller surface area from which to lose heat than small ones. The smallest marine mammal is many thousands of times larger than the smallest terrestrial mammal. As an insulator, a layer of subcutaneous blubber is more efficient than fur, but as its insulating properties are, in part, dependent on thickness, good insulation requires a big body – another reason for marine mammals to be large. The smaller marine mammals are those that rely, at least partially, on fur for insulation. The blubber of pinnipeds and cetaceans is of variable thickness and lipid content, its distribution optimising streamlining and the insulation of vital organs. Marine mammals also employ counter-current heat exchangers to minimise heat loss. Cetaceans also employ counter-current heat exchangers to cool the testes of the male, which are within the body cavity (for reasons of streamlining) and would otherwise overheat.

The other disadvantage of being a marine mammal is the breathing of air. Not only must the animals come to the surface to breathe – a procedure that might itself create problems for Arctic marine mammals because of the extent of ice cover – they must also store oxygen for relatively long periods if they are to feed successfully. The easy answer would appear to be large lungs – and large body size would appear to be just the thing to accommodate them. But storing a large supply of air has limitations. Large, air-filled lungs would act as buoyancy tanks making diving more difficult, and as pressure increases with depth the collapsing lungs would compress the air, with potentially lethal side effects. Water pressure increases by one atmosphere for each 10m of depth. At high concentrations oxygen is poisonous, while nitrogen is a narcotic. Bubbles of nitrogen and oxygen forming in the blood as the animal surfaced and the gases decompressed would also give rise to 'bends', a problem that can be fatal to human sub-aqua divers. Marine mammals dive to prodigious depths – Sperm Whales are known to dive to 3,000m (and to stay submerged for more than two hours) – and must therefore overcome this problem. The muscles of marine mammals are rich in myoglobin, which 'stores' oxygen (in much the same way as haemoglobin) and releases it gradually during a dive. Marine mammal blood is also rich in haemoglobin, the oxygen storage potential of the two compounds reducing the need to store air when diving. Pinnipeds exhale before diving, effectively eliminating air storage and, therefore, the potential for the bends. But cetaceans inhale before diving. They possess networks of blood vessels known as a *rete mirabilia* (literally 'wonderful network'), in the chest cavity (and other areas), which, it is believed, may act as a sink for nitrogen as the animal surfaces. At the huge pressures of deep dives, the collapsing lungs of cetaceans also force air into the nasal passages, where nitrogen absorption into the bloodstream is not possible.

## Pinnipeds

The Pinnipedia (the name means 'wing-footed') are divided into the Phocidae (true seals), and a superfamily comprising the Otariidae (eared seals – fur seals and sea lions) and the Odobenidae (the Walrus). Pinniped origins are shrouded in mystery. They are carnivores, relatives of cats, dogs and bears. Some authorities consider the pinnipeds represent two separate re-invasions of the sea, with an otter-like ancestor giving rise to the true seals and a bear-like ancestor evolving to the eared seals and walrus. However, more recent molecular research suggests a single re-invasion.

Pinnipeds are extremely well-adapted to the marine environment with spindle-shaped bodies, and limbs that have developed into flippers, but there are significant differences between the two groups. The phocids lack external ears (which would enhance drag), and have hind flippers closely akin to the tail flukes of whales in shape that provide the power for swimming. The front flippers are short and are held close to the body during swimming, though they can function as fins to aid steering. Insulation is by blubber, ancestral fur having been reduced to a sparse scattering of coarse hairs. On land the hind flippers are useless for locomotion, the seal using its front flippers to haul itself along, progress being an ungainly wriggle.

By contrast, the eared seals can rotate their hind flippers underneath their bodies and, using these and their long front flippers, the animals are reasonably mobile and able to move surprisingly quickly. Travellers familiar with the slow-moving, essentially sedentary phocids are in for an unpleasant surprise if they stray too close to an eared seal. Not only does the animal accelerate and move quickly, it has an array of business-like teeth: being chased by an irate eared seal is somewhat akin to being pursued by a large dog, though thankfully the seal usually gives up more readily. Eared seals also differ in their insulation, relying, in part, on fur: the underfur of the fur seals is luxuriantly thick, a fact that led to their near–extinction due to overhunting. Consequently, this different, and less effective, mode of insulation, means fur seals and sea lions are chiefly animals of cool temperate waters. The Walrus has blubber insulation similar to the phocids though it shares the reversible hind flippers of the eared seals. Eared seals and the walrus also differ in being gregarious, forming large, sometimes huge, colonies, whereas the phocids are more solitary.

One feature that all pinnipeds share is vibrissae, or whiskers. Studies have shown that these are sensitive to sound, which may be an advantage in avoiding predators. Studies with blindfolded seals have also shown that the seals are still able to catch fish, while if the vibrissae are removed the seals are much less efficient at fishing. The suggestion is that the vibrissae can detect vibrations in the water such as those caused by the wake of a swimming fish. Vibrissae may also detect hydrodynamic changes caused by fixed features, and so help a seal to navigate in murky waters.

The smaller pinnipeds exhibit countershading, being darker above and paler below. The dorsal colour is also disrupted to break up the animal's outline when it hauls out. Ribbon and Harp Seals have striking patterns, which are more definite in males and which develop with age: these are likely to be related to courtship. Young seals are, in general, born with a covering of white fur (lanugo), a contrast to Antarctic seal pups, which are dark. The white fur is assumed to be camouflage against sea ice for animals born in the land of the Polar Bear: those pups that are not entirely white are pale or partially white, and have a disrupted pattern.

Phocids moult their skin annually, but this is not equivalent to the moulting of fur-bearing mammals when losing or acquiring a winter pelt. For the seals, such a change is unnecessary, the moult representing the replacement of potentially damaged skin.

## True seals

Phocids live in both fresh and saltwater, though species such as the freshwater Baikal Seal *Pusa sibirica*, the inland (but saltwater) Caspian Seal *Pusa caspica* and the Arctic–breeding marine Ringed Seal all evolved from the same ancestor. Apart from the monk seals, all phocids are polar or sub-polar, with all northern species apart from the two inland seals being Arctic animals.

Ringed Seal, Kongsfjorden, Spitsbergen, Svalbard.

### Ringed Seal *Pusa hispida*

The most numerous of all Arctic mammals, with a population estimated at around 6 million. Ringed Seals can dive to 90m, though depths to 40m are more common, and they can stay submerged for 20 minutes, though 4–8 minutes is more usual. Ringed Seal pups are born in a birthing chamber, above the sea ice, usually excavated in a pressure ridge by the female. It is estimated that predation by bears and Arctic Foxes accounts for around 50% of each year's pups.

Found at the ice edge throughout the Arctic, the highest concentrations are in eastern Russia, the Bering Sea and the North American Arctic.

### Ribbon Seal *Phoca fasciata*

The 'ribbon' of the name is a series of pale bands on the body which circle the head, flippers and tail, these more prominent in males which have a darker background colour.

It breeds on pack ice far from land in the Bering Sea, southern Beaufort Sea and the Sea of Okhotsk.

### Harp Seal *Phoca groenlandica*

The 'harp' of the name is seen on the flanks, the two harps linked across the back so that the overall pattern is more saddle–shaped. Found from the Arctic islands of eastern Canada to the Laptev Sea.

### Harbour Seal *Phoca vitulina*

Found in coastal waters of southern Greenland, Iceland, northern Scandinavia and Svalbard (north to Prinz Karls Forland). Absent from northern Russia, but found in the northern Pacific and southern Bering Sea from Kamchatka to southern Alaska, and in the waters of eastern Canada, including Hudson Bay, northern Quebec and Labrador, and southern Baffin Island.

## Freshwater Ringed Seals

Despite being marine mammals, Ringed Seals are found at three freshwater locations – Lake Ladoga in Russia and Lake Saimaa in Finland, both close to the Gulf of Finland, and Lake Nettilling on Baffin Island's west coast. Both Lake Ladoga and Lake Saimaa are connected to the sea, but there is no evidence to suggest that the seals travel along the connecting waterways. It is assumed that ancestral populations either migrated to the lakes, or have been (partially) isolated by landscape changes. The Lake Saimaa population contains only about 200 animals, so the seals must have arrived recently, or be on the verge of extinction, as such a population is hardly sufficient to ensure genetic diversity and long term survival. The populations at the other two sites are much higher.

Ribbon Seal, Bering Sea.

Harbour Seals, Tracey Arm, Alaska.

Largha Seal, Bering Sea, off eastern Kamchatka.

### Largha Seal *Phoca larga*

The name *Largha* is that given to the seal by the Tungus people of the western Sea of Okhotsk. It was adopted because the North American name, Spotted Seal, though an accurate description, caused confusion with the Harbour Seal, which was occasionally called the Spotted Seal in Europe.

It is found in the North Pacific, the Sea of Okhotsk and the Bering Sea.

### Bearded Seal *Erignathus barbatus*

Unusually, females are slightly larger than males. The seal has a profusion of long vibrissae that curl when dry, giving the moustached look that (more or less) explains the name.

Found throughout northern waters, and also in Hudson Bay and the Sea of Okhotsk.

Bearded Seal, Hinloppen Strait, Svalbard.

**Hooded Seal** *Cystophora cristata*

Significantly sexual dimorphic, males being almost twice the weight of females (though only about 25% longer), this is the heaviest of the northern true seals (some males exceed 400kg) though not as long as the largest Grey Seals. The hood of the name is an enlarged extension of the nasal cavity that forms a proboscis which, in males, hangs over the mouth, but is much less pronounced or absent in females. The hood can be inflated to form a large black cushion or blister that spreads from the forehead over the mouth. Males can also extrude and inflate the internasal septum membrane. This extrudes from one nostril, usually the left, as a red balloon. The inflation mechanisms of hood and balloon are dissimilar, the hood requiring closed nostrils, the balloon an open nostril. Consequently, both hood and balloon cannot be inflated simultaneously. However, a 'half-hood' and balloon can be inflated, the effect being grotesque. Although the hood and balloon are used in mating displays, they are also inflated if the seal is surprised by an observer (in anxiety or as a threat) and, occasionally, by resting seals, seemingly just for the fun of it. The hood develops in males from the age of 4 years.

Found on the eastern seaboard of North America from Newfoundland to Lancaster Sound, but rarely west of Labrador's Cape Chidley or north into Smith Sound, they are also found in the north-western Atlantic around Iceland's north coast, Jan Mayen, Svalbard and Bear Island, but rarely as far east as Franz Josef Land.

Grey Seals *Halichoerus grypus* are the longest of the northern phocids. They also exhibit the most striking sexual dimorphism, with the male up to three times larger than the female. Sub-Arctic rather than Arctic, they breed on the Labrador coast, the southern and western coasts of Iceland, and northern Scandinavia eastwards to the White Sea.

## Eared seals

As noted above, most eared seals occur in cool temperate waters, but one sea lion and one fur seal are found at the edge of the Arctic boundary.

Steller's (or Northern) Sea Lion (*Eumetopias jubatus*) is significantly sexual dimorphic, males being almost twice the weight of females (though only about 25% longer). Recorded male weights have exceeded 1200kg. Seen

Steller's Sea Lions using a buoy in the Lynn Canal, Alaska, as a resting place.

in the western Sea of Okhotsk and across the southern Bering Sea from Kamchatka to southern Alaska, they are also found on the Kuril Islands and along the eastern seaboard of North America as far south as California. The Northern Fur Seal (*Callorhinus ursinus*) also exhibits extreme sexual dimorphism, with males being up to 5 times heavier than females and *c.*70% longer. It is found in the Sea of Okhotsk and across the Pacific from Japan to the Californian coast, but chiefly in the southern Bering Sea. The primary breeding grounds are on the Pribilof and Commander Islands, where some 90% of the animals breed. The luxurious pelt of the Northern Fur Seal is reflected in the scientific name, which derives from *kallos rhinos*, beautiful skin. It was for its pelt that the species was ruthlessly exploited in the 19th century when the original population on the Pribilof Islands was reduced from *c.*3million to *c.*300,000). So dense is the fur that water does not reach the skin even if the seal scratches itself under water. When hunting was regulated the population increased, but then dropped dramatically in the 20 years from the late 1960s to the late 1980s, possibly due to an increase of commercial fishing. Harvesting, which had continued, was also blamed, and was banned in 1988. Unfortunately the population is still declining, though slowly.

**Walrus** *Odobenus rosmarus*

Walruses split from the eared seals about 20 million years ago. The fossil record suggests they were once the dominant pinniped group, but only a single form now remains. Walruses are the largest of all Arctic pinnipeds, and second in size only to elephant seals in world terms.

The skin colour of adults varies with blood flow. In the water, or recently emerged, walruses can be very pale light grey or grey-brown. But when hauled out blood is pumped to the skin to aid cooling, and the animal becomes pink. Walrus skin is very thick (particularly around the neck, where it can be up to 4cm deep) and

Atlantic Walrus, Andretangen, Edgeoya, Svalbard.

Pacific Walrus, Chukotka's Bering Sea coast.

Atlantic Walrus, Fosterbukta, Hold-with-Hope, north-east Greenland.

tough, and was used by the Inuit to cover summer and winter houses because of its durability. Walrus blubber can be up to 15cm thick, though on average it is only half that thickness. The upper canine teeth are massively extended to form protruding tusks. The species exhibits extreme sexual dimorphism, with males being *c.*50% heavier (weight to 1900kg) and 20% longer (up to 3.5m) than females. Atlantic and Pacific Walrus are different subspecies and differ in the length and shape of their tusks. Male Atlantic tusks are *c.*75cm, with those of the females to *c.*60cm. Pacific Walrus tusks are longer, males to 100cm, females to *c.*75cm, and are also curved rather than straight (though there are exceptions). In general female tusks are circular in cross-section (whereas male tusks are elliptical) and more slender.

They are found in north-west and north-east Greenland, Svalbard, Franz Josef Land, Novaya Zemlya, north-east Siberia (the Laptev Sea walruses are considered a distinct subspecies), Wrangel, the west coast of Alaska, Baffin Island and the islands to the north of Hudson

## Tooth-walker and mollusc sucker.

The Walrus uses its tusks to make or maintain holes in the ice, and to help it haul out of the water, the latter task giving the animal its scientific name – *Odobenus* from *odontes baino* – tooth-walker. The second part of the name derives from *ros maris* – sea rose, a reference to the colour of the animal and its maritime habitat. The common name is from the Scandinavian *hvalross* –whale horse.

Walrus feed by standing on their heads and feeling for prey in the sediment with their highly sensitive vibrissae. The chief food is molluscs, the meat being extracted from the shell by suction. Walrus have a formidable ability to suck, the Inuit telling of animals coming up beneath swimming ducks and sucking them under. Walrus also eat young seals, and the Inuit maintain they occasionally kill Beluga. Adult Walrus have no enemies apart from humans. Polar Bears occasionally invade Walrus colonies, seeking to take a young animal. On land the Walrus is ponderous and no match for an agile bear, though bears ensure they stay well clear of adults as the tusks can inflict savage, potentially fatal, wounds. In water the tables are turned, the bears staying far away.

Bay. Walrus are famed for their haul-out areas, where hundreds of the animals bask on shore, but while such places are a joy for photographers that is not always the case for locals. In 2015 an estimated 35,000 walruses hauled out close to a small village (population 270) on Alaska's north-west coast, apparently arriving there because global warming was causing a loss of sea ice in the Chukchi Sea. In 2016 the villagers, who see the walrus as a useful source of food, appealed to visitors to stay away as the crowds, in boats and sightseeing planes, were causing mass panic in the animals, stampedes from the beach into the sea causing hundreds of deaths as young calves were trampled.

## Cetaceans

Although Aristotle recognised as early as the 4th century BC that whales breathed air, it was not until the work of Linnaeus in the 18th century that they were formally identified as mammals. The adaptations of the cetacean body for its marine environment took many more years to unravel, and even today aspects of the animals' physiology and lifestyle are still not completely understood.

The whale shape mirrors that of the pinnipeds in being streamlined for energy-efficient locomotion. The head merges with the body, there being no discernible neck and shoulders. The head cannot move independently of the body, except in the Beluga which has unfused cervical vertebrae that allow the head to turn and nod. The hind limbs are vestigial and within the body so that they do not interfere with streamlining, the power for swimming being provided by a large tail comprising twin flukes. The tail is powered by huge back muscles and moves vertically, the flukes staying parallel to the water surface: whales therefore differ in this respect from fish, the tails of which move from side-to-side. The front limbs have become flippers that are primarily used for steering, and many species have evolved a dorsal fin to aid stability. Unlike the hull of a ship, the whale's body is not rigid and so can move in response to water pressure. The skin also exudes a polymer that may assist the shedding of the outer layer (the Beluga is again an exception to the cetacean norm in undergoing an annual moult rather than continuously shedding the outer skin), and/or may assist in overcoming turbulence (and therefore drag).

The flippers and dorsal fin are the only protuberances on the body, the male penis being within the body cavity except during mating, and the teats of the female being set within slits close to the genital opening. Body hair is minimised, insulation being provided entirely by layers of sub-cutaneous blubber. In the case of the Bowhead blubber can be 50cm thick.

Whales are considered monophyletic, and to have evolved from a common ancestor in the Eocene, with two branches emerging in the Oligocene to form the two modern-day suborders, the toothed and baleen whales. Although the two groups have many common features, they differ markedly in feeding methods and, consequently, in the structure of the head. Toothed whales feed primarily on fish and squid. In general their jaws are extended into a beak-like snout (this being most pronounced in the beaked whale), the Sperm Whale being the major exception to the pattern. The jaws have an array of teeth for grasping prey (or tearing at it in the case of the Orca). The forehead is rounded, forming a 'melon' within which is a wax-like substance that is the basis of an echolocation system that is used to find prey in the deep, dark waters in which the whales tend to feed. The melon of the Sperm Whale, the deepest diver, is huge, the head accounting for 25–30% of the animal's total length. The wax it houses – spermaceti oil – was highly prized by early whalers. However, it is worth noting that the exact function of the spermaceti is still debated. While it is likely that the oil acts as a sonic lens to aid echolocation, it may also have a role in buoyancy regulation.

In the baleen whales the bones of the cranium and jaws have been extended and widened. The upper bones form the rostrum, from which the baleen plates hang. Although often called whalebone, baleen plates are not bone, neither are they modified teeth. Rather, they are keratinous plates emerging from the jaw bone. The plates are smooth, but the inner edges abrade to form 'bristles', the bristles of individual plates overlapping to form a sieve. Food is obtained when the tongue is pressed against the baleen, with engulfed water being squeezed out through the sieves and the trapped prey items then being swallowed. Although this action is common in the baleen whales, they have different strategies for engulfing. Some swim slowly forward and allow their sieves to extract food continuously, while others take huge gulps of water. Sievers include the Right and Bowhead Whales: they have huge heads to allow space for large baleen plates. Gulpers include the Blue and Humpback Whales: they have pleats or furrows of skin on the lower jaw, which allows the mouth to expand to engulf vast quantities of water at each gulp. These furrows give this group of whales their common name, *rorqual*, though the original meaning of this is disputed. It could be from the Scandinavian for pleated, or from 'red throat', a reference to the colour change of the skin when it is expanded, exposing the blood vessels. One species, the Sei Whale, feeds with a combination of sieving and gulping, while the Grey Whale differs in sieving bottom sediments. Some gulpers also employ lunge-feeding, rising from beneath a concentrated prey source. This method is used by Humpback Whales, which occasionally hunt collectively, driving a shoal of fish into a tight ball by producing strings of bubbles from their blowholes and then lunging through the ball. While gulpers tend to seek out prey-rich areas, there is no indication that they use sophisticated echolocation systems. Baleen whales do emit sounds, but these are usually at low frequencies and so would not be much use for echolocation (as the 'visible' target cannot be smaller than the wavelength of the emitted sound). Instead, the sound is primarily for communication, with low frequency sounds travelling huge distances underwater: the 'song' of the Humpback Whale, an evocative, ethereal noise, has become a famous example.

Toothed and baleen whales also differ in the number of nostrils. In toothed whales the two nasal passages combine to form a single, crescent-shaped blowhole. In some species it seems that only one passage is used for breathing, the second being part of the echolocation system. In baleen whales the two nasal passages form a blowhole of two parallel slits. When the whale exhales, water trapped in folds around the blowhole is also expelled, forming the characteristic 'blow', the pattern of which is a useful species guide. In the smaller toothed whales the blow may be minimal – and may even appear absent – and is a much less useful identification feature. The statistics of whale breathing are amazing, e.g. a Fin Whale exhales and inhales about 1,500 litres of air in just two seconds which is several thousand times the volume a human exchanges. Some of the smaller dolphins, with lung sizes more comparable to those of humans, can exhale and inhale in a tenth of a second. Whales also extract more oxygen from a volume of air than humans, 10–12% against about 4%. They also use a greater percentage of their lung capacity, exchanging about 90% of the volume each breath, compared to 10–15% in humans.

Cetacean eyes are relatively small, sight being much less useful in water. However, some whales 'spy hop', raising their heads out of the water, apparently to view the local area. Small eyes, acting as pinhole cameras, allow a greater depth of focus and may assist these whales – which for the most part stay close to shore – to locate land features. The larger rorquals do not do this, perhaps for biomechanical reasons, or alternatively it is possible that their low-frequency sounds are of use as coarse echolocation systems to help them identify underwater topography. However, the right whales do spy hop, as does the Minke.

Spy-hopping is just one of many distinctive whale behaviours. Others include lob-tailing, the waving of the tail flukes and subsequent slapping of the water surface with them, fin-waving, which may also involve water slapping, and breaching. These are all forms of communication, though they are more common in some species than in others. Humpback Whales are famous for breaching, but it comes as a surprise to discover that the huge, slow Bowhead and Northern Right whales also do it.

Cetaceans mate and give birth in water, and whale calves suckle underwater. Mating occurs horizontally and vertically in different species. Calves are born tail-first, as might be expected, and are quickly ushered to the surface by their mothers to take their first breath. However, head-first births have occasionally been observed. The problem of suckling underwater has been overcome by the female having contractile muscles in the mammary glands, which enable her to squirt milk into her calf's mouth, allowing the transfer of large quantities of milk in a short time to help the calf avoid spending protracted periods feeding underwater. Whale gestation is, as would be expected, long, and the period between births is also long, this hindering the recovery of populations that were overhunted in the past. While northern toothed whales usually give birth in northern waters, many of the baleen whales migrate long distances to breeding grounds. The reasons for this are not clear:

various suggestions have been made, but there is no consensus, particularly as any theory must explain why Fin Whales, for example, do not migrate.

The range limitations of terrestrial mammals and, to a lesser extent, birds, are not applicable to cetaceans so essentially southerly species may been seen in Arctic waters if the weather and food resources allow. In the descriptions that follow a pragmatic approach has therefore been adopted, and some species that may occasionally be seen in northern waters have been excluded as such sightings are akin to the sightings of vagrant birds.

## Intelligent whales?

While it is true that the brains of cetaceans are large – the 9kg brain of a Sperm Whale is the heaviest of any animal (compare the 1.5kg human brain) – the ratio of brain weight to body weight is considered more important than brain weight alone when comparing relative intelligence. On that basis, the Sperm Whale looks much less impressive, the brain/body ratio being 0.02%, compared to 2% for a human. However, the ratio for the Sperm Whale is low for a toothed whale (though larger than for most baleen whales). For the Bottlenose Dolphin the ratio is around 1.0% (close to that of chimpanzees), which implies a greater level of intelligence, and it is certainly true that studies on captive dolphins show that in terms of their response to spoken commands they exceed all non-primates.

## Toothed whales

Species of the suborder Odontoceti, the toothed whales, make up most cetaceans. Most are surface feeders and show similar countershading to that of pinnipeds.

### Orca (Killer Whale) *Orcinus orca*

The largest member of the dolphin family that is found to the ice edge in both the North Atlantic and North Pacific. Also found in all Earth's oceans.

### White-beaked Dolphin *Lagenorhynchus albirostris*

Found in the North Atlantic, it is the most northerly of the smaller dolphins, reaching the southern shores of Svalbard and the Barents Sea, though more southerly in the colder western Atlantic where it is rarely seen north of Labrador or south-west Greenland. Despite the northerly range the animals are poor ice travellers and many die after becoming entrapped in the pack ice.

The Long-finned Pilot Whale (*Globicephala melas*), the second largest dolphin after the Orca is found in the North Atlantic, as far north as Iceland and southern Greenland.

### Beluga *Delphinapterus leucas*

One of the species in the family Monodontidae, both of which are true Arctic dwellers. Found, rarely, in the Greenland Sea (and perhaps absent altogether), near Svalbard and east from there along the Russian Arctic coast to the Chukchi Sea. Also found in the Bering Sea and the Arctic waters of North America to eastern Greenland, and in Hudson Bay.

Orca pod, Lofoten Islands, Norway.

Young Beluga. Adult Beluga are snowy white, but youngsters are silver-grey, southern Hudson Bay.

### Narwhal *Monodon monoceros*

This whale, an ice-going species, has no dorsal fin. Narwhal skin (*muktuq*) is considered a great delicacy by the Inuit and is eaten as soon as a hunted animal is landed. It tastes, vaguely, of hazelnut.

Found, rarely from west Greenland to the New Siberian Islands, more commonly in the Canadian Arctic, from Banks Island to east Greenland.

### Northern Bottlenose Whale *Hyperoodon ampullatus*

Found in the north Atlantic to the ice edge. Rare in the Barents Sea and Hudson Strait.

### Sperm Whale *Physeter macrocephalus*

The largest toothed whale, with a legendary place in both the history of whaling, because of the ferocity of some whales towards their pursuers, and in literature as a result of Herman Melville's classic novel *Moby Dick*. It exhibits the most extreme sexual dimorphism of any cetacean, with males weighing up to three times more than the female. Found in all oceans, north and south, to the ice edge. However, it is rare in the far north.

## The Unicorn of the sea

Narwhal have only two teeth, both in the upper jaw. In males, the left tooth pushes through the lip to form a tusk. In some males, the right tooth also erupts, though these double-tusked narwhals are very rare. The tusk begins to grow when the animal is 2–3 years old (occasionally at one year) and grows continuously. Very old males may have tusks of 3m weighing 10kg. In females, the teeth often do not erupt during the entire life of the animal, though tusked females have been seen. The Narwhal's tusk is claimed to be the source of tales of the Unicorn. This cannot be conclusively proved, but it does appear that the true source of the tusks was suppressed to promote the Unicorn legend (and hence the price of the tusks). Despite occasional nonsense written on the subject, the tusk is not used to skewer fish. The tusk is a secondary sexual characteristic. It is sometimes used as a weapon: scarred males and broken tusks are seen. The tusks are also sometimes laid across the back of another animal in what appears to be a gentle, tactile gesture. The whale's curious name is from the Scandinavian *nár hvalr*, corpse whale, because the skin colour resembles that of a dead human.

Sperm Whale, Lofoten Islands, Norway.

## Ambergris

The Sperm Whale's main food is squid – it feeds on the Giant Squid, another creature of legend. The squids' horny beaks are indigestible and irritate the stomachs of the whale. To ease the irritation the whale secretes a resinous substance that coats the beaks. Coagulated masses of this material, normally excreted, were occasionally vomited out by harpooned whales. The masses floated and were collected. Known as ambergris, the masses were highly prized for their use in the perfume industry. It would be interesting to observe the faces of users when they discovered that one ingredient of their expensive perfume could be termed 'whale faeces'.

### Baleen whales

The huge baleen plates of the second, smaller suborder of cetaceans is the basis of their scientific name – the Mysticeti, or 'moustached whales'. The Mysticeti represent only about 10% of all whale species, though in Arctic waters the percentage rises to about 50%, the baleen whales trawling the rich polar waters for plankton and other prey. The Mysticeti comprises three families, the Balaenopteridae or rorquals, the Eschrictidae, which contains one species, the Grey Whale, and the Balaenidae or right whales.

### Blue Whale *Balaenoptera musculus*

The largest animal ever known – larger than any dinosaur. Found in all oceans, it has been seen in the Bering Sea, in Baffin Bay, and in the Barents Sea.

### Fin Whale *Balaenoptera physalus*

The second largest whale that is found in all oceans. It has a similar distribution to the Blue Whale, but tends to be more southerly.

### Minke Whale *Balaenoptera acutorostrata*

The smallest of the Arctic baleen whales. Found in the north Atlantic as far north as Davis Strait, and occasionally reaching Svalbard and the Barents Sea. In the north Pacific they are found throughout the Bering Sea and sometimes in the Chukchi Sea.

### Humpback Whale *Megaptera novaeangliae*

The whale's falcate dorsal fin is small and mounted on a hump, usually more easily visible in front of the fin. It is

Minke Whales in fog-shrouded waters off Spitsbergen's southern coast.

## Giants of the ocean

Recent analysis of whale fossils has suggested that the reason the baleen whales are so big is the growth of northern ice sheets some 4.5 million years ago. The ice altered ocean currents, which, in turn, altered nutrition levels allowing the populations of some species, for instance, krill, to multiply dramatically. It is suggested that this allowed the baleen whales to scoop up food more readily and their size increased so that they became the largest creatures ever to have existed on Earth.

Blue Whales of the Southern Ocean are larger than their northern cousins, with lengths to 33m and weights up to 190 tonnes, though the real giants that were encountered during the early days of rorqual hunting do not seem to have re-occurred, even though the species is now fully protected. Overhunting has drastically reduced the population of Blue Whales, though numbers are now slowly recovering. The current population is estimated at about 11,000 animals, about 15% of the pre-hunting population.

Interestingly, the second part of the species' scientific name means 'little mouse' which, given the whale's size, could be evidence of humour on the part of Linnaeus, or (and perhaps more likely) a mistranslation of 'muscle'.

Breaching Humpbacks in the Lynn Canal, Alaska.

this hump, which is prominent when the whale arches its back before diving, that gives the species its common name. As with Grey Whales, Humpback calves are hunted by Orcas, females occasionally turning over and lifting their calves out of the sea on their bellies to prevent them being attacked. A recent study in the Southern Ocean also found that while adult Humpbacks are famous for their 'singing' calls, which can travel great distances, the calves 'whisper', or may even soundlessly nudge their mothers when hungry so as to avoid giving their position away. In another recent study, it was found that adult Humpbacks also attempt to prevent Orca attacks on other species as well as their own calves, once being observed to take a seal thrown off an ice flow by Orcas onto its belly and cradling it in its fins. Experts advise against seeing this anthropomorphically as altruistic behaviour, suggesting that any attempt to persuade the Orcas to go and hunt elsewhere is of benefit to a Humpback female.

It is found in all the world's oceans. Seen in the Atlantic as far north as Svalbard, but rarely east of the Barents Sea. Also found in the Bering Sea.

## Grey Whale

The Grey Whale (*Eschrichtius robustus*) is the only representative of its family; once thought to be basal to other baleen whales, but DNA analysis has suggested a more recent evolution. The species possesses characteristics of both the rorquals and the Right Whales. Because of their feeding method, ploughing through bottom sediments, the whales stir up clouds of silt and food particles, so feeding whales are often located by noting the accompanying flocks of gulls and other seabirds.

Found in the North Pacific, as far north as the Chukchi and Beaufort seas, and in the Sea of Okhotsk. There was formerly an Atlantic population, the animals feeding off Iceland and Greenland and migrating as far south as the Bay of Biscay, but this population became extinct in the 18th century, probably due to overhunting by Basque whalers.

### Migrating Whales

Grey Whales migrate between the Bering Sea and the lagoons of Baja California and the Gulf of California and back each year, a distance of 12,000km or more. The whales move continuously on migration, day and night, and apparently without sleeping. Mating and birth take place in the same lagoons, implying a 12-month gestation. The single calf (5m long and weighing 650–700kg) accompanies its mother on the long migration north. On this journey the calves are vulnerable to Orca attack, the Orcas attempting to isolate the calf from its mother, then drowning it by lying on its back. Often the Orcas will eat only the tongue. Calves that survive the migration are weaned at 6–7 months and independent before undertaking the long migration south. Until recently the Grey Whale migration was considered the longest by any mammal, but in early 2007 a Humpback Whale travelling from the Antarctic Peninsula to waters off Costa Rica and Panama was found to have swum over 8,000km, implying a round trip of over 16,000km.

The Sei Whale (*Balaenoptera borealis*) – the name is pronounced 'sigh' rather than 'say' and derives from the Norwegian for pollock, which was believed to be a principal prey of the species – is found in all oceans. It tends to be more southerly than either Blue or Fin Whales, rarely being seen north of Jan Mayen, Labrador, south-east Greenland or the Aleutians.

## Right whales

Right whales acquired their name from the early whalers – these were the right (i.e. correct) whales to kill: they were slow and so could be easily overhauled by a rowed boat; they were passive, and so did not turn every killing into a battle in which the whalers were at equal risk to the whale; they floated when dead, making them easier to transport to ships or shore; and they yielded huge amounts of baleen and oil. The mass of blubber, which yielded the oil, explains the reason dead whales floated. This combination resulted in the three species of large right whales (two Arctic and one Antarctic species – the Southern Right Whale was once thought to be conspecific with the Northern Right, but is now considered a separate species) being hunted almost to extinction. Despite full protection (though the Bowhead is still hunted by native peoples in both Alaska and Siberia) the populations do not appear to be recovering, possibly due to inbreeding.

There are some key anatomical differences between right whales and rorquals. The right whales have an arched nostrum, which results in a bow-shaped mouth rather than the straight mouth of the rorquals – their baleen plates are consequently much longer. Right whales have no throat pleats, and so feed by sieving rather than gulping. Also, there is no dorsal fin.

**Bowhead Whale** *Balaena mysticetus*

Originally known as the Greenland Right Whale, this is the most Arctic of all whales. Bowheads can break through ice of considerable thickness, up to 60cm thick. They are circumpolar, though the population in the Atlantic Arctic is small.

## The age of Bowheads

Some Bowheads taken by native hunters in recent years have been found to have ancient harpoon heads embedded in them. These heads have been dated to well over 100 years old. This discovery encouraged scientists to investigate other methods for determining the life expectancy of Bowheads. One determination, analysing the ratio of two stereoscopic isomers of aspartic acid, a constituent of the lens protein of the Bowhead's eye, indicates ages of up to 200 years. That means that the whales rival the Giant Tortoises as the longest-lived animals.

A rare photograph of a Bowhead calf. The calf was asleep or resting at the floe edge in Foxe Basin, Canada. The lower white jaw and flipper can be clearly seen. After a short period of observation, hydrophones picked up the call of a mother whale and the calf submerged and disappeared.

Fin Whale blow, Jan Mayen Island.

The Northern Right Whale, one of the world's rarest cetaceans, is not a true Arctic dweller. *Eubalaena glacialis* is extremely rare in the area between Labrador and Maine, while a second species as subspecies, *Eubalaena japonica* is found from Japan to Kamchatka and in the Sea of Okhotsk, and, rarely, near the Aleutians and southern Alaska. The population of *Eubalaena glacialis* is thought to number no more than 300 animals and is classified as Critically Endangered by the IUCN: an eastern Atlantic population is now believed to be extinct due to hunting. In June 2017 a carcass was found floating off Canada's coast, with another 9 carcasses being found floating, or being washed ashore on Newfoundland in the weeks to early August. These deaths represent up to 3% of the population and are double the probable annual birth rate. The deaths are unexplained, but may have been due to ship collisions or crab net entanglement. Some scientists believe the population may now be below the viability limit, with the species doomed to extinction.

The population of *Eubalaena japonica* is believed to be 1000, perhaps a little higher, the species being classified as Endangered. However, the same concern over the long-term viability of the species has been expressed as noted above for the Atlantic Northern Right.

Kulusuk, south-east Greenland.

# 17 A traveller's guide to the Arctic

Arctic travel is one of the growth industries of international tourism. In a sense that is sad as there is an underlying pessimism about much of it – visit now, a last chance to see the Arctic's wonders. A case could also be made that tourism itself represents a threat to the area. Air travel is a significant producer of $CO_2$, and the effect on wildlife should not be understated. Research on penguins in Antarctica has shown that the arrival of visitors at breeding colonies often stresses the birds, even if the close approach that is allowed appears to suggest otherwise. Presumably the arrival of humans at Arctic bird breeding sites is equally stressful. Yet in those, and other, threats lie the hope that those who visit will return home entranced and willing to make efforts to ensure the region's survival, celebrating its beauty and so encouraging others to do likewise.

It is with that hope that this section explores the options available for travellers. The fact that tourism is on the rise means that options are also increasing, so as with all guides this one will be out of date on a timescale as likely measured in months as years. With that in mind only very brief details are given.

## Iceland

Iceland lies between 63.5°N and 66.5°N, just below the Arctic Circle, which passes through the island of Grimsey off the northern coast. Yet for many, the fact that the island's history is tied to the exploration of the Arctic, that some Arctic species may be seen there, and that travel and accommodation is distinctly easier and more 'civilised' makes it an ideal destination. The

The 125m Haifoss in southern Iceland is the country's highest, and is a double falls, but is not the easiest to reach.

Landmannalaugar, Iceland.

weather is not only variable, but variable on a short timescale, and temperatures are invariably warmer than more northern destinations.

Most visitors will be tempted by the 'Golden Triangle' of Þingvellir, site of the world's oldest parliament (the *Althing* – General Assembly, set up in 930, though it had little to do with democracy as it is now understood), Strokkur, as the sole remaining active geyser at Geysir is called, and the beautiful Gullfoss waterfall. At Þingvellir the visitor can also walk a path between the American and Eurasian landmasses.

Þingvellir and Geysir are aspects of Iceland's volcanic landscapes, but there are others, the island having a textbook array of cinder cones, explosion craters, fissure eruptions, spatter cones and shield volcanoes. The resultant lavas offer superb examples of ropy and scoria lavas. There are also classic volcanic features, such as the bubbling mud pits of Krisuvik and Hveragerði, the thermal springs that feed the Blue Lagoon close to the international airport of Keflavik, and the basaltic columns behind the Svartifoss waterfall in the Skaftafell National Park. Volcanic distortions of the landscape have also created some of the finest waterfalls in Europe, not

only Gullfoss, but the 'pony-tail' fall of Seljalandsfoss, and Skógafoss, all in the south of the island, and the northern falls of Dettifoss and Goðafoss. Iceland also has several glacial ice caps, one of which, Vatnajökull, is by far Europe's biggest glacier.

### Wildlife

Iceland has only one indigenous terrestrial mammal, the Arctic Fox. American Mink have also become established due to escapes from fur farms. Brown and Black Rats, House and Wood Mice have been accidentally introduced; Reindeer have been deliberately introduced, though numbers remain low. Grey and Harbour Seals live in the waters around the island and many species of whales are found close to the island. But the main wildlife interest is its avifauna. Iceland is the only place in Europe where Barrow's Goldeneye and Harlequin Duck breed. Mývatn – midge lake – is a good place for these and other wildfowl. The island also has a small population of grey-morph Gyrfalcons, and is regularly visited by migrating Greenlandic white Gyrs. The sea cliffs of Látrabjerg are one of the largest seabird nesting areas in the North Atlantic, and is particularly good for auks.

## Jan Mayen

Jan Mayen lies at 71°N, 8°30'W, about 600km north-west of Iceland. The climate is kept relatively mild by the ocean, but the island is often cloud- or mist-shrouded because of the high relative humidity, with some years seeing as few as five clear days. Jan Mayen is also frequently windswept. The island is dominated by the 2277m volcano of Beerenberg. Jan Mayen is composed of volcanic lavas the sombre colours of these, from black to red-brown are a complete contrast to the vivid green of mosses and liverworts which have colonised what would be assumed to be a sterile landscape. The island flora also includes about 70 flowering plants, these including members of the saxifrages. The island was famed for its blue Arctic Foxes, but these were hunted to extinction. Northern Fulmars occupy the island throughout the year, these being joined by various waders, auks, skuas and gulls that breed there.

## Scandinavia

The presence of the North Atlantic Drift eases the Arctic northward in Scandinavia, particularly in Norway. Consequently, agriculture is possible north of the Arctic Circle though in winter the region becomes much more Arctic, the combination of the long northern night, high snowfall and the rugged terrain (of much of northern Sweden in particular) combining to provide a true polar experience.

Visitors interested in the Sámi, true Arctic dwellers, will head for the numerous museums and cultural centres near places such as Kautokeino in Norway, Jokkmokk in Sweden and Rovaniemi in Finland. One of the best museums is at Inari, in Finland (the *Siida*, the name is a Sámi local community, a group of families occupying a defined area of land). Northern Scandinavia has small and endangered populations of Wolves and Brown Bears, and other large mammals such as Elk, Lynx and Wolverine, as well as numerous smaller animals. It also has excellent birdlife. The highlights, as would be expected, are the various National Parks and wilderness areas, but Varangerfjord in northern Norway though not a park, is one of the best spots for the birdwatcher for both its breeding and migratory species.

One site which delights many visitors is the 'tri-point' near Kilpisjärvi, the place where Norway, Sweden and Finland meet. There is another such site, where Norway, Finland and Russia meet, close to the southern tip of the Øvre Pasvik National Park, but that is much less accessible.

Because of Jan Mayen's position and the height of Beerenberg, the top of the volcano is usually shrouded in cloud, appearing only a handful of times each year. From the ice cap summit, tidewater glaciers flow down to calve icebergs into the north Atlantic.

Dog-sledging in Norway's Øvre Dividal National Park.

## Bear Island

Bear Island lies at 74°30'N between the northern coast of Norway and Svalbard. An alternative name, Island of Mists, reflects the prevalence of cloud cover. It is occasionally visited by Arctic cruise ships. It is estimated that more than 1,000,000 seabirds are present during the breeding period, though a long-term decrease in the numbers of guillemots has been identified, almost certainly due to consistent over-fishing of local waters. The island is also a staging post for the Svalbard populations of Barnacle, Brent and Pink-footed Geese on their autumn migration.

## Svalbard

Svalbard is the name given to the archipelago of islands lying between 74°N and 81°N to the north of Scandinavia. The archipelago includes Spitsbergen, the largest and most widely known of the islands, and the name frequently used to cover the whole island group. The archipelago has Norwegian sovereignty, and is administered by a Norwegian governor. However, the Svalbard Treaty of 1920 affords equal rights of access and residence, and commercial exploitation (by fishing and mining) to both the initial and subsequent signatories. Article Nine of the Treaty declares that the archipelago must not be used for 'warlike purposes' but there is little doubt that the strategic importance of the islands which explains the Soviet, now Russian and Norwegian presence: coal mining on Svalbard is hardly economic, but each nation is there because the other one is there. The islands are also a superb research base, being far to the north yet comfortably accessible.

Scheduled flights link Norway to Longyearbyen on Spitsbergen. In summer ship tours of the archipelago offer visits to historically interesting sites such as the remains of the whaling station on Amsterdamøya, and scenic highlights such as Magdalenafjorden. The trips also offer a good chance to see Polar Bear, Arctic Fox, Svalbard Reindeer (a smaller cousin on the mainland species) and Walrus as well, perhaps, as whales. Birdlife includes gulls (the lucky might see the wonderful, but elusive Ivory Gull) and auks. Recently winter touring on snow scooters has been restricted, but is still scenically worthwhile.

Bear Island is of interest both to scientists, who can study the rich geology, and birds who populate the steep cliffs. The photograph is of the southern end of the island. Off-shore is the Sylen rock needle.

## Russia

The Russian Arctic extends from its border with Finland in the west to the Bering Sea in the east, just a few kilometres from Alaska. Yet for much of this vast distance the mainland does not lie in the Arctic as defined here, the deep cold of the Siberian winter being moderated by warm summers. Consequently, in general only a narrow section of the mainland is truly Arctic, though all of the country's northern islands are. Siberia is a wonderful land but it has had a bad press, the sad history of exploitation of its native peoples and animals having given way to exploitation of the landscape itself, when it was used as a repository for criminals and dissidents. Today, exploitation of the area's mineral wealth is threatening the landscape again, but now in a very real sense.

Despite *perestroika*, travel within Russia is still difficult, the invitations and permits required for journeys making genuinely independent travel difficult, though the persistent will manage to reach most worthwhile places. Purpose-built trips are also possible as new companies shake off the rigidity that characterised Intourist. The better Russian Arctic destinations, the islands, still require a visitor to join an organised cruise, but these cruises – through the North-East Passage – are becoming increasingly frequent (though expensive). There are nuisances that can try the patience of the traveller, delays without explanation and changes likewise, and occasionally the bureaucracy can be exceedingly trying. I once spent an uncomfortable 12 hours at the entrance to Petropavlosk Bay, Kamchatka *en route* from Starichkov Island, with limited food, water and shelter on a nowhere-nearly big enough boat when the bay was arbitrarily closed by the naval base commander.

Because of the ever-changing nature of Russian travel it is not considered worthwhile to go into details, but a few places are worthy of consideration. The Kola Peninsula is among the easier places to reach in northern Russia, as the infrastructure is reasonable. There are several *Zapovedniks* (Nature Preserves) with numerous plants and birdlife. The Yamal Peninsula is also relatively easy to reach and it is possible to organise a trip to and with the Nenet Reindeer herders. The Yamal Peninsula, and the nearby Gydan Peninsula are breeding areas for Red-breasted Geese.

The Taimyr Peninsula is much more difficult to reach and explore, though exploitation for oil and gas may make it easier in the future. It has Red-breasted Geese and other, rare, northern birds. It also has the largest remaining Eurasian population of wild Reindeer. The eastern river deltas are superb for birdlife. Easiest to reach is the Lena where the visitor may see Ross' Gull amongst other rarities.

Chukotka is often visited by cruise ships, allowing visitors to meet Yupik sea mammal hunters. To the south, Kamchatka is easier to reach and offers the chance to meet the Koryak people and to visit spectacular volcanic scenery, e.g. the Valley of the Geysers and the Uzon Caldera, as well as seeing Steller's Sea Eagle and other rare and localised species – Snow Sheep, Largha Seal, Black-billed Capercaillie, Far Eastern Curlew, Slaty-backed Gull, Spectacled Guillemot, Siberian Rubythroat, and Middendorf's Grasshopper Warbler. Trips can also be made to the historically interesting Commander Islands, though the weather there means a great deal of flexibility is required, as is a willingness to suffer long delays (outward and return). The islands' breeding birds include many Bering Sea auks plus Red-legged Kittiwakes, Red-faced Cormorants and a subspecies of either Arctic or Asian Rosy-finch.

Of Russia's Arctic islands, only Franz Josef Land and, occasionally Wrangel Island, are on cruise itineraries

Zhupanovsky Volcano rises above the morning mist in this photograph from the Zhupanova River, Kamchatka, Russia.

Interesting glacial architecture on Jackson Island, Franz Josef Land, Russia. In the foreground is the sea beyond which a narrow strip of land with two prominent hummocks borders a (not visible) inlet of the sea. The blue glacial face beyond calves icebergs into the sea. The white band above is the glacier, two arms of which encompass a mountain mass.

other than NE Passage transits. The Franz Josef archipelago lies between 80°N and 82°N, from 45°E to 65°E and comprises almost 200 islands, the majority very small. Historically the islands are fascinating, and the wildlife is excellent with Polar Bears, many seal species, and excellent bird life. Wrangel escaped glaciation during the last Ice Age (when it acted as a refuge), and having formed part of Beringia, the island has one of the most extensive and interesting floras of any of Russia's Arctic islands. More than 380 plants have been identified to date including several endemics. Excavations have revealed the existence of a Palaeo-Eskimo human culture. The people hunted sea mammals, but may also have been at least partly responsible for the extinction of dwarf Mammoths, which are known to have inhabited Wrangel until about 4,000BCE, up to 6,000 years after Mammoth disappeared from the rest of Siberia. Wrangel is famous for its Snowy Owls, and has an endemic Lemming.

## Alaska

With an area of more than 1,500,000km² Alaska represents about 16% of the total area of the United States. Travel to and around the state is relatively easy, but some wilderness areas are difficult to reach. The state is historically very interesting – the gold rush town of Nome for instance, and the Russian churches at various sites – and has several worthwhile wildlife hotspots.

To the north, Barrow, is easily reached. It is famous for Snowy Owls – I was once able to combine enjoyment of a restaurant meal with photography of a Snowy

### The sale of Alaska

The US purchase of Alaska – for $7.2 million, a price that, famously, worked out at 2 cents per acre – was not seen, at the time, as the bargain it is now considered. The Russians wanted the United States to have Alaska as they feared British ownership, Britain at the time being the predominant European power; if Canada expanded west to the Bering Sea, Russia would be 'surrounded' by Britain. The United States also feared British expansion to its north, but despite that the enthusiasm of William Seward, Secretary of State in 1867, for the purchase was not shared by most Americans. After the purchase, Congress, which had been reluctantly convinced, lost interest and the state remained a lawless place for two decades until the discovery of gold.

Owl perched on the tundra, with sea ice forming a background. The local area is the breeding ground of both Steller's and Spectacled Eiders, as well as other wildfowl and a fine collection of northern waders. Breeding birds from Russia are also seen frequently, these including Curlew Sandpiper, Sharp-tailed Sandpiper and Red-necked Stint.

The Dalton Highway to Prudhoe Bay is also now open to visitors, though care is needed by drivers as the massive trucks speeding along this 'haul road' take no prisoners. The road offers limited facilities along the way so it is best to be self-sufficient. The journey crosses the Arctic Circle, goes over the Brooks Range, and edges the Gates of the Arctic National Park to the west and the Arctic National Wildlife Refuge to the east, though in each case a long walk or a chartered aircraft is required for access.

Further south the Denali National Park is scenically magnificent – with views of Mount McKinley – and offers easy access to significant wildlife opportunities. The Seward Peninsula to the west is home to the Bristle-thighed Curlew, Aleutian Terns, New World sparrows and warblers. Nome, the largest town, is also the jump off point for trips to St Lawrence Island.

Southern Alaska has so many possibilities for the independent traveller that to mention them all is beyond the scope of this book, but most visitors will spend some time in Anchorage, from where flights leave for the Pribilof and Aleutian Islands.

## Canada

Canada is the second largest country on Earth, and its northern islands represent a large fraction of the land area of the Arctic. Because of the vagaries of the world's climate, the 'Arctic' also extends much further south into mainland Canada than it does into mainland Russia. In general travel to the Canadian Arctic is straightforward, and many of the destinations reachable by scheduled flights are interesting, both historically and for the wildlife. There are also frequent cruise ships transiting the North-West Passage which land passengers at most of these sites.

The Yukon is famous for a gold rush, and Dawson City is a delightful place to enjoy its history. Yukon has several National Parks, though access to these is not straightforward. Visitors could also travel the Dempster

Old gold dredger, Nome, Alaska.

The twin summits of Mount McKinley, Alaska. The photograph was taken from the site where, at the time of shooting, the new Eielson Visitor Centre was being constructed.

Summer on Alaska's Kenai Peninsula. Pools such as this are a favourite haunt of Moose.

An aerial shot of Glacierfjord, Axel Heiberg Island, Nunavut, Canada.

Highway to the mouth of the Mackenzie River, though wildlife spotting is limited for much of the journey as it threads through dense conifer forest. North-West Territories, now reduced in size after the creation of Nunavut, also has several National Parks, though again these are not easy to access: that on Banks Island may be briefly visited by NW Passage cruise ships.

Nunavut (meaning 'our land' in Inuktitut, the language of the Inuit) represents the bulk of the Canadian Arctic, including the major Arctic islands of Baffin and Ellesmere, as well as a large fraction of Victoria Island. Consequently, it also includes many sites important to the history of the search for the North-West Passage, such as Beechey Island with its graves of three of the crewmen of Franklin's final expedition and Gjøahaven where Amundsen overwintered during the first transit of the Passage.

While Passage cruise ships visit the southern islands, such trips rarely go north. However, there are many local outfitters who will help the visitor used to expedition-style trips to visit places such as northern Baffin and Bylot. The entirely self-sufficient northern traveller can also charter an aircraft and land on Ellesmere and other remote spots. But it must be emphasised that such trips should be undertaken with appropriate caution. Personal experience includes bad weather preventing charter pick-ups which could have caused severe problems if contingency plans had not been made.

The rewards of trying are considerable though with white wolves and white falcons, narwhal and bowhead in the High Arctic.

To the south, in Manitoba, the town of Churchill is one of the best and most easily accessible wildlife destinations in the Canadian Arctic. In spring and early summer several thousand Beluga congregate in the Churchill River to moult, while during the autumn Polar Bears, which have been marooned on the southern shore of Hudson Bay, mass to await the winter freeze. These events have led to the town styling itself as 'Polar Bear Capital of the World' and the 'Beluga Capital of the World' at the relevant times. Churchill is also famous for its remarkable birdlife. From the town, a series of roads head off into habitat that varies from shoreline to tundra (some dry, but mostly wet) and boreal forest. The numbers of species that might be seen is staggering, and the area is renowned for rarities.

Close to Churchill, the Wapusk National Park has recently been set up to protect the bears of this part of southern Hudson Bay. The park encompasses a coastal strip from west of Cape Churchill to the Nelson River, and includes one of the most important bear-denning sites in the world. Access to the Park is controlled, but there are tour operators in Churchill that offer visits. As well as the wildlife opportunities, the town of Churchill is also interesting historically. Across the river mouth from Cape Merry is Prince of Wales Fort, built by the Hudson's Bay Company in the 18th century. Further down river on the western side is Sloop Cove, the natural dry dock where the Company's ships were repaired.

In Ontario James Bay is home to the world's most southerly population of Polar Bears. The bay can be reached by train from Toronto via Cochrane (the train being called – and this will come as a surprise to no one – the Polar Bear Express), which reaches Moosonee. From the town, a short boat trip, often by freighter canoe, reaches the historically interesting Moose Factory, a Hudson's Bay Company establishment.

On the eastern side of Hudson Bay lies the final section of the mainland Canadian Arctic. That part of Quebec lying above the 55th parallel is known as Nunavik and includes Inuit coastal villages scattered along the shores of the Ungava Peninsula. The larger Nunavik settlements are reached by scheduled flights.

## Greenland

To the indigenous Greenlanders this is Kalaallit Nunaat, the Land of the Kalaallit. For the traveller, Greenland offers an odd mix of almost pure Inuit culture and modern western ways. In the north-west the 'Polar Eskimo' culture encountered by John Ross continues almost untouched, while further south profitable industry has brought modernity. On the east coast, Greenlanders still hunt Polar Bear from dog sleds, but use the income from selling the skins to North America

Rugged, but beautiful, landscape between Kulusuk and Ammassalik, east Greenland.

to buy the latest electronic gizmos. Modern weapons and medicines (aiding a rise in population) have led to a worrying destruction of island wildlife, particularly on the more heavily populated west coast. Yet the island remains in large part a magnificent, and truly Arctic, wilderness, one to be savoured.

With such a vast area, no in-depth consideration is possible. Outside the villages and townships there are no roads (and few tracks) so walking is the only way to explore. In winter snow scooters can be used, though hire options are limited and expensive. In winter the wildlife is also much scarcer. Getting around within the country requires air or sea travel. For the more experienced and determined traveller charter flights are an option. These are available from Iceland for the east coast, and Canada for the west. However, such flights are difficult to arrange, as Greenland is a foreign country to the pilots of both Canada and Iceland and

## Greenland statistics

The largest Arctic island, Greenland has a surface area of 2,167,000km² including the offshore islands. For comparison, the next largest Arctic island is Baffin, with an area of 476,070km². Greenland also has the highest Arctic mountains (if Mount McKinley and the other Denali peaks are assumed to be non-Arctic). The highest peak is Gunnbjörn Fjeld at 3,700m (at 68°54'N, 29°48'W). Mount Forel, 3360m, at 67°00'N, 37°00'W is the only other Arctic mountain rising to more than 3,000m.

so international regulations apply, and may limit the willingness of some operators. If chartering is not an option, Greenland's main towns are still worth visiting as the scenery close to them is excellent, though wildlife may be in short supply.

The Hart Glacier outlet from the Inland Ice, near Thule, north-west Greenland.

Spider-shaped rock/ice formation, Dickson Land, Svalbard.

On the west coast Kangerlussuaq is reached by scheduled flights from Copenhagen, and has air links to the major towns (including the capital Nuuk). Narsarsuaq is currently also served by seasonal flights from Iceland and Denmark (but please check, history suggests the companies operating these flights may be forced to limit them). The town has very limited facilities, but is one of the few places where the Inland Ice can be visited in a (long) day's walk. The walk passes the site of a now-demolished US military hospital and goes through the beautiful Valley of Flowers. From Narsarsuaq boats cross to Brattahlíð where the ruins of Eirik the Red's farm can be seen. At Qaqortoq a half-day trip can be made to the Norse church at Hvalsø, the most complete structure of the period in Greenland.

Kangerlussuaq has better facilities than Narsarsuaq. From it visitors can join a trip to the inland ice (rather too far for a day's walk) with a good chance of spotting Musk Ox. Flights from the town reach Ilulissat, from where the iceberg-choked fjord fed by the Jakobshavn Glacier is a short walk away. Ilulissat can also be used as a base for boat trips around Disko Bay. Northward flights continue to Uummannaq, with its excellent seabird cliffs, and Qannaq. Qannaq, in north-west Greenland, is the most culturally unspoilt part of Greenland.

There are far fewer townships on the east coast, a more rugged and inhospitable land. Kulusuk is reached by scheduled flights from Reykjavik (and also has cross-ice flights to and from Nuuk). Kulusuk is a beautifully positioned and very picturesque town, and has onward flights to Ammassalik, one of the few places from which walking tours can be made. Further north Constable Point is reached by seasonal flights from Reykjavik and Akureyri. Cruise ships heading for Scoresbysund, the longest fjord on Earth at more than 300km, and the North-East National Park also visit. The Park is, at 700,000 km², the largest national park in the world. It is visually magnificent, and home to Polar Bear, Musk Ox, Wolf, Arctic Hare and Northern Collared Lemming, all the North Atlantic pinnipeds, including Walrus, numerous waterfowl include Barnacle Geese, and to Gyrfalcons.

Bowhead Whale rib, Spitsbergen, Svalbard.

# 18 The fragile Arctic

Earth's ecosystems are finely balanced. Changes to a habitat or a change to the number or diversity of species, by the introduction of an 'alien' organism or the elimination of an existing one, can have profound effects. The Arctic is no exception to this rule, but it is a special case for several reasons. Firstly, the Arctic ecosystem is young, having developed only since the retreat of the ice at the end of the last Ice Age: such systems can be especially unstable. Secondly, the Arctic is an unforgiving and hostile environment, one in which climatic effects can be sudden and devastating, and recovery times can be lengthy. The Snow Goose population of Wrangel represents a good example. During the late 1960s and early 1970s a series of late springs prevented the birds from raising chicks: the population crashed by over 90%. In more southerly latitudes a late spring means that birds may raise just one clutch rather than two. If the following spring is early the population can soon recover. In the Arctic a late spring means no clutch, and even if the spring is early two clutches are very rare. The habitat itself is also slow to recover. In southerly latitudes, a ploughed field will be seeded, the crop will ripen and be harvested and by the following spring the field will look much the same again. In the Arctic, soil denuded of its plant life may stay barren for years, the tracks of vehicles on the tundra may remain for decades.

Because Arctic species are continuously stressed by their environment, any additional stresses imposed by external, man-made changes can cause major, and rapid, disruption. In this chapter we explore the current threats to the Arctic and its species. But we begin with a history of the direct exploitation of Arctic animals, continuing with the exploitation of mineral resources and the subtle effects of pollution on the Arctic ecosystem.

The 'Boneyard' at Gambell, St Lawrence Island, Alaska. The yard is actually a centuries-old midden into which the remains of hunted sea mammals were thrown. The inhabitants of Gambell, who still hunt sea animals, now excavate the yard. It is a treasure trove of old harpoon heads, carvings and other finds. These are sold to mainland collectors to supplement the villagers' meagre incomes.

## The exploitation of Arctic animals by native peoples

Although the native peoples of the Eurasian mainland exploited the terrestrial animals of the Arctic, hunting Reindeer for food and clothing and other species primarily for food, only the Pomores of the White Sea area and the Bering Sea Yuppiat ventured into the far north to hunt seals, Walrus and bears. But although these peoples clearly had Arctic survival skills, it is the Inuit of Canada and north-western Greenland who are the true Arctic dwellers, and it is instructive to understand the way in which they used the northern fauna to survive. The Inuit used the skins of Walrus for tents. Walrus hide also made a strong rope, so strong that when the British navy met Inuit for the first time they were amazed to discover that the ropes of these supposedly inferior people, made by braiding strips of smoked Walrus hide, were stronger than their own. Sealskin was used as clothing because it was water-repellant: Caribou skin was a good all-purpose clothing material; Polar Bear skins were warm, as were those of the Arctic Fox. Interestingly, the use of Polar Bear fur was mostly confined to Greenlandic Inuit as in Canada seal and Caribou were sufficiently abundant to provide an easier alternative. Qiviut, the fine underwool of the Musk Ox, is one of the lightest and warmest wools known, while the longer guard hairs of the ox's coat were made into caps that were claimed to be an effective mosquito deterrent (though personal observation suggests that the caps are no better than any other protection against the pest). The Inuit also used the skins of birds to produce warm undergarments. Several of the 15th century mummies discovered at Qilakitsoq, west Greenland, wore inner parkas made of the skins of auks, divers and other birds. As the Greenland Inuit had no access to Caribou, the use of birds to supplement the skins of seal and Polar Bear was essential.

The Inuit ate the meat of all these and other animals. Occasionally they ate meat raw, but the meat of the Polar Bear was always cooked as they knew it often contained parasites. Polar Bear fat is a good source of Vitamin A, but the liver was avoided (and not even fed to dogs) as the concentration of Vitamin A in it is lethal. Arctic Char are a good source of Vitamin B, while the soft bones of the fish are rich in calcium. The blubber of most Arctic animals is a rich source of omega-3 fatty

The Arctic is occasionally no place for the squeamish as this discarded group of heads from skinned seals indicates. But is it really so much worse than would be seen (but is more carefully hidden) at a slaughterhouse?

A dead walrus hauled ashore by Yuppiat hunters in Chukotka.

acids. The skin of Beluga and, especially, of Narwhal are rich in Vitamin C. Bowhead Whales provided meat and blubber, the latter prized as a cooking and lighting oil as it does not leave the sooty black residue of seal oil. The baleen of the Bowhead was used to make kayak frames. The uses other parts of prey species were put to reflect the understanding the native peoples acquired both of their prey and their own environment. For example, the Inuit realised that the temperature at which the fats in a Caribou's leg went solid reduced as distance from the body increased, so they chose fats from the lower leg for greasing bow strings and other tasks that they needed to work at very low temperatures.

The achievements of other Arctic peoples should not, however, be ignored. The skills of the Asian Arctic peoples were legendary among the Russians who had to deal with, and occasionally fight, them. The Koryaks had sledges of wood bound together with strips of animal skin that could bend almost double and carry more than 100 times their own weight without breaking. Not only were they good for trade, but as the Russians discovered to their cost, the sledges could be used as war chariots, one man guiding the dogs while another let fly with a bow and arrow. The bows were made of strips of birch and cedar, the plies held together with vegetable glues and cords of nettles. They were incredibly strong, and the arrows, tipped with bone or rock crystal barbs, were lethal at great distances. The archers also devised a method of killing ducks that was remarkable, firing their arrows so close to the water's surface that the flight feathers dabbled the surface. Mother ducks, fooled into believing that the dabbling was the struggling of a swimming duckling would move towards the disturbance – and into the path of the arrow.

## Exploitation of Arctic animals by outsiders

### Fur trappers

Writing in 551AD, the Roman Jordanes noted that his countrymen were wearing furs that they had obtained from the *Suehans* (Swedes), who had themselves acquired the furs from the *Screrefennae* (presumably the Sámi). Other references point to a well-established fur trade from Fennoscandia in place since antiquity. It is probable that the Sámi were not only trading furs, but supplying them as tribute to the Norsemen of southern Scandinavia,

The village dump at Gambell, St Lawrence Island, Alaska. This is the modern equivalent of the 'Boneyard' illustrated on p319. It is relatively easy to persuade ships and planes to carry high-price consumer goods to out-of-the-way northen settlements, but almost impossible to get them to remove scrap. So quad bikes, snow scooters, TV sets etc. come in – and stay. The dump therefore grows annually. One advantage it does offer is acting as a makeshift hardware store. With spare parts hard to come by, the locals forage, extracting likely looking bits which can be turned into spares. The ingenuity occasionally shown is impressive.

Fox and wolf pelts hanging outside a trapper's hut in north-east Greenland. It is believed the photograph dates from about 1928.

figures as are known suggest killing on a far more massive scale: in 1595 when the Holy Roman Emperor Rudolf II demanded soldiers from Tsar Boris Gudanov to help a crusade against the Turks, Gudanov feared repercussions on his valuable trade with Constantinople if he complied and so he declined, sending a consignment of furs by way of compensation. The consignment comprised the pelts of 3,000 beaver, 40,360 Sable, 20,760 other martens, 337,235 squirrels and 1,000 wolves. The furs filled 20 rooms of the Emperor's palace, with numerous wagons parked outside still loaded with the less-valuable squirrel furs. When the consignment was valued by merchants in Prague they declined to put a price on 120 Sable pelts, so rare and beautiful were the colours and quality. The 400,000 furs the consignment included are very unlikely to have represented the accumulation of seven years, as official figures suggest – even a gift to an Emperor would not go that far.

who traded them to the countries of early medieval Europe. Furs were valuable not only because work and travel meant being out in the elements, and heating systems were much less efficient than today, but because of their status. Fur was the preferred clothing of royalty. Although there were fur-bearing animals throughout Europe, it was to northern Scandinavia and Russia that the crowned heads and aristocracy of Europe looked, and when trapping reduced animal numbers and threatened the economy of Russia, the Tsar sent men across the Urals to discover new sources. The men 'discovered' Siberia, a country that seemed limitless, with fur-bearing animals whose numbers were apparently equally limitless.

Using boats to navigate Siberia's rivers, groups of *promyshlenniki* (hunters and trappers) explored the country and extracted tribute (*yasak*) and *pominki*, a 'voluntary' gift, for the Tsar from the native tribes encountered on an eastward push. Annual yasak was high, up to 20 Sables per person at first, though this figure had fallen to three within a century, such was the destruction of the animal population. The numbers of animals slaughtered is difficult to estimate as the official figures for pelts does not include the vast numbers smuggled past unknowing or bribed officials: the corruption and other abuses by officials far to the east of central government gave rise to the still-heard Russian comment 'God is high above and the Tsar a long way off', to express the unfairness of a life without checks and balances. Official estimates suggest 60,000 pelts annually, but that seems much too low as such accurate

Fur trapper's hut at Antarctichvn, north-east Greenland. The beautiful locations and the isolation of such sites suggest a romantic lifestyle, but trapping was a hard and dangerous life, without even considering the brutal nature of the capture and dispatch of the animals.

Sable was the chief prey of the trappers, the animals sought in winter (as were all the fur-bearers) when their pelts were at their thickest. Immense ingenuity went into the trapping methods as any damage to the pelt reduced its value. So good at their job were the trappers that within a just a few years a location could become devoid of animals. When the Russians reached the Yenisey they were amazed to discover that Sable were so numerous that the local people used pelts as comfort padding on skis: within 30 years there were so few Sable left that trapping in the area ceased.

Production of furs was such that supply exceeded demand in Europe, the position being exacerbated by the rising trade in North America. The Russian trade (and economy) was saved by an increase in trade with China that underwrote further Russian expansion, into Chukotka and Kamchatka, though that was not without its problems, the Chukchis and Koryaks being particularly aggressive in their response to Russian incursions and Moscow's demands for yasak. This belligerence was, in part, responsible for Bering's expeditions which, it was hoped, would lead to the discovery of yet more sources of furs in less hostile areas. It did.

## Trapping on the Commander, Aleutian and Pribilof Islands

When Bering's second expedition struggled to what are now called the Commander Islands, his men found them inhabited by vast numbers of Arctic Foxes and Sea Otters. They killed and ate the otters (though the meat was disagreeable and they quickly found the Steller's sea cows to be much more palatable: this animal, a huge manatee, was rapidly hunted to extinction) and noted the luxuriousness of the animal's pelt. The discovery led to an eastward 'fur rush'. It is estimated that over the next 20 years 70,000 otters were taken, as well as huge numbers of foxes. In the context of the total fur trade at the time the numbers were massive as the rest of Siberia had been largely hunted out. Foxes were also introduced onto all but a handful of Aleutian Islands to aid future hunting: as there had never been terrestrial predators before, the effect on indigenous bird populations was disastrous, though fox eradication programmes have aided some island populations to recover.

Male Northern Fur Seal, St Paul, Pribilofs, Alaska.

Then, in 1768, Gerassim Pribilof spotted the islands that now bear his name: he sailed home with 40,000 Northern Fur Seal pelts and 2,000 Sea Otter pelts, as well as over seven tons of Walrus ivory. Russia's new fur trade with China was driven mainly by desire for Sea Otter pelts and within a few years the Pribilof numbers of Sea Otters and fur seals had been greatly reduced.

The killing of fur bearing animals continued when America purchased Alaska, but by now rifles had become commonplace, and in the rush to acquire pelts less care seems to have been taken in ensuring an undamaged skin. Otters were shot at sea, with many dying and sinking before they could be retrieved. Others were injured, escaping to die later. The losses could not be sustained, and by the early years of the 20th century a hunter might return from a season with only 20 pelts. Estimates of the original population of Sea Otters vary from 100,000 to 200,000 animals: by 1925 the take had shrunk to zero and it was widely assumed the animal had been hunted to extinction. Then, in 1931, a remnant

Sea Otter, Resurrection Bay, Alaska.

Northern Fur Seals, St Paul, Pribilofs, Alaska.

colony was discovered. With full protection, numbers increased spectacularly so that today's Arctic traveller can again enjoy the sight of this most beautiful animal.

Although Sea Otters were quickly exterminated on the Pribilofs, the population of Northern Fur Seals allowed hunting to continue for much longer. However, by the late 19th century numbers had reduced to the point where fears were being expressed for the health of the species. It is estimated that by the purchase date of 1867 the Russians had taken almost 2.5 million seals.

The Americans probably took another million by the early 20th century. By then returns for hunters were diminishing markedly. The seals were hunted both on land and at sea and the leading pelagic sealing nations – Canada and Britain – resented the suggestion that pelagic hunting should be limited, a call lead by the United States, whose land-based sealing industry would have been unaffected by any limit. Negotiations took place between interested parties, but these became so heated and irrational that one man suggested, apparently without irony, that the ideal solution would be the hasty extinction of the species to avoid further conflict. Ultimately a grudging agreement was reached and the seal population increased again. Today, however, the population is once more in decline, though the reason for this is still debated.

## Trapping in Siberia

In Siberia, the invention of the snow mobile and the use of leg-hold traps made trapping easier, and the creation of the USSR pushed fur hunting to new extremes. As the communist regime maintained a monopoly on currency exchange and the rouble rate was artificially frozen, the government was continuously short of valuable exportable commodities required to maintain the trade balance. Such commodities included precious metals, oil and furs (so-called 'soft gold'). By 1980 it is estimated that there was one Arctic Fox trap for every 10km$^2$ of tundra, with corresponding numbers of traps for other animals, particularly Sable. The Arctic Fox kill exceeded 100,000 animals in many years during the 1970s and 80s, representing as much as 60% of the total population. The fecundity of the fox could sustain such losses in years when the birth rate was high because of high prey density, but not in poor prey years and not surprisingly the trade eventually caused a significant fall in fox numbers. Added to the sheer destruction of the species was the effect on an individual animal. The leg trap causes terrible suffering as the animal rarely dies quickly: some bite through their own lower leg to escape (this cruelty made the European Union ban the import of fur caught using leg-hold traps in 1991). Leg-hold traps can also kill other species such as Gyrfalcons and Snowy Owls attempting to take the bait or, occasionally, by perching on the trap. One estimate, based on capture records, gave a figure of one Gyrfalcon killed for every 50 traps, a mortality highly significant in terms of the overall population. In Russia, an attempt to reduce the death rate of Arctic birds has led to the re-introduction of the *pasti*, the traditional log-fall trap that at least has the advantage of killing the animal quickly.

As a digression, those Gyrfalcons which avoid death in a leg-hold trap may, if they are the white morph of the species, be illegally trapped to be sold to falconers. Drugged and then smuggled abroad, the birds are sold in areas where white birds are in constant demand. Such incidents have also happened in North America, though the greater financial standing of protection agencies on the continent has reduced the trade. In Russia, agencies have much less spending power and a vast area to police, and from the poaching incidents known to have occurred there seems to be a consistent trade. Such poaching can have a significant impact on the wild population in more accessible areas, as the density of the falcons is never high. As roads are built to service increased industrial activity, access to sensitive areas improves and the poacher's task is made easier.

## Trapping in North America

Though the capture rate in North America, in terms of sheer numbers, never matched that of Siberia, the continent's fur trade also had a significant effect on fur-bearing animal populations. The beginnings of commercial hunting can be dated to French trade in North American beaver in the early 17th century. Beaver fur was waterproof, easily shaped and very durable, ideal material for hats, the Canadian animals being a lucky replacement for the European species that had been almost hunted to extinction to satisfy the trade. For half a century the French controlled the Canadian beaver trade, but then the English joined in, the Hudson's Bay Company being, initially, granted a monopoly of fur trade: ultimately the French were ousted to prevent competition. The rights of native peoples were also ignored even though they did most of the actual trapping. The history of fur trapping in Canada is largely the history of the Hudson's Bay Company. Indeed, it could be said that the early history of Canada is largely the history of the Company as it was given jurisdiction over the country by the British crown. The Company's

Old leg-hold traps, Victoria Island, Canada.

It is often claimed that the idea of a fox (or other animal) caught in a leg-hold trap gnawing its foot off to escape is a myth. Here is the proof that it is not. The sprung leg-hold, still attached to a Caribou head which was the bait, grasps the paw and lower limb of an Arctic Fox. The fox would not have survived long with a severed foot and would have died a painful death. Victoria Island, Canada.

factors (managers) occasionally used this authority to act as rulers rather than traders, the initials HBC being eventually said to mean 'Here Before Canada', reflecting the view outsiders took of the Company's methods. The Company traded guns, ammunition and other useful products of 'civilised' Europe for the furs of beaver and

Without realising its significance, a young and curious Arctic Fox examines an old fox trap outside the trapper's hut at Myggbukta, north-east Greenland.

not fully embracing the ethical issues of the west and fur still being the traditional way of combating winter's frightening cold. Trapping continues there, and to the surprise of many, in North America though with much reduced takes: with modern, lightweight and warm materials now cheaply available there really is no need to wear fur. And it has been said many times, the fur always looks better on the original wearer.

## Whaling

Although the number of animals killed in the fur trade far exceeds the number of whales slaughtered during the era of 'industrial' whaling, the effect of whaling on populations of individual species has been much more dramatic. It is assumed from the existence of Stone Age rock carvings of whales that Neolithic folk knew of them, though it is not clear if their knowledge was from animals washed up or stranded on beaches, or whether they hunted whales, using boats to drive them close to shore for killing. That technique, still used in the Faroes today, was certainly in use in 9th century Norway. The construction of Whale Alley on Yttygran Island, which dates from a later, though still early, period, indicates that the ancestors of today's Inuit were also hunting whales many centuries ago. By the 16th century Basque fishermen, who had hunted whales in the Bay of Biscay, began to hunt off Newfoundland. From written accounts and archaeological evidence, it seems they killed around 450 whales annually, rendering them to oil at shore bases. The Basques were clearly efficient whalers, because by the late 16th century the number of whales taken each season had reduced to such an extent that whale oil – the primary reason for taking the animals – had become an expensive luxury in Europe. The whale the Basques took was chiefly the species now called the Northern Right Whale, though the Basques also killed the Greenland Right (now known as the Bowhead).

With the discovery of vast numbers of whales in Svalbard's waters, the trade shifted west. It is estimated that there were at least 25,000 Bowheads when the whalers arrived, often so densely packed in some bays they collided with ships and anchor cables. The whaling involved large ships from which rowing boats were launched to kill the whale. A successful boat towed its kill to a shore station where the blubber was

other Canadian animals. The Company motto was *Pro pelle cutem*, which translates roughly as 'a skin for its equivalent'. When the native hunters became a little more astute, the factors were more inclined to suggest the motto meant 'we skin you before you skin us', though the native trappers preferred 'we risk our skins for your pelts'. Later still, when alcohol increasingly became the trading tool of choice, the motto was said to mean 'a skin for a skinful'.

Because the Hudson's Bay Company trade was better controlled than its equivalents in Siberia, and was also less open to corruption, the take of fur-bearing animals is better quantified: in the century from 1769 the Company exported to London almost 5 million beaver furs, together with 1.5 million mink, 1.25 million marten, over 1 million lynx, and almost 900,000 fox, 500,000 Wolf, 288,000 bear and 275,000 badger.

The economic crash of the 1930s, the 1939–45 War and then the development of central heating reduced the need for furs in the west, while fur farming reduced the need for wild capture. Changes in the public's attitude towards fur also helped, but although this led to a reduction in the slaughter of animals in Siberia as well, there is still a market within Russia itself, the population rendered and poured into barrels that were loaded back on to the ship. Some of the shore stations became seasonally permanent, the most famous being the Dutch Smeerenberg (Blubber Town), whose remains can still be seen on Amsterdamøya, off Spitsbergen's north-western shore. It is still common to see in the literature descriptions of Smeerenberg stating that it had a church, bakery, gambling hall, dance hall and brothel, serving a population of up to 10,000. But the archaeological evidence does not support this colorful view, suggesting a population of 200 at most, housed in barrack-like buildings, and an absence of clergy and women.

The Danes, French and Germans all sent ships to Svalbard to join in the killing, though it was the British and Dutch who made up the bulk of the fleet, and soon whale numbers were badly depleted. By the 1640s the catch barely covered the costs of the voyage: Smeerenberg was abandoned in the late 17th century, perhaps as early as the 1650s. The station, and a Dutch one on Jan Mayen, had to be set up and manned each season (attempts at overwintering failed because of scurvy), which was expensive and, once numbers were depleted, uneconomic. Whales could still be hunted close to the ice edge, a long way from shore, but the carcass then had to be dismembered in the water and hauled on deck in large chunks for rendering. The ice, hunting and processing was dangerous and as, again, numbers dropped was also abandoned.

Both the Dutch and the British then moved to the west Atlantic. There hunting lasted longer because the season was shorter. Conditions were far more dangerous, so ship losses were much higher, but again ended only when whale numbers made the trade uneconomic. Bowheads were also found in the Pacific, where they had sustained Yuppiat populations in both Siberia and Alaska for centuries. Soon the Americans, who were already hunting Sperm Whales in both the Atlantic and Pacific, were pursuing Right Whales in the Bering Sea and Sea of Okhotsk, and Grey Whales near Baja California.

The losses of men and ships, particularly after the hunt switched to the ice-filled waters of the Beaufort and Chukchi seas, coupled with the development of a petroleum industry that offered a cheap alternative to whale oil, meant respite for the whales. But it was temporary, the demand for baleen – called whalebone

A pile of Beluga bones on a beach of Bellsund, Spitsbergen, Svalbard. Whalers were indiscriminate when it came to species, Beluga might not be a large whale, but they form very large pods.

Debris from an old whaling station on Bear Island.

by the fashion industry, which preferred a catchy name to a correct one – and the development of steamships that allowed the whalers to push through ice that would have imprisoned and sunk sailing ships, allowed hunting to increase again.

It is difficult to be exact about the number of Bowheads killed by the whalers, as in many cases a struck whale would escape, only to die later. Various figures have been suggested, but the most likely seems to be 120,000–150,000 in the whale's eastern range (the Atlantic and Hudson Bay), with 20,000 in the Sea of Okhotsk and at least the same number in the Bering Sea. The effect on the species was disastrous. When whaling ceased, the Atlantic population was numbered in hundreds, though Pacific numbers were higher. Today it is estimated that the Atlantic Bowhead population is still no more than 500–600 animals (perhaps 450 on the western side, no more than 100 on the east). In the Pacific there are thought to be around 6,000–8,000 whales. For the Northern Right Whale the numbers are even more depressing, the present population being estimated as no more than a few hundred animals in the Atlantic and perhaps only 100 in the Pacific: the population is

also failing to recover. The Bowhead population may be recovering slowly, though some experts doubt this. Inbreeding and consequent depressed breeding fitness resulting from loss of genetic diversity are acting against recovery. Another consideration is the 'Allee effect', which suggests that for some species that habitually schooled (or flocked), as the population level declines individuals suffer increased stress consequent on the absence of group members, reducing lifetimes and fecundity rates. Given the vastness of the oceans and the enormous reduction in the right whale population it is possible that such stresses may also be affecting the whales, and such effects would be even more pronounced in the Northern Right Whale: some experts believe that this species is doomed to extinction.

But it is not just a question of numbers. Death by harpooning was usually agonizing and drawn-out. Some whales towed row boats for days before dying. There are stories of the men in the boats being covered in gore as the whale spouted blood, and one appalling tale of a Bowhead diving so fast that it crashed into the seabed, burying itself almost 3m into the sediment and breaking its neck in the process.

[Above and opposite page] The Siberian Yuppiat also have a quota of Bowheads. For yuppiat on both sides of the Bering Strait the hunting is a tradition centuries old, though their ancestors did not have the luxury of a giant tractor to haul a killed whale on to the beach. In the photograph above, chunks of blubbers are being cut off the whale and hauled up the beach. They will be allocated to villagers according to traditional rules on the make-up and outcome of hunting trips.

The Arctic species were also not unique in the plunder, the pursuit of whales in other waters being equally relentless. For the rorquals, coveted because of their huge size, the technique of using rowed boats that had been so deadly against the right whales was useless as the whales can reach speeds of 20 knots. But that situation changed in the 1860s with the introduction of steamships and the invention of the explosive harpoon gun. The great whales of the Arctic fringe were then hunted mercilessly, though the slaughter of rorquals was greatest in southern waters. During the period of operation of the Grytviken whaling station on South Georgia (1904–1965), almost 90,000 Fin Whales, over 40,000 Blues, about 27,000 Humpbacks, some 15,000 Sei and almost 4,000 Sperm whales were killed, figures which are in addition to the catches of Southern Right Whales that had brought the species to the edge of extinction in the first phase of Antarctic whaling. And the catches there are dwarfed by those of all species in Antarctica: it is estimated that overall at least 350,000 Blue Whales were killed over the period of hunting, about 90% of these in Antarctica. Fin Whales were killed in even greater numbers, perhaps 420,000 animals in total. The populations of the other great

whales, Sei and Humpback, were also dramatically reduced, with some estimates putting the reduction of Humpbacks at 95%. Finally, with many species on the edge of extinction, the slaughter of these magnificent creatures was halted in 1986 by the International Whaling Commission (IWC), though Japan objected and continued to hunt, largely for 'scientific' purposes. Iceland and Norway have also hunted, though their take has been considerably smaller.

Fortunately, the populations of Humpback, Fin and Sei Whales show definite signs of recovery (the Humpback population is now thought to be about 35% of the pre-whaling figure), but that of the Blue Whale appears to be static, with some authorities concerned that the species may never recover and may even be on a slow road to extinction. Iceland, Japan and Norway claim that stocks of some species are sufficient to allow harvesting without endangering populations, but opponents point out that population numbers are disputed. The method of calculation, using aerial photographs of the ocean and counting the number of whales seen, does not allow for the known behaviour of the animals which tend to form groups. Opponents also point out that one justification for a resumption of whaling, that the whales take a huge volume of fish and so endanger commercial fisheries, is not supported by scientific studies. Nevertheless, Iceland began whaling again in 2006 chiefly catching Fin Whales, the meat from which was exported to Japan. The take of this endangered species was apparently almost 150 annually by 2010 when the USA formally protested the continued hunting. Despite this, and protests from other nations and environmental groups, the hunt continues: 155 Fin Whales were taken in 2015, 1,700t of meat being exported to Japan. The sole company involved loses money on the whaling, but is intent on continuing. In 2018 the intention is to take as many as 200 Fin Whales, despite the species now being listed as 'endangered' on the IUCN Red Data List. Concerned that whale watching is a significant tourist industry, and that the sight of a whaler entering Reykjavik harbor with a huge whale attached to the side is a major 'own-goal', the Chairman of the Icelandic Tourist Board worried about the consequences of this 'bloody-mindedness' in the face of national and public disapproval, noting it was 'like peeing in your shoe: you get warm for a little while, but then you get really cold.' Iceland also hunts Minke Whales for domestic consumption. The catch has been diminishing, reducing by around 65% over the last five year (while the number of people on the watching tours had increased five-fold). The lack of whales has implications for whale watching tour operators, though it must be said that scientific opinion contends that the reduction in local whales might be as much to do with a crash in the population of sand eels, the Minke's main food source, which have forced the whales to move elsewhere.

Japan has also continued to hunt whales (not only the Minke, the major 'headline' catch, but Sperm, Sei and Bryde's: Japan also takes numbers of smaller whales, for instance Baird's Beaked Whale, which are not included in the IWC moratorium). The Japanese position is aided by the fact that many small, non-whaling countries (some land-locked, some tiny Pacific and Caribbean islands) had joined the IWC and sided with Japan. The fact that these countries had received Japanese aid packages and had joined the IWC because of Japanese pressure is stated to be a coincidence. In 2010 the claim that Japanese whaling was for 'scientific' purposes, so outraged the Australian government that they took Japan to the International Court of Justice. In 2014 the Court ordered Japan to desist as it was clear the whaling was not for scientific purposes, but despite the ruling, in November 2015 Japan announced it would continue whaling 'for scientific purposes'. During the 2016/2017 austral summer the Japanese took 333 Minke Whales, noting that since the majority were mature adults this showed the population ws healthy. The Japanese intend to take another 4000 whales over the next 12 years to continue their research: the IWC responded to the news by rejecting the research idea, stating that there was no justification for the killing. Meanwhile, during the northern summer Norway has announced it will take 999 Minkes, and makes no apology for saying that they are for meat not research, and accepts that many may be pregnant females.

## Native Arctic whale hunting

Before the advent of commercial whaling it is thought that the Inuit of the eastern Arctic took perhaps ten Bowhead annually, while the take in the Bering Sea was higher, perhaps 15–20 animals in total for the Siberian and Alaskan sides. European whalers in the Atlantic reduced the Bowhead take by native folk effectively to zero by the early 19th century. In the Bering Sea, commercial whaling had a curious effect, increasing the take initially as native people acquired new technology that made their kills easier and less dangerous, but eventually also reducing the take to zero. By the 1970s there was no native whaling for Bowheads in the Bering Sea, though the hunting of Grey Whales on the Russian side continued. However, the development of the Prudhoe Bay oil fields pumped money into the local economy, allowing people to buy boats and modern equipment, and they began whaling again. But the native peoples had lost their basic skills and many more whales were struck and escaped to die elsewhere than were landed. These 'kills' contravened the zero-quota agreed by the IWC for the Bowhead and caused concern among environmentalists. When the IWC asked the US government to intervene to stop native hunting the Yuppiat were outraged, considering the Bowhead hunt to be part of their heritage and pointing out that it was the southerners who had reduced the stock to dangerous levels. They also disputed the IWC figure for the Bowhead population and set up the Alaska Eskimo Whaling Commission, which used hydrophones to listen for whales, claiming that this was a better method of assessing numbers than visual counts. The study gave a population estimate of 6,000, four times the IWC's estimate. Consequently, the IWC agreed a quota of 280 landed whales for the five-year period 1998–2002 and, subsequently, a quota of 255 for the five years to 2007. In the late 1990s the Yuppiat killing efficiency rose to over 75% so the number of whales struck and not landed decreased substantially. The quota for the years 2013-2018 is 336 Bowhead in total for Alaskan and Chukotkan native peoples. There is also a quota of 140/year over the same period for Grey Whales. In East Greenland native people can take 12 Minke annually until 2018, while on the west coast 164 Minke, 19 Fin and 10 Humpback may be taken annually until 2018. There are also limits on the number of catches that can be carried forward if quotas are not met.

Norway also continues to hunt whales, despite objections from the IWC, with catches in 2014 and 2015 be around 700 Minke, up from kills of 450-550 in the years 2008-2012, but comparably to those in the early years of the new century. Some of the meat is exported to Japan, a trade which continues despite the Japanese dumping meat imported in April 2015 because it contained levels of aldrin, dieldrin and chlordane which were above approved Japanese safety limits.

In all three whaling nations the position seems to have become entrenched more because of nationalism, a simmering resentment at being told what to do, than pure commercialism. The problem is exacerbated because the opponents of whaling are equally entrenched. Whales have become a totem, both for environmentalists and the public, particularly with the latter as whale watching has become big business and has resulted in considerable sympathy for the animals. That whaling is also undoubtedly cruel has added to public sympathy.

## Fishing

The northern native peoples have always taken fish as a supplement to their diet of terrestrial or marine mammals. The civilisations of northern Europe commercialised fishing, taking cod and herring in the seas off Norway and European Russia, as well as seas further south. Over time, as both vessels and fishing equipment became more efficient, and people realised the potential of the sea for food and profit, catches increased. By the second half of the 20th century the apparently limitless fish stocks of the Barents Sea and northern Norwegian Sea were showing clear signs of overfishing. The normal migration patterns of herring in the north-east Atlantic were also disrupted, and the stock collapsed dramatically. The fishing fleets of Norway and Russia therefore transferred their attention to capelin and, predictably, stocks of that species also collapsed within a few years. Conservation measures have seen fishing activity reduced, though there is evidence that illicit fishing is being carried out and that the Barents cod stocks are being seriously depleted.

Elsewhere a similar story has unfolded. Cod fishing in the north Atlantic increased to the point where by the 1970s the Icelandic economy was based almost exclusively on the fish. To maintain stocks, Iceland extended its coastal limit to 200 miles (320km) in 1977 and fought 'cod wars', particularly with Britain, as the new limit was considered illegal. The limit has allowed cod fishing to continue in Icelandic waters. However, Icelandic fisherman have overfished local herring, and now pursue capelin. Icelandic and foreign fishermen have also fished the waters off eastern Greenland, catching Greenland halibut and cod, and some capelin. Off eastern Canada the history of the cod fishery centred on the Grand Banks deserves a book of its own, the story being one of astonishing riches, greed and incompetence. The fishery not only saw the virtual extinction of the cod, but also played a significant part in the extinction of the Great Auk. Fishing for Greenland halibut and shrimp continues.

Fishing in the Bering Sea has come under intense scrutiny in recent years because of its possible (some would say probable) involvement in the decline in the number of Sea Otters and pinnipeds. Historically, fishing for salmon species, herring and halibut has been important and remains so, salmon being a sport as well as a commercial species. In the 1970s and 1980s the fishing of Walleye Pollock increased dramatically, with annual catches of up to 20 million tons eventually being recorded. However, catches then dropped sharply, as did the populations of both Sea Otters and Steller's Sea Lions. Pollock decline was probably due to a combination of overfishing and a rise in water temperature, though experts are divided on which of these was the more significant. One likely scenario sees warming water pushing temperature-sensitive organisms at the bottom of the food chain north, and crustaceans dying as they are unable to respond fast enough. Young fish also die from lack of food, and with fishing reducing adult fish numbers, populations decline dramatically. Next the sea lion population slumps because of a lack of fish. Now, perhaps, Orcas, deprived of the mainstays of their diet – fish and sea lions – take otters instead. The otters are the main predators of the sea urchins that graze on kelp. The subsequent increase in urchin numbers then causes the kelp forests around the Aleutians to thin or disappear. Many authorities believe that the ecology of the southern Bering Sea has been irreparably altered: as this sea is one of the wonders of the Arctic fringe, the loss of this ecosystem, which supports, among other things, millions of seabirds, would be a tragedy.

Two other effects of fishing cannot be ignored. Bycatch is the name given to the accidental taking of sea mammals and birds caught in, and killed by, fishing tackle. Birds and mammals can be entangled in nets and drowned, and birds may also take bait from long-line fishing and so become hooked. The latter is a real threat to albatross species in the North Pacific, particularly the endangered Short-tailed Albatross. The other effect is damage to the seabed by trawl nets. In shallow waters with soft seabeds it is likely that storms cause more disruption to sediments than trawling, and that benthic creatures have adapted to occasional disturbance. In deeper waters where the effects of storms are negligible, disturbance from the dragging trawl nets will have more impact. This impact increases on harder seabeds, with considerable damage being caused to corals and to rocky environments.

## The exploitation of Arctic minerals

### Oil

The use of 'Arctic' oil has a long history as it is known that the Inuit of northern Alaska used the oil that seeped to the surface near Cape Simpson as fuel, compressing oil-soaked earth into bricks that could be burnt. However, not until the last half of the 20th century was commercial exploitation begun, after geophysical studies indicated that large areas of the Arctic had sedimentary basins that were likely to be oil-bearing, these including northern Alaska, the Mackenzie delta and Beaufort Sea, the northern Canadian Arctic islands, offshore western Greenland, and Yamal, Yakutia and Sakhalin Island in Russia.

The potential importance of Alaska's North Slope was recognised as early as the 1920s, and in 1923 the National Petroleum Reserve, Alaska (NPRA), to the west of Prudhoe Bay, was designated as an area within which oil production could proceed, but only in the case of a national emergency. Exactly what constitutes an emergency, particularly given the lead time of extraction, is a moot point. Oil was discovered at Prudhoe in 1968: in 1977 oil from the bay's wells was first loaded onto tankers at Valdez on Alaska's southern coast. The oil had been transferred along the Trans-Alaska Pipeline, construction of which involved about 70,000 workers and cost $8,000,000,000. The pipeline is 1250km long,

Well-heads Prudhoe Bay, seen across driftwood on the shore of the Beaufort sea. The tree trunks in the driftwood probably originated in the Mackenzie River, reaching the Arctic Ocean then being carried to the Bay by the Beaufort Gyre.

the pipe itself having an outside diameter of 1.22m and a wall thickness of 1.3cm. About 52% of this pipeline is above ground to prevent thawing of sensitive areas of permafrost. In general, buried sections are in rock or dry gravel, though even there it is well-insulated from the substrate. Where burial was essential in permafrost areas, circulating refrigerated brine chills the pipeline. The pipeline does not form a continuous straight line, the zig-zag pattern allowing for seasonal thermal expansion and contraction, catering for assumed extremes of -55°C (with the pipe assumed empty in winter) to +63°C (with the pipe assumed to be filled with oil at near-extraction temperature). In areas of potential seismic activity, the pipeline sits on Teflon sliders, and is designed to allow a movement of 60cm. Oil reaches well-heads at a temperature of about 70°C and enters the pipeline at about 45°C. The oil flows at 6–7km/h, taking 8–9 days to reach Valdez, by which time its temperature has fallen to about 18°C. At any given time, when in operation, the pipeline holds about 9 million barrels of oil. Pumping is provided by five pump stations along the pipeline. On 14 January 1988 the throughput was 2,145,297 barrels, though the daily average is now usually about 500,000 barrels. The Valdez oil terminal has 18 storage tanks, each capable of holding 500,000 barrels. Production has not been without problems, the worst being on 24 March 1989 when the *Exxon Valdez* tanker went aground on Bligh Reef and about 250,000 barrels were released into Prince William Sound. The disaster was then exacerbated by the lack of emergency preparedness, which meant that little effort was made to control the spread of the oil for many hours, by which time a storm had dispersed the slick. Ultimately the oil contaminated 2,500km of shoreline. At least 2,000 Sea Otters were killed (a more exact number has never been established, but figures as high as 5,000 have been suggested: the current best estimate is 2,800). Over 200 Harbour Seals and 20 Orca are also known to have died, while the total number of dead birds exceeded 250,000. The effect on lower trophic levels of the marine ecosystem are difficult to assess, but it thought that the combination of oil and chemical residues from the spill and the clean-up, which made their way into the food chain, would have been considerable. The use of high-pressure hoses to clean up areas of contaminated beach also caused environmental damage, destroying organisms at the sea margin. The populations of some species, such as Harbour Seals and Harlequin Ducks, have not recovered and it is feared they never will. In early 2007 a study of Prince William Sound and the Gulf of Alaska found that the oil was still a threat to wildlife as its decay was much slower than had been anticipated. Exxon, which has funded several hundred studies that indicate no significant long-term effects from the release, suggested that the study told us nothing that had not been anticipated. The study was released the day before Exxon Mobil revealed that its 2006 profit had been $39.5 billion dollars, the largest annual profit ever posted by a US company. Following the disaster, the tanker's captain, Joseph Hazlewood was sentenced to 1,000 hours of community service for a Class B misdemeanour. Hazlewood was said to have been in his cabin at the time of the grounding sleeping off the previous night's heavy drinking. A Third Mate was in charge of the ship, and the radar that would have warned him of the reef was switched off: it is claimed it had not worked for a year as it was considered too expensive to fix. The Exxon Corporation (now Exxon Mobile) were fined $1 billion for a violation of various Acts, with a further $5 billion in punitive damages. The company contested the level of damages awarded in various courts for over 20 years, though damages were paid to certain companies or company groups. A payment in 2009 appears to have been the final settlement for punitive damages (at an apparent level of 10% of the original award). One year later, in response to a report that 60% of workers involved in the oil clean-up had become sick, Exxon said that there was no evidence that the clean-up had resulted in any ill-health.

A section of the TransAlaska Pipeline which carried oil from Prudhoe Bay to Valdez.

The pipeline itself is protected by four independent leak detectors. This sounds excellent, but the systems do not protect the oil fields and feeder systems, and in March 2006 a spillage of over 6,000 barrels was discovered. The spill was from a corroded pipe, BP being fined a total of $20 million for the incident. BP was also required to carry out inspection of the feeder pipelines as well, as it had not done so for 14 years. Corrosion levels of up to 80% of the pipe thickness were found, and the feeders were closed for several months, causing a rise in world oil prices.

Although Prudhoe Bay is the best known of Arctic oil fields, there are now several more, and many others are planned for the future. In 2008 275 drilling licences were granted in the Chukchi Sea, in the US Arctic sector, and in 2015 President Obama gave clearance for Shell to commence drilling. Environmentalists were outraged, particularly as the decision came soon after Obama had announced that climate change was his priority. Later Obama blocked further licences and licence extensions in the Beaufort and Chukchi seas (the current licences expire in 2017–2020). Coupled with Shell's decision

to cease their exploration as yields were much lower than expected will reduce Alaskan operations for the foreseeable future: Shell had spent $7 billion on the search and were under pressure from shareholders at a time when low oil prices were making operations uneconomic, and were also aware that continued work was not helping the company frame the debate on tackling climate change as it was attempting to do. While environmentalists welcomed the news of the licence blocking, one of Alaska's senators, Republican Liza Murkowski, condemned it as threatening US energy security. However, Shell, who held licences in the area, announced they were pulling out of the area as their first well had found only traces of oil rather than the anticipated significant finds. President Obama also proposed to make a large section of the Arctic National Wildlife Reserve (to the south and east of Prudhoe Bay) a wilderness area to protect it from future oil exploration. He could have used his power to make it a National Monument and so preclude drilling permanently, but chose not to, leaving the door open for future administrations to alter the designation.

The Arctic National Wildlife Refuge, Alaska.
Conservationists fear that oil exploration in the refuge
will irreparably damage its fragile ecosystem.

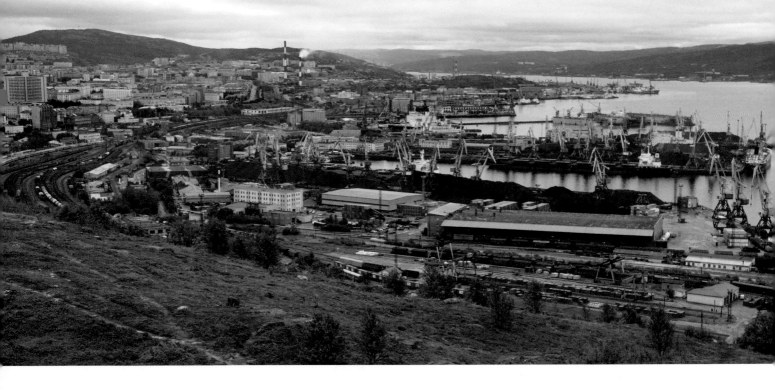

Murmansk harbour, Russia.

Elsewhere in the Arctic, oil exploration continues in Canada, and in the western Arctic particularly in the Barents Sea (Norwegian and Russian fields), as well as further east in the Russian Arctic and on the Russian mainland. One Barents field (it is in what is occasionally called the Pechora Sea, between Novaya Zemlya's southern island and the mainland), Gazprom's Prirazlomnaya, has been in the headlines recently. Production was delayed for several years with rumours of 'safety concerns', and the field was then visited by the Greenpeace ship *Arctic Sunrise* in September 2013, three activists attempting to board the platform. The ship was seized by Russian coastguards and sailed to Murmansk where the 30 occupants were imprisoned and charged with piracy, which carried a potential 15-year prison term. The international community reacted with outrage, but it is likely that President Putin's decision to free the 30 under an amnesty law had more to do with concerns over disruption of the Sochi Winter Olympics in 2014. Further east, production of a field in the Kara Sea has been delayed at Univeritetskaya-1 by sanctions imposed on Russia following the Ukraine crisis. Elsewhere in Russia, the main interest is in the Khanty-Mansi Autonomous Region, north-east of Novosibirsk, the Yamal-Nenet Autonomous Region, the latter an important Arctic area, and in the Sea of Okhotsk.

Drilling operations and pipelines beneath the northern seas are in danger from moving ice, which occasionally gouges trenches in the seabed, and, perhaps, from sub-sea pingos. Oil production in the Arctic has unique environmental problems because of the long recovery times of the landscape in the event of damage. Although improving technology has reduced the footprint of drill heads (though the drive to do so was as much to do with economics as environmental concerns), oil fields still do considerable, if localised, damage to the environment. Wheeled vehicles create ruts that deepen and widen, construction causes changes in local hydrology (the construction of a 7km section of road near Prudhoe Bay caused the subsequent flooding of more than 130ha of surrounding land), while oil production releases hydrocarbons and other pollutants, including heavy metals into the environment. On Alaska's North Slope 180 species of breeding birds have been identified, with a further 60 species having been noted. The total number of birds is estimated at 10 million. There are also 15 terrestrial mammals, while six species of marine mammals are regularly seen offshore. Studies of these animals and North Slope vegetation have revealed both costs and benefits from oil extraction. In summer the Dalton Highway (the 'haul road' beside the pipeline) is very dusty: the dust has killed roadside vegetation, though

Nenet Reindeer herder.

the effect reduces relatively quickly with distance from the road. For bird populations, the effect of oil extraction has been essentially neutral. Road and site construction work has, of course, reduced nest site availability, but there appears to have been no instance in which a habitat made unavailable does not exist elsewhere, so that the effect is short-term only. Nests have even been discovered on the pipeline structures. For Caribou, the oil field and pipeline has also been basically neutral. Some sections of the elevated pipeline were constructed less than 1.5m above the ground, making it difficult for the animals to pass beneath, but newer sections have greater ground clearance. Brown Bears that scavenge on the oil field dumps show lower cub mortality than those whose diets are not supplemented in this way. The reason appears to be less to do with the condition of the mother because of enhanced food intake (though heavier mother bears tend to have healthier cubs) and more to do with the relatively sedentary life of these animals. Cubs of peripatetic mothers may be drowned crossing rivers, injured in falls or become separated,

and are more likely to meet predators. A similar effect has been seen in Arctic Foxes, which also scavenge at the dumps. However, there is a downside for the prey of these larger, healthier bears: their numbers are not enhanced by scavenging, and the bears kill more of them. Arctic Ground Squirrels and some nesting birds have shown population decreases in areas where bears and foxes hold territories that include the dumps. While these studies are encouraging to a degree, the fact remains that poorly considered oil and gas pipelines can and do interfere with local animals. As an example, a railway built to service a Taimyr gas field, and the pipeline running beside it, affected local Reindeer, which declined to cross the lines and so were isolated from their traditional calving grounds. Though the problem was ultimately solved by creating a fenced migration corridor, the initial impact was severe. Shipping can also have a deleterious effect on wildlife, not only because of the possibility of oil slicks, but engine noise and traffic volume affecting species which use sonar for hunting or communication. Environmentalists have warned that the proposal to ship oil from Albertan bitumen via a pipeline to ports on the Salish Sea (between Vancouver Island and the mainland) will likely cause the extinction of a pod of 80 Orcas already stressed by the reduction in salmon numbers because of sea warming.

Oil spills in the sea have a devastating effect on all levels of the marine ecosystem (a fact reinforced by the devastating release after the Deepwater Horizon disaster in April 2010). Potential drilling in the Chukchi Sea is at greater depths than was Deepwater Horizon, and at the latter it was found that oil discharged on the ocean floor behaves differently from surface spills, forming suspended plumes rather than rising to the surface and so is difficult to either control or remove. Oil under sea ice may be impossible to reach, and it also persists for longer in the Arctic than in more southerly areas, since the oil may become trapped under the ice. On land, spillages tend to be more localised, but surface water can spread the oil. Oil does considerable damage to Arctic plants which, being shallow-rooted, cannot avoid the pollutants. Terrestrial birds and mammals are in less obvious danger than their marine counterparts, but nonetheless can become oiled, with similarly fatal consequences. It is worth noting that the oil released at Deepwater Horizon, which caused so much damage, is about half that released annually into the environment by Russian oil production operations, according to the Russian Natural Resources and Environment Department.

## Gas

Natural gas is often a by-product of oil production. In the case of the Prudhoe Bay fields, oil extraction results in millions of cubic metres of gas being brought to the surface each day. The gas is used to power the oil production process, or it is pumped back into the oil reservoir. However, gas can be the primary reason for drilling, particularly as its value as a fuel has become increasingly important in recent years. The potential for much greater quantities of gas production on the North Slope has led to the proposal of a second pipeline running beside the oil line, then beside the Alaska Highway and eventually to the rest of the continental United States. Although the gas could be liquefied for tanker transport, liquefaction and regassing is energy-intensive, consuming a significant fraction of

the energy content of the gas, so compressor-driven pipelines are preferred. An alternative suggestion is the so-called 'over-the-top' route east to the Mackenzie delta, where the Canadians are exploiting gas reserves: delta gas is already used to power Inuvik and other communities along the northern section of the river. However, a proposed pipeline from the delta village of Tuktoyaktuk, through Norman Wells to Fort Simpson, planning for which had begun in 2004, was finally shelved at the end of 2017 when energy prices made it uneconomic.

Russia has vast reserves of natural gas. The combined oil and gas reserves in Russia's West Siberia Basin – in the Yamal-Nenets Autonomous Region – make it one of the largest reserves in the world, and comfortably the world's largest natural gas field. The field produces 90% of Russian (and 22% of world) gas a significant fraction of which is piped to western Europe. As with oil production, gas extraction results in environmental damage. The drill sites and pipelines have their own footprints, but other damage occurs from site construction: at the Bovanenkovo Gas Field on the Yamal Peninsula the quarrying of sand for construction work destroyed over 1,200km² of tundra.

## Mining

Mining within the Arctic is not a new concept. The Klondike and Alaskan Gold Rushes saw men exploiting previously pristine wilderness areas, and gold is still worked in the Dawson City area. Although Nome is no longer the gold town it was, gold is still mined in Alaska, though most of the identified potential sites have yet to be exploited. Gold is also mined in Yakutia and Chukotka, with other mines active near Magadan and Krasnoyarsk. There are also known gold reserves in Greenland.

Historically, cryolite was mined in Greenland, particularly at Ivittuut. Cryolite was critical to the production of aluminium during the first decades of the 20th century, the mineral being used as a flux, reducing the melting temperature of the ore bauxite and making the otherwise uneconomic electrolysis process viable. During the 1939–45 War the United States essentially annexed Ivittuut for the war effort, but in the post-war period a synthetic alternative was developed and cryolite

The abandoned Giant gold mine, Yellowknife, Canada.

The Russian town of Barentsburg, Spitsbergen, Svalbard, which was built to house coal miners.

Norwegian coal mining at Ny Ålesund, Spitsbergen involved the construction of the world's most northerly railway. The mine has closed, but the railway still attracts visitors.

production slowed, ceasing altogether in 1987. During the period of cryolite production it is estimated that almost 4 million tonnes of the mineral was extracted.

Coal is another historically important mineral mined in the Arctic, both Norway and Russia operating mines on Svalbard. The Russian mine at Barentsburg was closed in 2006 when a fire was detected. Although the fire was extinguished mining did not restart until 2010. Norway has mines at Longyearbyen (which feed the local power station) and at Svea, the largest of the three sites.

Iron ore has been mined in the Swedish Arctic for many years, the country's iron reserves being the foundation of the highly successful Volvo and Saab companies. Iron is still mined at Kiruna, Gällivare and Malmberget. In 2014 it was decided to move the town of Kiruna eastwards by 3km as the LKAB mine had extended beneath the town centre and there was clear evidence of subsidence. In 2013 the Beowulf mining company applied for permission to develop an iron ore mine at Kallak near Jokkmokk in Sweden. The mine was opposed by local Sámi reindeer herders and initial permission was refused on the grounds it would damage herding, but passed to the Swedish Government for a final decision. At the time of writing, although discussions had taken place between the government and Beowulf no final decision had been made.

Elsewhere in the Arctic, the search for new mineral sources, both 'old-fashioned' minerals such as gemstones and precious metals, but 'modern' materials required by new technologies, these including rare earth metals is gathering pace. In Russia finds of molybdenum, niobium, platinum, titanium, vanadium and rare earths have been made, as well as chromium, copper, iron, manganese, nickel and diamonds. Similar resources have also been identified in Alaska and Canada. In Greenland, there are sources of rare earths, but also uranium and thorium, the latter splitting local opinions into factions who welcome the possibility of mines as they will bring prosperity and those who say they will move from the area if such dangerous elements are mined, as they are fearful of potential health effects. While radioactive materials bring special problems, the potential conflict between prosperity and protection of the environment has been raised almost everywhere mining has been proposed.

Coal carriers at the mine in Longyearbyen, Spitsbergen.

Though heavy metals are a problem because of their toxic nature when released into soils and water, other forms of mining may affect the fragile Arctic ecology, as no mine can be entirely self-contained and all must therefore have some environmental impact. Mining and smelting activities also result in the emission of sulphur dioxide and nitrogen compounds, which have caused acid-rain damage to trees downwind of the plumes. Arctic waters are especially vulnerable to acid precipitation, as the acid burden may be contained in snow. Snow melt then results in acid pulses to local aquatic vegetation: such pulses are far more damaging than the year-averaged acidic concentration might imply. On the western Kola Peninsula at Nikel, acids in the furnace plume from the nickel smelters have created a large dead footprint downwind. At Monchegorsk, about 120km south of Murmansk, there is an even bigger smelting complex where nickel, copper and other heavy metals are produced, while south-west of the Taimyr Peninsula, close to Noril'sk, the largest city in the Arctic with a permanent population of about 180,000, is a former gulag which is now the largest producer of non-ferrous metals in Russia, the smelting of nickel, copper and cobalt, and smaller quantities of other rare metals (including platinum and gold), being fired by gas from local gas fields. Noxious furnace gases and heavy metals have contaminated Reindeer grazing areas and poisoned local waters – emissions from Noril'sk have been detected in Alaska and northern Canada. Some estimates suggest that 50% of all heavy metal contamination detected in the Arctic comes from the metal smelters of northern Russia: anthropogenic heavy-metal pollution far outweighs that from natural sources such as volcanoes and forest fires. In an ironic twist, much of the nickel produced at Nikel is used in the manufacture of catalytic converters, to reduce pollution from motor vehicles.

Old lead mine on Bear Island.

But while it is easy to categorise the environmental damage at Noril'sk and Nikel as the product of a system of government that considers such issues unimportant in comparison to competitive industrialisation – an argument that applies equally to the economies of emergent Asian countries – damage is not confined to such nations. The Red Dog Mine near Kotzebue in northern Alaska is the world's largest zinc mine currently in operation, the output of which represents a significant

The smelting town of Nikel, Russia.

fraction of global production of the metal. Mining began here in 1980, the mills powered by local natural gas and oil. A haul road connects the mine to a purpose-built port from which the milled ore is transported to smelters around the world. Studies of the heavy-metal pollution close to the haul road showed high levels of contamination due to spillages, and infringements of regulations on the contamination of local water.

While the ecological footprint of many current Arctic mines remains reassuringly small, the ability to overcome the problems of mining in permafrost areas is improving and the costs of transportation are falling, so mining is likely to become more significant. This will not only increase the total area of mines, but the roads constructed to service them will further fragment the ranges of species, particularly the larger and migratory species such as Reindeer – as an example, surface and underground diamond mining in Canada's North-West Territories (north-east of Yellowknife) lie at the centre of the range of the Bathurst Caribou herd. Herd fragmentation could lead to a reduction in genetic diversity within the population. At the time of writing there is also concern that mining for heavy metals, copper and zinc, near a major tributary of the Chilkat River in Alaska, could result in run-off which threatens the salmon and eagles at the Bald Eagle Preserve which annually sees hundreds of eagles congregating to feed on spawning fish. As elsewhere, the inhabitants of Haines, close to the Preserve comprise those dependent on tourism and those seeking better-paid employment.

## Pollution in the Arctic

### Chemical pollutants

The Arctic is vulnerable to pollution from sources other than metal mining and smelting, and the drilling for fossil fuels. Arctic rivers carry run-off from these industries to the Arctic Ocean. Although the huge flows of these rivers dilute the pollutants, the seasonal flow resulting from annual freezing and thawing also tends to concentrate them. Pollutants in rivers from more temperate regions also carry pesticides and other agricultural chemicals, the run-off from southern farmlands, ocean and atmospheric currents carrying these north along temperature gradients, so that the

Arctic can in some ways be considered a litmus test for the planet's health.

Of the transport mechanisms, atmospheric currents are the fastest, carrying pollutants north in days or weeks, and of the pollutants carried perhaps the most important are the persistent organic pollutants, the organochlorides. The story of DDT and its disastrous effect on wildlife, particularly birds where breeding failed due to egg-shell thinning, is well known. However, although DDT was banned in North America and Europe in the 1970s, it is still used to combat malaria-transmitting insects, and significant stocks are held worldwide. DDT (specifically DDE – Dichlorodiphenyldichloroethylene – formed by the dehydrohalogenation of DDT) is therefore still making its way to the Arctic, as is HCB (Hexachlorobenzene) an organochloride fungicide. More significant are the PCBs (polychlorinated biphenyls), of which there are some 200 variants. PCBs do not easily dissolve in water, but dissolve readily in lipids – fats and oils such as blubber and mammalian milk (the latter including that of nursing native women). The adaptations of Arctic wildlife and its peoples to their environment make them especially vulnerable to these pollutants, this being particularly true for animals that rely on fat reserves to survive the winter. Absorption transfers the PCBs to the animal's internal organs, the new spring's blubber growth taking up yet more chemicals. The Arctic's climate also makes it more susceptible to PCB pollution, the lack of photodegradation of the chemicals during the long Arctic winter allows them to persist longer than in southerly regions, while low temperatures inhibit the natural biodegradation of the compounds. Pollutant particles also attach readily to snowflakes, which have a large surface area relative to rain drops, and so are easily brought to the ground or ocean surface.

Low levels of PCB contamination in humans cause skin problems such as acne, as well as stomach, thyroid and liver damage. Higher levels can cause severe liver and immune system problems, and reductions in reproductive capacity. Similar effects must also occur in contaminated Arctic animals. Because of their propensity for dissolution in fat and oil, the PCBs concentrate in seal blubber. Consequently, Polar Bears, sitting at the top of the food chain, show very high concentrations. In relative terms, assuming the concentration in sea water is 1, the concentration in Arctic zooplankton in the worst affected areas would be 12,000, in fish 200,000–500,000, in seal blubber 500,000–1,500,000, and in Polar Bears, Beluga and Narwhal, up to 30,000,000. Such huge differences arise because of the ratio of predators to prey: it has been calculated that before it is full grown a Polar Bear will eat around 2.5t of Ringed Seal. Similar scale differences are seen in birds, with predatory birds (such as Glaucous Gulls) exhibiting concentrations of 5–10 times those of ducks and auks. Although the chemicals are not linked to birth defects in humans, this does not seem to be the case in bears, where there is a direct correlation between maternal organochloride concentration and an increased birth rate of hermaphrodite cubs. With levels of the contaminant 100 times the agreed safety levels in adult bears, and 1000 times higher in cubs, some experts believe contamination is a greater threat to the species' survival than global warming.

Because traditional food still forms a significant fraction of their diet, Inuit also show elevated PCB levels, this being particularly worrying in nursing mothers. The startling conclusion of one study was that some of the tested Inuit had such high concentrations of PCBs that their bodies would be classified as hazardous waste if they had to be disposed of as 'non-human' material. Studies also indicate that the ratio of girls to boys born to northern mothers has fallen from just under 1, consistent with the worldwide average, to 2:1. In some northern communities, only girls are being born. Pollutants such as PCBs and flame-retardants which mimic the human hormones are believed to be the cause.

And, of course, if it is bad for the Inuit it is worse for Polar Bears, which also have other pressures and no access to medical facilities. Production and use of PCBs is now limited or banned by international convention, but although that is good news, history suggests caution; a succession of compounds has been synthesised, used, found to be toxic, banned and, soon, replaced by other compounds, so that the cycle begins again. Severe doubts are already being expressed about replacements such as polybrominated diphenyl ethers (PDBEs), and more recently fears have been raised about neonicotinoids, particularly with regard to their effect on bees and songbirds.

One final aspect of pollution must also be mentioned. The beaches of Svalbard and other Arctic coasts are now littered with plastic detritus and the remnants of fishing gear, as are the beaches of the rest of the world. As elsewhere this pollution can kill local wildlife, with animals swallowing the plastic or becoming entangled in discarded nylon fishing line and trawl nets. The entanglement of plastics in the baleen plates of whales and the subsequent inability of the whale to seal its mouth have been implicated in whale deaths. Much of the plastic is in the form of 'nurdles', the lentil-sized pellets that are the basis of most plastic products. In a survey in early 2017, over 70% of UK beaches were contaminated by nurdles: on one beach more than 125,000 were found in a 100m stretch. At much the same time, a survey of the ocean floor between Greenland and Svalbard noted that there, at a depth of 2500m, the quantity of plastic debris had increased twenty-fold in a decade. It was a sobering finding, given that at the same time it was found that the world's consumption of plastic bottles had increased to about 20,000/second and that scientists are predicting that by 2050 the total tonnage of plastic in the world's oceans might be as much as the tonnage of fish.

### Ozone depletion

Ozone ($O_3$), a molecule of oxygen with three atoms compared to the more usual two, is present in the upper atmosphere, where it is formed by the interaction of free oxygen atoms with molecules of 'standard' oxygen. The free oxygen is produced by the interaction of ultraviolet (UV) radiation with oxygen molecules high in the atmosphere, the free atoms combining with other oxygen molecules at lower altitudes. About 90% of ozone is found within the stratosphere, at an altitude of 15–30km. Ozone is poisonous, and is therefore a hazard at sea level (where it is produced when the gases from petrol engines are broken down by sunlight), but in the upper atmosphere it acts as a protective shield to life on Earth by absorbing UV, particularly UV-B (that part of the UV spectrum with wavelengths in the range 290–320nm), which can cause cellular damage.

Despite there being only three molecules of ozone to every 10 million molecules of oxygen in the atmosphere, ozone's role as a shield is crucial, so there was immediate concern when, some 50 years ago, it was noticed that the ozone layer above Antarctica was thinning, and that a hole was appearing in the ozone cover. Similar, but reduced, thinning was then discovered above the Arctic. Data from balloon flights above Antarctica showed that the thinning was due largely to the presence of chlorine, which was destroying the ozone molecules (chemicals other than chlorine, such as bromine, are also involved, but chlorine is by far the most important). The amount of chlorine required to have a significant effect was beyond that which could arise from natural sources, and it was soon established that the source was compounds such as chlorofluorocarbons (CFCs), used as coolants in refrigerators, in aerosol spray cans and other industrial products. When these compounds are broken down by UV radiation they release free chlorine atoms that react with ozone molecules, stripping off a single oxygen atom to form chlorine monoxide (ClO). This reacts with another free oxygen atom to form an oxygen molecule ($O_2$), with the chlorine atom being released to destroy another ozone molecule. It is estimated that a single chlorine atom can destroy up to 100,000 ozone molecules before it is absorbed by nitrogen compounds, bringing an end to the trail of destruction.

As with global warming, the initial reaction of those with a vested interest in CFC production was that more research was needed and that the effects of reduced ozone levels had been exaggerated. (One American government official claimed that as people did not habitually stand in the sun there was no problem and, in any case, they could buy a hat and sunglasses.) However, in response to growing concern, in 1985 20 nations signed the Vienna Convention for the Protection of the Ozone Layer and accepted the need to control anthropogenic activities that caused the release of ozone-depleting chemicals. In 1987 these measures were incorporated into the Montreal Protocol on Substances that Deplete the Ozone Layer. The Protocol, and it subsequent revisions has been ratified by 197 parties, the first ever 'universal ratification' in the history of the United Nations. The Protocol called for the phasing out of CFCs and other chemicals linked to ozone depletion. However, the replacement chemicals, hydroflourocarbons (HFCs), while solving the ozone depletion problem were found to be potent greenhouse gases, requiring further UN conferences and, in October 2016, an amendment to the Montreal Protocol was agreed, phasing out use of HFCs. However, recently there has been a sudden rise in the level of atmospheric CFC-11, one of the most damaging of the various CFCs. It is not yet clear if this a deliberate use of the chemical or is arising from 'back street' recycling of refrigerators which is accidentally releasing the gas. And, in another instance of the unexpected consequences of apparently good intentions, it has recently been discovered that the newest replacement, dichloromethane, used as a propellant and in other processes, is also destroying ozone. Scientists hope that the chemical will soon be brought within the Protocol, but at the time of writing that has not happened, and the chemical is now feared to be responsible for the depletion of the ozone layer away from the poles.

### Nuclear weapons

In 1955, having forcibly removed the local native population, the Soviets exploded a nuclear weapon beneath the sea of Chernaya Bay, on the western side of the southern island of Novaya Zemlya: the sediments of the bay still contain high levels of radioactive caesium and plutonium. Between 1955 and 1990 the Soviet Union conducted a total of 132 nuclear tests, mainly at three land sites on Novaya Zemlya. These included 87 atmospheric tests (one of which was the largest bomb ever exploded, a 58-megaton hydrogen bomb on 23 October 1961), the tests resulting in the uncontained spread of radioactive material. In 28 of 42 underground explosions the resulting radioactive material was not completely contained (the remaining tests were underwater). At the test sites radiation levels remain high, but officially the rest of the islands are 'only marginally above background'. In addition to the weapons testing, as many as 11,000 containers of highly radioactive material are thought to have been dumped in the Barents and Kara seas, to each side of the islands. Radioactive debris in the seas apparently includes reactor units from the Soviet nuclear icebreaker *Lenin*. Unconfirmed reports suggest that the *Lenin* had two nuclear accidents in the mid-1960s. In one (or perhaps both), the reactor cores melted. Lives (maybe as many as 30) were lost, and on at least one occasion it was decided to dump the crippled core into the Kara Sea.

In April 1989, the Soviet nuclear submarine *Komsomolets* sank, about 120km south-west of Bear Island after a fire had forced the submarine to the surface and efforts to contain the incident failed. Only

The US Thule airbase, north-west Greenland.

one of a skeleton crew of five survived the sinking. No attempt has been made to salvage the boat, its reactor power system or the two nuclear-tipped torpedoes it is rumoured to have been carrying. Monitoring expeditions have detected increased levels of radioactive materials (Sr90, Cs137 and Co60) in benthic animals, but not at dangerous levels. There has been no indication, as yet, of massive increases in plutonium or nuclear fuel-related isotopes. Aged, redundant nuclear submarines have also been left to decay in bays along the northern Kola coast. (Nuclear submarine activity in the area is a consequence of Murmansk, an ice-free harbour, being the principal port for the Soviet Atlantic navy.) There are storage facilities for spent nuclear fuel at nearby sites on the Kola Peninsula. Petropavlosk, in southern Kamchatka, is one of the headquarters of the Russian Pacific Fleet, in particular its nuclear submarines, and it is known that radioactive leakage from waste disposal sites has occurred there.

US bomb testing chiefly took place in the Nevada desert, but three tests were carried out on Amchitka Island in the Aleutians (in 1965, 1969 and 1971). The tests were carried out to a background of strong local protest. The 1964 earthquake was fresh in the minds of local people and they feared that the explosions might precipitate a further shock. The groundswell of feeling led one group of protesters to acquire a ship in 1971 with the intention of sailing to Amchitka. A combination of bad weather and the US coastguard stopped them, but despite the failure of their mission the protest spawned a worldwide environmental movement. Although the reasons for its choice of name are lost in history, the protest group called themselves 'Greenpeace'.

During the 1939–45 War the United States created 17 military sites on Greenland. After the war the US continued to consider Greenland strategically important. Several stations of the Distant Early Warning (DEW) system were established on the island (and at other sites across North America). Most have been abandoned, but they leave a legacy of spilled oil and toxic waste that have been implicated in higher cancer rates of villagers close to some sites. In 1951 the Americans built an air base at Thule, the construction of which required the forced resettlement of local Inuit to Qannaq, 140km to the north. In January 1968, a B52 bomber crashed onto the

Norwegian border guards keeping an eye on Russia.

Although there has been a decline in overt military activity in the Bering Sea the border between Norway and Russia near the Øvre Pasvik National Park is still closely watched. In the upper photograph Norwegian border guards peer into Russia, while the photograph above shows a Russian watch tower, from which vantage point the Russians survey Norway. Behind the tower the town of Nikel is just discernible through the haze and smoke.

sea ice about 12km west of the base. A clean-up operation followed, though it was not immediately announced that the plane had been carrying nuclear weapons. It is claimed that the cancer rate in those involved in the clean-up was higher than normal, and the Danish government later paid compensation to workers who exhibited health

problems. Eventually it was admitted that the plane had been carrying four hydrogen bombs, and there had been an escape of plutonium into the environment (the nuclear fusion of a hydrogen bomb is triggered by a fission explosion, in this case plutonium-based). Secrecy over the four bombs has been maintained. The fusion elements of the device present minimal risk to the environment, but that is not the case with the plutonium. The Americans admit that not all the plutonium was recovered, with an amount variously stated to be 0.5–1.8kg having gone missing. Workers at the base claim that the clean up involved the scooping up of contaminated ice and snow into barrels, which were then stored at the base, but that ultimately the barrels rotted and the contaminated water they contained leaked into the ground. An environmental study was carried out, but the report – apparently running to 4,000 pages – has never been published and remains secret. Inuit hunters claim to have seen Musk Ox and seals with physical and internal deformities, but these are unconfirmed.

### Nuclear power

Russia has four nuclear reactors at the Polyarnye power station, near the city of Polyarnye Zori close to Murmansk. The reactors are VVER-type, not the RBMK design involved in the accident at Chernobyl. However, the two older reactors (which began commercial operations in 1973 and 1975) at the site are considered by western nuclear experts to be as dangerous as the Chernobyl type. The older reactors are due for final shutdown in 2018 and 2019, the newer plants (commissioned in 1981 and 1984) are planned for shutdown in 2026 and 2029.

There are also four 12MW reactors at Bilibino in Chukotka, built to power local gold and tin mines. The Bilibino reactors were due to be decommissioned in 2007, but their working life has been extended to 2018. The Russian plants are the world's only nuclear power stations built on permafrost. Russia is also planning to build floating nuclear power stations in part to provide power to oil and gas installations in Arctic waters. Sites close to the Kola, Yamal, Taimyr, Chukotka and Kamchatka peninsulas have been identified for the stations. Russia has three nuclear fuel reprocessing plants, at Kranoyarsk on the Yenisey River, and at Mayak and Tomsk on tributaries of the Ob River. Both Finland

The Russian nuclear-powered icebreaker *50Let Pobedy*.

A nuclear-powered Victor III class Russian attack submarine.

## Radioactive statistics

In all there have been more than 2,000 nuclear weapon tests, around half of these performed by the United States. It is estimated that the atmospheric testing of nuclear weapons and the leakage of radioactivity from underground tests has resulted in the release of around 4.2 tonnes of plutonium into the atmosphere and a further 3.8 tonnes into the ground. The total release of radiation is about $2.5 \times 10^{20}$Bq. For comparison, the Chernobyl accident, by far the worst accident in a civil nuclear reactor, released about $2 \times 10^{18}$Bq, about half of which comprised inert noble gases. The most recent sampling of the Kara Sea suggests that the sea's radioactive burden from weapons testing fall-out and dumped nuclear material amounts to a total of more than $3 \times 10^{15}$Bq. Some of the radioactive material in both the Barents and Kara seas originated at the UK's Sellafield nuclear fuel reprocessing plant: levels of Cs137 in Minke Whales, for example, show a direct correlation with releases from the plant.

## Climate change

As discussed earlier (see Chapter 12), the ratio of two isotopes of oxygen in ice cores, together with the distribution of plant life as indicated by the fossil record, enable a good estimate of the Earth's climate over the last several million years to be established. As well as investigating oxygen isotopes, ice cores also allow the variation of the concentration of other gases in the atmosphere to be measured, bubbles of air trapped in the ice being analysed for gas composition. Ice cores from Antarctica have allowed the concentrations of carbon dioxide ($CO_2$) and methane ($CH_4$), two important 'greenhouse' gases, to be studied over the last 400,000 years or so. As both these gases and oxygen can be sampled at the same time, the relationship of the Earth's temperature to the concentration of greenhouse gases can also be inferred. Pushing the relationship back further in time is more complicated as other natural processes must be considered, but the studies indicate a correlation between the concentration of greenhouse gases and the Earth's atmospheric temperature. Particularly significant is the fact that atmospheric $CO_2$ concentration during the Carboniferous/Permian Ice Age (some 300 million years ago) was almost the same

and Sweden also have nuclear power stations, though these are situated in the south of each country.

In 1986, Reactor 4 at the Chernobyl power station in the Ukraine exploded after a major reactor fault. The ejected radioactive material contaminated large areas of the northern hemisphere. In Scandinavia, the Sámi suffered a double problem as they were advised not to eat the meat of the Reindeer herds, but they could not sell it either as it was banned from public consumption (despite the Norwegian government raising the allowable level of Cs137 to 10 times that of the European Union – 6000Bq/kg rather than 600). It is estimated that 20,000 Reindeer were slaughtered and dumped. Scandinavian meat still has Cs137 levels which are above the EU limit.

## Greenhouse Gases (GHGs)

The main constituents responsible for the atmospheric greenhouse effect are water vapour, $CO_2$, $CH_4$, nitrous oxide ($N_2O$), ozone and chlorofluorocarbons (CFCs). Apart from the CFCs, the gases are produced by natural as well as anthropogenic processes: $CO_2$ from volcanoes and the metabolism of animals (including humans); methane from coal mines, by the decay of organic materials and the flatulence of ruminant mammals (which is no laughing matter when the number of domestic ruminants on Earth is considered, though the main source of methane is actually the animal's mouth rather than its rear even if the latter has more comic potential); and nitrous oxide from soil bacteria. But it is the man-made production that has caused the most alarm over recent years: $N_2O$ as an anaesthetic and aerosol propellant; $CH_4$ is released in oil and natural gas production; and $CO_2$ from the burning of fossil fuels. It is $CO_2$ that has become the 'shorthand' for all greenhouse gases.

Although the concentration of methane is much lower than that of $CO_2$, methane is 23 times more effective as a greenhouse gas, the fugitive releases in the production of natural gas outweighing many of the claimed advantages for switching from coal to gas for electricity production (without considering whether it is sensible to use a primary fuel to produce a secondary fuel at low conversion efficiencies): as we shall see below, temperature rise in the Arctic are also causing permafrost to thaw with consequent increases in methane release . CFC-12 ($CCl_2F_2$) is about 7,000 times more effective than $CO_2$. Its concentration is 700,000 times lower, which would seem to make it a negligible greenhouse gas, but there are many forms of CFCs, so when their individual contributions are added together the total effect is significant.

as today's levels (when the Earth is in a similarly cold phase).

As already noted, the Earth's atmosphere is essentially transparent to incident, short-wave solar radiation, but opaque to the long-wave radiation re-emitted from the Earth's surface. The long-wave radiation is absorbed by constituents of the atmosphere, its energy then being re-emitted and re-absorbed, resulting in warming. The incoming and outgoing radiations are balanced, and in the absence of an atmosphere the average temperature of the Earth would be about -20°C. In practice, it is about +15°C. This raising of the Earth's temperature is analogous to that within a greenhouse, so the process has become known as the 'greenhouse effect' and the atmospheric constituents which contribute to it as 'greenhouse gases (GHGs)'. In practice greenhouses work chiefly by preventing convective heat losses rather than absorption and re-emission, but the analogy is not unreasonable.

Some 30 years ago scientists began to report that the temperature of the Earth was rising. The temperature rise accelerated in the 1990s as a debate raged between those (chiefly scientists) who claimed it was a result of increases in GHGs in the atmosphere, these being the result of man-made emissions, chiefly from the burning of fossil fuels, and those who either denied the existence of the rise or claimed it was a natural occurrence, caused by changes in the Sun's output or other processes. Then

the rise in temperature slowed, the Earth's temperature apparently stabilising. The 'climate change deniers' as those who claimed there was either no rise, or that is was a natural event were now known, pointed to this as evidence of their claims. Scientists were not convinced, though at first the reason for the pause was unclear. Then, in 2010 the temperature rose again. It fell slightly in 2011, but this was followed by a series of rises. In 2014 the temperature of the Earth was the highest that had ever been recorded. But 2015 was hotter still and in 2016, the temperature was pushed up by a strong El Niño event (though it is considered that El Niño contributed only about 20% to the overall temperature rise). Figure 18.1 shows the variation of the Earth's temperature over the period 1880–2017, with an assumed 'zero' temperature for the period 1880–1951. The figures note a temperature rise to the end of 2015 of about 1°C. However, in early 2016 the January temperature was 1.15°C higher, and in February 1.35°C higher. Later months confirmed the sharp increase in the Earth's temperature. 2016 broke the record high of 2015: it was the hottest year on record with a land temperature 1.23°C above the pre-industrial value. (The ocean temperature was 0.99°C above the pre-industrial value: overall the Earth's temperature had risen by 1.1°C.) 2017 was cooler than 2016, but was still the third hottest on record, after 2015 and 2016, and was the hottest year not boosted by an El Niño event. 17 of

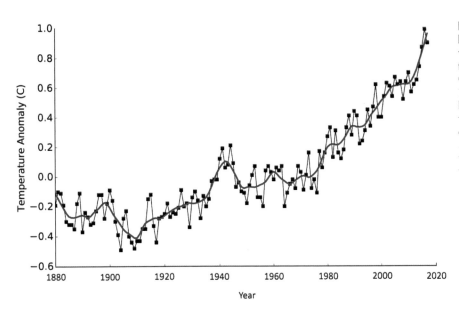

Figure 18.1 The increase in Earth's total (i.e. land and ocean) temperature since 1880. Drawn from data of the NASA Earth Observatory. The sharp increase in 2016 was boosted by a strong El Niño event, though it is considered this contributed only about 20% of the temperature rise. The black squares are the calculated values, the red curve being the 'best-fit' to the points.

the 18 hottest years ever recorded have now occurred in the last 18 years.

In 2016 scientists discovered that the missing heat in the decade of stable Earth temperature had been stored in the oceans. The increase in sea temperature had led to a massive 'bleaching' event with around 40% of the world's coral reefs having been affected to some degree: some reefs had died completely. Following two consecutive bleaching events in the austral summers of 2015/16 and 2016/17 almost two-thirds of the Great Barrier Reef has been damaged, with some Australian scientists having given up efforts to control water quality by combating farm and industrial pollution run-off and dumping because it seems pointless when climate change is causing such havoc.

Almost all experts now believe the increase in the Earth's temperature is a consequence of human activities, and that global warming has resulted in disruption of climate patterns, with more extreme weather events being seen. In the UK, the catastrophic floods seen in recent years are almost certainly due to warming of the Atlantic Ocean: the ocean warms the air above it, warmer air holds more moisture and so UK rainfall is higher. Scientists had also noted that the temperature rise in the Arctic was higher than the Earth's average rise. Therefore, before considering the response of the world's political structures to global warming the reasons the Arctic is warming faster need to be considered. It is also worth noting here that the oceans are not only becoming warmer, but more acidic

as a result of the absorption of $CO_2$ from the atmosphere: seawater is now 30% more acidic than it was at the end of the 19th century. Acidification is a threat to marine life: dissolved $CO_2$ breaks down calcite and aragonite which are the building blocks of shells and external skeletons. The acid also alters water chemistry and as many fish use the 'smell' of the water to find or avoid prey, acidification has been shown to alter their behaviour, something which could, eventually, also alter other brain functions and lead to entire food webs being disrupted.

## A warming Arctic

While the Earth has, to date, warmed by about 1°C, the Arctic has warmed by almost double that, with both summer and winter temperatures significantly warmer (and winter temperatures even warmer in some areas).

On Svalbard, a Global Seed Vault was opened in 2008. Buried in a mountain, the vault is a depository for around one million packets of seeds from plants vital to human food supplies and medicines. Svalbard was chosen because the permafrost underlying the island was considered impregnable. In the winter of 2016/2017, with local temperatures 7°C above normal, rain and melt water entered the vault. No seeds were damaged, but 24-hour monitoring has been instigated to avoid what might be a catastrophic loss.

The reason for the apparent anomaly in Arctic temperature rise in comparison to the Earth average lies in feedback mechanisms which are at work in the far north.

# Feedback mechanisms in the Arctic

## Reduction in albedo

The primary reason is a reduction in albedo, the reflection of incident solar radiation by ice and snow, the white surfaces of which act (as all white surfaces do) as mirrors. Once ice or snow is lost what lies beneath is exposed, and is invariably dark. Typically, sea ice reflects 85% of solar radiation, while ice-free ocean reflects only 10%. Because the dark ocean absorbs more of the incident solar radiation (as do all dark surfaces), the surface layer of the sea therefore warms, and the warmer water then laps at the ice edge, both reducing the coverage of the ice and thinning it: the average August 2016 summer surface layer temperature of the Arctic Ocean was 5°C higher than the 1980–2010 mean. But although the surface layer warms, the sea temperature is much lower than that of water in the tropics, and so there is much less evaporation in northern latitudes, the heat absorbed by the atmosphere tending to stay there, raising ambient temperature further. Loss of sea ice thus boosts the effect of the atmospheric $CO_2$ increase (by about 25%), the effect being the main reason the Arctic is warming faster than the planet as a whole. Warmer water and air then generate a subtle alteration to the pattern of oceanic and air currents, each of which tends to draw warmer water and air towards the region. Sea ice loss is accelerated, generating with, again, positive feedback raising Arctic temperatures still further

The melting of snow and ice on the land (snow cover and glacial ice) also reveals a darker surface, tundra, rock and glacial till. Though the albdeo of dark land surfaces is double that of the ocean (about 20%) it is still significantly less than that of sea ice, and similar warming is seen. The average June snow cover of the Arctic is now almost 60% lower than it was in 1979. The northern hemisphere has a greater proportion of the Earth's land mass than the southern hemisphere, and the Arctic Ocean is surrounded by land. Consequently, loss of snow cover is another positive feedback mechanism serving to raise Arctic temperatures. The heating of soils once held at lower temperatures beneath a blanket of snow also causes a rise in temperature of meltwater streams and, consequently, river temperatures. As some of the world's largest rivers discharge into the Arctic Ocean this enhances the sea temperature, with a consequent effect on sea ice thickness and area.

The resultant temperature rises have also caused the shrinkage of glaciers, not only in the Arctic, but in all alpine areas: Mount Kenya's glaciers have shrunk by 75% in the last 100 years, while Alaska's Glacier National Park has lost 90% of its ice over the same period. A study of 39 of Greenland's widest glaciers over the period 2000–2010 showed that those retreating (based on area change) were nine times the number advancing: the total ice loss was 1368km². The highest retreats were seen in the most northerly glaciers, with the Humboldt Glacier (in north-east Greenland) losing 311 km². In terms of summer end points, the nearby Petermann Glacier retreated 17 km between 2009 and 2010. (As a digression, some climate change deniers point to the advance of some glaciers as evidence that change is not occurring, but as we shall see below, one effect of climate change is to increase precipitation, and if this falls as snow on a glacier, it may, with sufficient volume, cause an advance even if glacial thinning processes are at work. As a further digression, use of the name 'denier' for most of those on the opposite side to climate scientists, appears incorrect, implying a genuine scepticism – but sceptics question evidence and seek to examine it. Most 'climate deniers' are actually 'climate cynics', questioning the motives of those who suggest climate change rather than addressing the evidence.)

## The northern migration of plants

As the Earth warms, trees will expand their range northward. Forests have even lower albedo than tundra, so again local energy absorption will be higher. It is an irony that the idea of planting trees to offset carbon use (an idea that appeals to some in developed countries) works only in the tropics where $CO_2$ absorption is higher – but where logging is intensive, negating the effect. That is not, of course, to say that planting trees is a bad thing. It is not: forests are an important part of the Earth's biosphere and loss of them has diminished both wildlife habitat and recreational opportunities. But the 'feelgood' factor relative to climate change should not be overstated, particularly in the Arctic. One other effect of global warming has been that some Arctic plants now bloom early, upsetting the local ecology if they then outcompete later bloomers.

## Water vapour concentration and permafrost thawing

Water vapour is itself a significant GHG, and as warmer air holds more water vapour, the increase represents a positive feedback mechanism for the Earth. But as the Arctic temperature is rising faster than the Earth average due to other feedbacks, it is a more effective feedback mechanism in the Arctic.

In addition, as snow cover reduces and air temperature rises, the active layer of the permafrost that underlies the Arctic tundra and boreal fringe deepens each summer, and a deeper active layer provides less insulation for the underlying permafrost: borehole data from northern Alaska indicate that the temperature of the permafrost has risen by as much as 4°C over the last 80–100 years. Organic material trapped in the permafrost warms, leading to increased microbial activity, and increased $CO_2$ and methane release into the atmosphere. Though the magnitude of the release of these GHGs as the permafrost thaws is debated, some estimates suggest that the upper (and therefore vulnerable) layer of the northern permafrost holds around 30% of all the carbon stored in the world's soils. Several years ago, Siberian scientists reported a record thaw in the West Siberian Bog, the world's largest peat bog: it is estimated that the bog holds 70,000 million tons of methane. Recent surveys of permafrost areas indicated that at the present rate of loss, some 75% of the total permafrost will have thawed to a depth of 3m within the next 20–30 years. One concern is that once it has thawed the peat will dry, and then burn. In autumn 2017 a huge fire burned 15km² of peat in west Greenland, adding its GHGs to the atmosphere. A continuing temperature rise may also result in the release of the methane held in gas hydrates. Gas hydrates are crystalline solids trapped in molecular water 'cages'. Methane chlorate, one such hydrate, is found in deep ocean floors, but also in Arctic continental shelves (e.g. the Mackenzie delta) and in permafrost. They are a potential energy source, but rising Arctic temperatures could result in a release of the trapped methane. Such a release could be catastrophic in terms of GHG concentration, and there is already evidence that methane from submarine permafrost thawing is bubbling to the surface in several places in the Arctic Ocean. Inevitably as sea ice coverage reduces and the sea temperature rises such releases will increase.

Changes in the active permafrost layer also have the potential to affect local hydrology. Predicting actual effects is difficult as they depend on the distribution of water and on water chemistry, but concerns have been raised that lakes might shrink or disappear with a consequent loss of wildlife habitat. In 2016 a further dramatic effect of permafrost loss was noted when there was an anthrax outbreak among Reindeer and herders in Russia's Yamal region, at least 24 people and several thousand animals dying. Anthrax can lie dormant in frozen carcasses for centuries, spores being released when the permafrost and the carcass within it, thaws. In the case of the Yamal outbreak permafrost thaw was exacerbated by the number of Reindeer. The area is considered capable of sustaining only half the number of animals present, overgrazing eliminating plants which had been acting as a protective blanket to the permafrost. Loss of the blanket allowed high ambient temperatures and sunlight to reach the soil. While culling of this, and other herds, would reduce the anthrax risk in the short term, permafrost thawing will inevitably occur as ambient temperatures increase.

Interestingly, permafrost thaw has given a new purpose to the idea of recreating the Mammoth by taking DNA from remains long frozen in Arctic permafrost and using it in the egg cell of an Asian Elephant to produce, initially, a hybrid animal. Once seen as a scientific experiment of limited value and questionable ethics, proponents now envisage the advantage of Mammoth herds again wandering the tundra, their feet punching holes in the snow cover to allow freezing air to reach the soil, maintaining the permafrost. Whether true Mammoths should, or can be created, and in sufficient numbers to test this hypothesis is a very different matter.

## Particulates on ice and snow

It has recently been discovered that a potential feedback effect initially dismissed as irrelevant is, in fact, significant – the deposition of dark particulates, chiefly the soot from diesel burning and industrial processes, but also from forest burning (either natural or the 'slash and burn' clearance which has accompanied palm oil production). Such deposition reduces albedo, and creates local thawing, with bacteria and vegetation collecting in the meltwater causing the meltwater pool to expand, causing further albedo reduction.

# The impact of temperature rise in Arctic

## *Sea ice loss*

In 1979 the first year in which satellite imagery allowed a reasonably accurate value for Arctic sea ice coverage, the winter maximum was approximately 16.3 million km². While it was more difficult to measure the coverage prior to 1979 it is believed that the winter maximum had been stable (though with annual variations resulting from natural fluctuations in temperature). By 2016 coverage had fallen to 14.4 million km², a fall of almost 12%; the winter coverage was the lowest ever recorded. The reduction is shown in Figure 18.2. While the fall in winter ice coverage had been significant, the fall in summer ice coverage has been even more pronounced. The coloured lines on Figure 18.2 are the sea ice overage for the years from 2010. While 2012 saw an anomalously low summer ice coverage the trend over the last five years has been downward. Very high air temperatures during autumn 2016 meant that 2016/17 sea ice coverage was reduced: the winter ice coverage of 14.42 million square kilometres on 7 March 2017 was the lowest recorded since satellite records began almost 40 years ago. More significantly,

the reduction in winter sea ice formation will lead to a reduced summer sea ice coverage in 2017; on the basis of late 2016 data it is probable that the 2017 summer sea ice cover will be lower than that seen in 2012.

The sea ice is also thinning, with a volume reduction of 20% being seen in winter 2016/2017 relative to the mean of winters 1980–2016, and by about 50% in summer, again 2016 relative to 1980–2015. The average ice thickness was 3.6m in 1975: in 2012 it had reduced to 1.25m. Much of the thinning is at the continental shores of the Arctic Ocean, i.e. away from the pole, where first-year ice is freezing later and thawing earlier so that the multi-year ice which used to exist (and could be 5m thick) no longer forms. Figure 18.3 shows the reduction in Artic sea ice volume from 1980 to 2017. Were the ice loss seen to date (3,000km³/decade) to continue at the most recently measured rate the Arctic Ocean may be ice-free by 2035, but if the decrease of 2012 were to be replicated continuously, then an ice-free summer might before 2025; even if that were not to happen, the maintenance of present $CO_2$ emission rates means there will be no summer sea ice by the mid-2040s at the latest. Fig. 18.4 shows the variation in average ice thickness through the year, noting the year-on-year change during the new

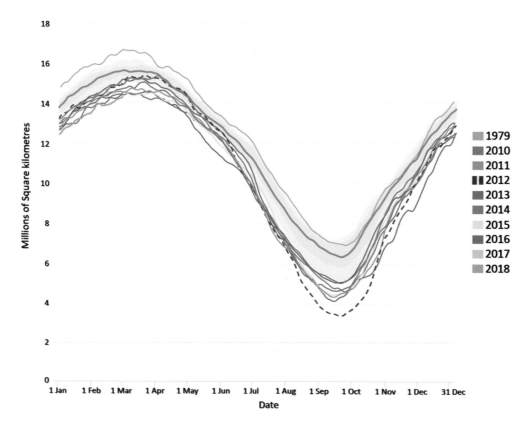

Figure 18.2 Variation of ice coverage through the year. The left scale is the sea area with at least 15% ice coverage. The black line is the average coverage for the years 1981–2010, with the grey shaded area representing ±2 standard deviations. The orange line is the coverage in 1979. As can be seen in that year the coverage was significantly higher in all months than the 1981–2010 average. The years from 2010 are represent by different coloured lines. The green dotted line is 2012: in that year the summer ice coverage was the minimum seen since records began in 2012. The 2018 line covers January and February only. Data from National Snow and Ice Data Centre, University of Colorado, Boulder, USA.

Legend: 1979, 2010, 2011, 2012, 2013, 2014, 2015, 2016, 2017, 2018

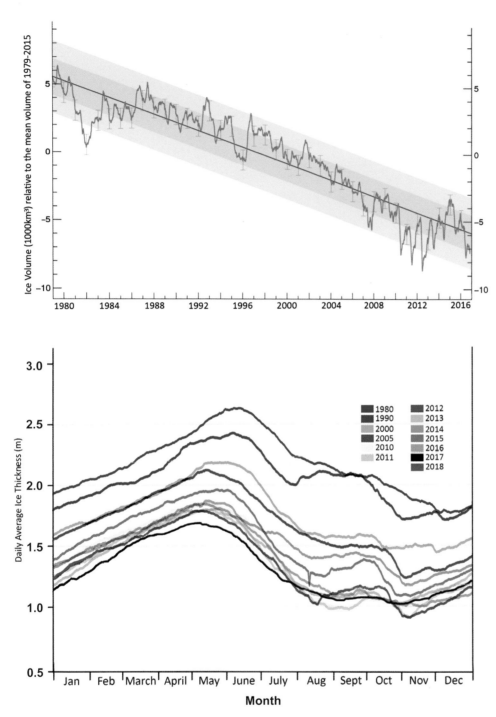

Figure 18.3 Variation of the volume of mean volume of sea ice on the Arctic Ocean during the period 1979-2017. The zero point on the vertical axis is 20.3 thousand km³, the average annual volume for the period 1979-2015 (i.e. there was c.25,000km³ in 1979 and c.15,000km³ in 2015). The rate of decline based on the line shown (the best fit to the data) is 2,800km³/decade. The shaded areas represent ±1 and ±2 standard deviations (1sd = 1,000 km³/decade) with respect to the best fit. Data from the Pan-Arctic Ice Ocean Modelling and Assimilation System (PIOMAS) developed at the Applied Physics Laboratory/Polar Science Centre of the University of Washington, Seattle, USA.

Figure 18.4 Daily average ice thickness across the Arctic, plotted for 1980, 1990 and certain years of the new millennium. Data from the Pan-Arctic Ice Ocean Modelling and Assimilation System (PIOMAS) with thanks to Dr Axel Schweiger.

millennium relative to the values in 1980 and 1990. One aspect of the reduction in ice thickness which should be noted is that thin ice, too thin to hold the weight of a seal or Polar Bear, is registered as ice by a satellite, so the data for sea ice coverage is likely to overestimate the true extent of 'real' (i.e. solid, substantial) ice. An example of this was seen in the summer of 2018 when, following an unusually, and remarkably, warm Arctic winter which both thinned sea ice and altered the normal wind

patterns, the sea ice along Greenland's northern coast began to break up. This area had always been assumed to hold the most substantial and persistent of all Arctic ice, and its break-up was unexpected. When asked his opinion on the possible implications for both the Arctic and the rest of world's climate, one expert felt only able to say they were 'scary'.

Global warming has implications for all of Earth's species, including humans. A study in 2016 noted that

Radar domes at Ny Ålesund, Spitsbergen, Svalbard.

of almost 1000 species which occupied the 'warm edge' of their range 47% had disappeared from those areas as the planet's temperature increased. For Arctic species, the problem is amplified by the fact that the area is warming at twice the rate of the Earth as a whole. The loss of sea ice has implications for Arctic species. Algal growth on the underside of the ice is the basis of a food chain. The algae are eaten by zooplankton which are consumed by fish which feed seals which are hunted by Polar Bears. But as the ice shrinks, the algae reduce, and the primary zooplankton are being replaced by warmer water species which have less fat deposits, a loss which is transmitted along the food chain. Thinner ice is also a hazard for seal pups which may fall through the ice and either drown or die of hypothermia.

Ironically, sea ice loss will also make the exploration for oil in the Arctic easier, though it might also accelerate the territory-grab in the area. In 2015 Norway began to look for oil in the sea around Svalbard. Although in 2010 Norway and Russia had agreed a frontier line in the Barents Sea, ending a dispute over ownership that had lasted 40 years, Russia objected to the Svalbard oil search. While Norway claims Svalbard lies on the Scandinavian continental shelf, a claim which reinforces its claim of sovereignty, Russia claims Svalbard is separate, and has a shelf of its own which comes under the Svalbard Treaty. A visit by the Russian Deputy Prime Minister to Svalbard in 2015 therefore made Norway nervous, and as a sign of their desire not to antagonize a powerful neighbour, late that year the editor of a local Norwegian newspaper was apparently sacked after demands by the Russians following the paper's unfavourable coverage of Russian oil

exploration in the Barents Sea. The dispute over ownership of the seas around Svalbard rumbles on. But such actions are hardly likely to placate a nation determined to stake its claim in the Arctic, and over recent years Russia has extended the area of the Arctic it considers to be Russian as well as planting a flag on the ocean floor below the North Pole in what was seen as a statement of ownership. If the world continues to burn fossil fuels and the Arctic becomes one of the last places where they are available it is not difficult to perceive a time when ownership claims on the area become a source of conflict.

The thinning of sea ice also makes hunting more hazardous for the Inuit, while thinning of freshwater ice also causes problems for other native Arctic dwellers. In the Yukon during the last two northern winters, the Gwich'in people who had travelled on frozen rivers in pursuit of Caribou for centuries found that they were no longer able to trust the ice coverage: there and elsewhere, people using snow scooters on river ice died when the ice failed.

### Rain, snow and ice

Global warming will inevitably lead to increased evaporation from the Earth's oceans and, consequently, an increase in both cloud cover and precipitation. Increased cloud cover in the Arctic will lead to warmer winters because of back-reflection of infrared radiation from the clouds and a reduced number of cloud-free skies, but tend to reduce summer temperatures because of albedo from the clouds, though this reduction is very unlikely to have a significant effect on the area's overall temperature rise. Current best estimates are that Arctic precipitation will increase by 20% by 2100, with precipitation increasing faster in winter than summer. Much of the summer increase will fall as rain, but winter rainfall is also likely to increase as winter temperatures increase. Nothing clears lying snow faster than rain: the effect of increased rainfall will be to further reduce snow cover. And even if the enhanced precipitation falls as snow the news is not necessarily good, as the albedo of wet snow is much lower than that of dry snow. Either way, the consequence is yet another positive feedback mechanism to drive Arctic warming.

If the increased precipitation falls as snow on the upper reaches of glaciers, glacial thickness will increase. However, if it falls as rain, ice thickness will quickly reduce. The effect of winter rain can also be devastating on wildlife. If rain is followed by hard and continuous frosts, the ground can disappear under a coating of ice as tough to penetrate as sheet metal. Already in

Reindeer use their hooves to scrape through lying snow to reach food. If it rains and then freezes, the iron-hard coating of ice can be impossible to break through.

## A warmer Earth and a colder northern Europe?

The likely increase in precipitation in a warmer world will lead to enhanced flow in the rivers that drain northern Russia and northern North America into the Arctic Ocean. This influx of freshwater might influence the flow of the North Atlantic Drift, which keeps winters in northern Europe mild and wet by reducing the Atlantic conveyor (see Chapter 4). The conveyor is a thermohaline current i.e. salinity- and temperature-dependent. An increase in freshwater flow will reduce salinity, and the Earth's oceans are warming. Greenlandic ice cores suggest that at the end of the last Ice Age there was an abrupt cooling on the island of about 5°C, followed by an equally abrupt warming. Such rapid changes are the signature of natural systems that have more than one stable configuration, and the rapid change indicated by the ice cores could have been a switching off and on of the Drift. In a report published in 2015 scientists noted that the Atlantic conveyor had weakened significantly after the 1970s, and that although there had been a partial recovery in recent years, the probability of enhanced melting of the Greenland ice sheet due to global warming might cause a future reduction or even a cessation of the conveyor. If the conveyor were to stop it is likely that northern Europe, particularly Britain and Norway, would see very much colder winters.

Scandinavia there have been such incidents, Reindeer herds suffering as they were unable to break through the ice cover to reach forage. These incidents are thought to have been related to the North Atlantic Oscillation (NAO). When the NAO Index is high, the resulting warmer, wetter winters lead to high snowfall and freeze-thaw conditions that make life difficult for female deer: their condition deteriorates and calf mortality increases. With a warming climate such incidents may increase. Some experts attribute the fall in the population of Peary Caribou in Canada from about 26,000 animals in the early 1960s to around just 1,000 by the end of the 20th century to warm, wet winters. Others are more cautious, though it is clear that at least some of the reduction has been due to autumn/winter rain: in both 1973–74 and 1995–96, Peary Caribou and Musk Oxen died in large numbers on Bathurst Island when frozen rain sheathed the ground. More recently (2016) researchers in Svalbard found that the weight of island Reindeers had decreased by 12% over the last 20 years as a result of such 'rain-on-snow' incidents, while on the Yamal Peninsula similar incidents have resulted in high mortality in Nenet Reindeer herds. An increase in winter rain, and a related increase in rain-on-snow incidents could also seriously affect mainland North America's Caribou herds, and other foraging animals.

An aerial photograph of surface water accumulating on Greenland's inland ice.

## Storms and sea levels

Coupled with an increase in precipitation, it is predicted that the Earth's climate will experience an increase in storminess. Indeed, such an increase is already being seen with high hurricane and typhoon counts, and more locally in the succession of high rainfall storms in the UK which have led to winter flooding. Storms produce or magnify waves in open water, and waves crashing against the sea ice edge may disrupt it, the fragmented ice being easier to melt than the bulk ice, so that overall ice shrinkage increases. Storms also increase coastal erosion in much the same way, with open water waves crashing against and disrupting the land edge (a process amplified in permafrost areas by permafrost thawing), and as a large fraction of the Earth's population, and a significant fraction of the Arctic's native dwellers, lives in coastal areas this will be significant. Coastal dwellers will also be affected by sea level rise: it is estimated that a 3°C rise in Earth's temperature would result in Shanghai, Osaka and Miami, among others, being under water.

The loss of sea ice does not, of itself, cause a rise in sea level, though a point often made by climate change deniers that there would be no effect misses a crucial point. It is true that if an ice cube is placed in a glass and it is then filled to the brim, the effect of melting the cube is minimal (as the cube displaces its mass of water when floating, but adds its volume once melted, and the density of ice is less than that of water there would be a minimal overflow), but as an analogy of the Earth the exercise is deeply flawed. The density of water is temperature dependent. For a small volume of water

over a limited temperature range the change is indeed small – but there is a vast amount of water in the world's oceans and so even a small temperature change can make a huge difference, with 33% of currently observed sea level rise being due to the temperature increase in seawater temperature. Loss of land ice is, however, critical to sea level rise, and the largest land ice mases are on Antarctica (by far the largest) and Greenland.

Currently, sea levels are rising by about 3mm/year, but computer models including projected global temperature increases predict rises by up to 80cm by 2100. Since 1996 the melt area of the Greenland ice sheet had been increasing year by year, with a maximum in 2012. In 2015 the melt area was 85,000km² greater than the average for the period 1981–2010, aided, in part, by an alteration in the position of the northern polar jet stream (jet streams are narrow corridors of high speed air flow situated close to the tropopause caused by heat in the atmosphere coupled with the Earth's rotation and, ironically, the shift in jet stream is almost certainly a consequence of sea ice loss) which caused not only the succession of storms which have troubled the UK in recent years, but created a 'heat dome' of warm, southern air to form over Greenland. Were the sheet to melt completely, sea levels across the Earth would rise by about 7m. Such a rise would be catastrophic for coastal dwellers, but is minimal in comparison to the rise if the Antarctic ice sheet were to melt: if that were to melt sea levels would rise by over 60m. However, it is widely believed that this is extremely unlikely as the predicted increase in precipitation resulting from global warming would fall as snow on the Antarctic plateau, where altitude enhances the extreme cold of the southern continent. The increase in snow depth on the plateau outweighs the loss of ice sheet at the coast, so that overall the effect on sea level is negative, though small. Sea level rise would be catastrophic for, as the most obvious example, the inhabitants of Bangladesh where much of the agricultural land is barely above sea level, as well as for islands such as the Maldives where the economy is based on tourism. Within the Arctic it has already had an impact, the inhabitants of Shishmaref, a village of 600 Inupiat Eskimos on a land bar off the northern, Chukchi Sea, coast of Alaska have voted to relocate to the mainland as rising sea levels and increased erosion are threatening their houses. Sea level rise could also be deadly for some animal species: in 2016 the Bramble Cay Melomys (*Melomys rubicola*) was the first definite extinction of a species due to human-induced global warming. First noted in 1870, the rodent inhabited the 4ha Bramble Cay, an island off northern Australia whose highest point was no more than 3m above sea level. Not definitely recorded since 2009 a search in June 2016 failed to find any evidence of the rodent, and noted that 97% of its habitat had been lost to rising sea level in 10 years. It is considered the rodent is now extinct.

While it might be argued that the loss of one species, though regrettable, is not catastrophic, a recent investigation (late 2016) noted that across almost 1000 species studied, local extinctions as a consequence of climate change had occurred in 47%. Ultimately, of course, enough local extinctions results in total extinction.

### The northern movement of vegetation and species

As Earth's temperature rises, southern species of plants and animals will expand their ranges northward. The effect of forests invading the tundra has already been mentioned as a positive feedback mechanism in the Arctic, but all northward range expansions will impact Arctic ecology. Shrinkage of the tundra is a reduction in habitat for both resident species and the migrant birds that breed there. The living space of residents will also be squeezed as southern species head north: for example,

### A warming Arctic: benefits and threats

The reduction of sea-ice coverage will allow easier shipping in Arctic seas. It may allow both the North-West and North-East Passages to become available to conventional ships, as opposed to the ice breakers and ice-strengthened ships that currently ply these waters. These possibilities were brought into sharpfocus when in 2016 the *Crystal Serenity*, owned by Crystal Cruises, carried 1000 passengers through the North-West Passage (from Seward) in late 2016. The ship used low sulphur fuel and had tight controls on the discharge of water and rubbish, though this did little to allay the fears of environmentalists. The cruise brought both benefits and threats: coming ashore in groups of 150 to the town of Cambridge Bay (population 1500), the passengers brought C$100,000 to the economy, but the drummers, dancers etc. laid on for their entertainment were left exhausted and another (conventional, smaller) ship arrived the following day. Local leaders wished there had been a day's rest, but other communities were concerned that the centralisation of the wealth meant smaller villages were missing out and that the return from sales of carvings etc. were too small when measured against effort required to entertain the visitors. A few months after the transit, another ship, the Finnish ice-breaker *Nordica*, completed the earliest-ever transit: the Passage is now opening earlier and staying open longer. Then, in August 2017 a new Russian tanker, the *Christophe de Margerie*, made the first transit of the North-East Passage without icebreaker escort, carrying liquid natural gas from a gas field near the Yamal Peninsula.

Lack of sea ice may also allow easier exploitation of the oil and gas reserves below the seas of the northern continental shelf. Each of these, which many view as benefits, will bring with them the threat of increased pollution. There may also be sovereignty problems as land previously thought too difficult to exploit becomes valuable.

Reduced sea-ice cover may allow the opening of northern fisheries, while increased Arctic temperatures may allow the northward expansion of agriculture. Each is a benefit to an overpopulated, hungry world, but the farming of former tundra will result in a reduction of habitat for northern wildlife, while increased fishing will apply further pressure to an already over-exploited resource. Finally, perhaps the most startling benefit of a warming world was the news that in a blind tasting Finnish wine fought a draw with French vintners.

as already noted, the northern limit of the Arctic Fox's distribution is limited by food and den sites, while the southern limit is defined by the northern limit of the Red Fox, which outcompetes (and predates) its smaller cousin where the two overlap. Ultimately the Arctic Fox might be left with nowhere to go: for terrestrial animals pushed northward by competitors better suited to the changing habitat, the Arctic Ocean represents a final border. The same is true for trees. White Spruce is the most widespread boreal species in North America and is a valuable timber tree. But it is sensitive to temperature. As temperature increases the tree's growth rate decreases. If the temperature rises high enough the tree dies. As the ocean limits its northward spread, the tree's range - expanding northwards, but shrinking to the south - may eventually decrease dramatically.

More immediately devastating is the effect on breeding seabirds of the movement of fish. Northern species of fish are moving further north as sea temperatures increase, the fish heading for cooler waters in pursuit of prey species also seeking a cooler environment. In the Bering Sea, the increase of El Niño activity (coupled ocean-atmosphere warming events in the Pacific Ocean) has contributed to already rising sea temperatures. Many fish species can respond quickly to such changes (though some use traditional spawning grounds and so may starve when traditional prey abundance is reduced: this can be a real threat to populations already at risk due to overfishing). But birds cannot respond as fast as zooplankton and fish. Breeding areas are in relatively short supply and consequently have been used for millennia. The birds are 'programmed' to return to these traditional breeding grounds. If the fish depart the birds will need to fly further to find enough food to feed their chicks. Fewer chicks are then raised and the species becomes endangered. Reductions in the number of chicks raised and in breeding populations are already being seen. In Britain, seabirds dying of starvation have been seen in recent years, along with reductions in breeding populations. One reason is the northern movement and overfishing of sandeels, but there has also been an explosion in the number of pipefish (Syngnathidae: relatives of the seahorses), a formerly southern group. Pipefish are longer than sandeels and are unsuitable food for chicks, with instances of birds being unable to either swallow or regurgitate the fish,

## El Niño

El Niño (the boy child), was named by Peruvian fishermen, who had known of it for many years before it attracted the attention of scientists, because it is usually seen around Christmas. The fishermen noted the reduction in their catches when the phenomenon occurred. It occurs every few years and is a warm phase of the natural Southern Oscillation, akin to the North Atlantic Oscillation, with warm Pacific waters flowing east towards South America. El Niño may be relatively benign, but the increased temperature in eastern Pacific waters and a resulting cooling of the western Pacific can disrupt world weather patterns, sometimes very significantly.

and choking to death. Fears have been expressed that prey losses, coupled with deaths due to offshore wind farms, may result in the loss of the Atlantic Puffin in the not-so-distant future.

Rising sea temperatures not only cause fish species to migrate. The temperature gradient induced in the sea causes turbulence, bringing cooler waters and nutrients towards the surface. Plankton blooms were seen in the Kara Sea in 2013 when the surface temperature rose by 5°C above the 'pre-warming' average. While exceptional at the time, the Kara's surface temperature is now being seen across the Arctic where surface temperatures of 4°C are commonplace. Large areas of warm open water drive ambient air temperatures up, with local temperatures of 20°C above normal now being seen frequently. The combination of warm surface water and air temperatures make ice formation impossible, adding a further positive feedback mechanism to a warming Arctic as well as influencing the local seawater ecology in unpredictable ways.

Climate change also allows agriculture to move north, and this will inevitably bring northern species into conflict with humans. Despite protective legislation, wolves in Fennoscandia are frequently killed by livestock farmers, and recently both farmers and foresters in Sweden have been calling for a reduction in the Elk (Moose) population as young saplings and crops are eaten or trampled and fences damaged so that livestock escapes. At the same time,

Norwegian Reindeer herders and sheep farmers had called on the government to allow a cull of 25% of local Golden Eagles which, they claimed, were taking calves and lambs.

## The response to climate change

A recent UN Environment Programme report (Global Environmental Outlook (GEO) 6) published in 2016, followed an earlier version (GEO 5, 2012) in concluding that degradation of Earth's natural resources by humans is rapidly outpacing the planet's ability to absorb the damage. Data clearly shows that since the 1980s mankind's demands have exceeded the planet's capacity to regenerate: the best estimate being that mankind now requires 1.2 Earths to continue to live at the present rate of demand. Water resources, air pollution, marine ecosystem damage, land degradation by modern agriculture, as well as the obvious depletion of non-renewable resources by mining all indicate this phenomenon. All these are exacerbated by climate change.

In 1988 the Intergovernmental Panel on Climate Change (IPCC) was established by the United Nations to provide advice on climate change. The IPCC does not carry out original research, but bases its findings on published work. It produced its first assessment in 1990 and has subsequently produced four further assessments, the latest in 2014. The various IPCC assessments have made increasingly sobering reading, so much so that there have been clear attempts by interested parties to criticise or downplay the findings, with numerous tales (some being rumours, but others based on facts) that companies (chiefly oil companies) were paying scientists to produce contradictory reports, and governments were pressuring scientists to 'water down' reports, and even banning some from speaking at meetings and threatening others with loss of grants or jobs. The IPCC assessments have led to the development of worldwide attempts to counter the effects of man-made climate change. The first was the Kyoto Protocol, adopted in 1997, which called for reductions in emissions by developed countries, on the grounds these were mainly responsible for the increased levels of GHGs. The Protocol was signed by many nations, but not the USA (which also attempted to persuade European business and political leaders to abandon their support) or by China and India, which were both developing rapidly.

The years since Kyoto have seen a succession of international meetings organised with the intention of tackling climate change and/or building on the Protocol, the latest being the 21st Conference of Parties (COP) held in Paris just a few days after the terrorist attacks of 13 November 2015. Prior to COP21 important figures were released. The best estimate of the concentration of $CO_2$ in the atmosphere before the Industrial Revolution is 0.028% by volume (280ppmv, or parts per million by volume). At the start of the Paris talks the concentration was 398 with the years to 2013 having seen the highest annual rates of rise ever recorded: experts believe that catastrophic, irreversible climate change will occur at a concentration of 450. Equally sobering was the news that 2016 was the hottest year since records began (in the mid-19th century) and that during the Arctic winter temperatures at the North Pole were reaching 5°C, which was 35°C above average. (As a digression, when temperatures of -8°C to -12°C hit the UK in late February/early March 2018, Britain was actually several degrees colder than the still dark Pole.) A report published to coincide with the Paris talks noted that while Chinese emissions had fallen, India's decision to build coal-fired power stations to aid its industrial developments had caused an increase. The Paris meeting lasted an exhausting fortnight. As expected, India's Prime Minister Narendra Modi took a hard line, signing an alliance with 120 other countries to push ahead with solar power, but declining to accept any curb on development. But ultimately Paris did achieve an agreement, 195 countries stating their intention of limiting Earth's temperature rise to 'well below 2°C above pre-industrial level' and 'pursu(ing) efforts to limit the rise to 1.5°C' with an accelerated reduction in fossil-fuel burning, increase in renewable energy sources and a carbon market for the trading of emissions. The agreement's aim is for zero emissions by 'the second half of the century' and a budget of $100 billion to aid poorer countries adapt. Agreement was made on 12 December, though it will not come into effect until signed by all parties during the period 22 April 2016 – 21 April 2017 (at the time of writing almost 200 countries had signed). The agreement allows for stocktaking every five years to ensure compliance. However, countries would be allowed to set their own targets (Nationally Determined Contributions (NDCs)) of emission cuts within a general framework. The NDCs should be 'ambitious' but sticking to them will not be legally binding. There was also no mechanism for forcing countries to comply though there will be a 'name and encourage' policy. Naysayers immediately noted that China, the US, India, Brazil, Canada, Russia, Indonesia and Australia which between them are responsible for more than 50% of GHGs, will be subject only to voluntary cuts without any penalty or fiscal pressure to comply. The agreement is also a fine example of circular logic in which the premise is contained within the conclusion – we will limit the temperature rise to 1.5°C because we will pursue efforts to limit the rise to 1.5°C. It is also worth noting that in a report in late 2017 one leading group of climate scientists noted that there was only a 5% chance of maintaining Earth's temperature rise below 2°C by 2100, and only a 1% chance of it being just 1.5°C. Even if the 1.5°C were to be realised, positive feedback effects would probably mean a loss of summer sea ice in the Arctic, with little hope of its being re-established.

In the aftermath of the Paris talks American president Barack Obama hailed the agreement, claiming it would create more jobs and economic growth, missing the rather obvious point that the pursuit of unfettered growth is what has brought us to where we are – an irony not lost on some experts dismayed by the lack of teeth in the agreement. In 2015 the UK government granted a further £4.8 billion in tax breaks for North Sea oil and gas extraction (for the period 2015–2020) and halved the tax to be paid by fracking companies. The 2015 Infrastructure Act also legally binds the present and future governments to maximise fossil fuel extraction despite the 2008 Climate Change Act requiring a reduction in fossil fuel burning. In the same year the International Monetary Fund noted that worldwide the hidden subsidies of governments to the oil and gas industries were, annually, about £26 billion. Only a supreme optimist would believe that with all the efforts going into discovering new fossil fuel sources the outcome would be other than the extraction and burning of them once found, particularly with the human population increasing by 1.2% annually and the demand of people in developing countries for the lifestyle of those in the developed world.

## The future

While it was easy to be pessimistic about the future in the wake of the Paris talks, the agreement did represent an acknowledgement that a problem existed and did set a framework for action. Sadly, it was followed by an announcement that Earth's temperature will probably reach 2.7°C before it peaks and declines unless the NDCs are drastic and the 5-year stocktakes are equally drastic – as drastic as a cessation of GHG emissions altogether by 2020. If the agreement is to be adhered to then there must be no drilling in the Arctic – but also no offshore drilling, no fracking (because of methane release) and no more subsidies for oil and gas exploration. Looking at that list it is not hard to envisage that if one of the signatories decides it cannot replace fossil fuels with renewables until 2050 because of pricing, then others will follow suit to avoid being left behind in the economic race and Earth's temperature will rise towards 4°C and catastrophe, not only for the Arctic, but for the world. It is worth noting here that while the IPCC suggestion is that the earth's temperature will perhaps reach 4.8°C by 2100 in a 'business as usual' world, new research suggests 'at least 4.8°C and perhaps 7.4°C' for that scenario, while other work suggests that if the lowest figure for the remaining fossil fuel resources on Earth were to be burnt that would release a further 5 million million tonnes of carbon to the atmosphere resulting in a temperature rise of 6.4–9.5°C, and an Arctic temperature rise of 14.7–19.5°C. Such increases would also make large parts of Africa and the Indian sub-continent uninhabitable as neither plants nor humans could survive in temperatures in excess of 60°C. The result would be mass human migration northward on an unprecedented, and unsustainable, scale.

In 2016 it was reported that for the second year in succession $CO_2$ emissions had risen by only 0.2% (chiefly because China was burning less coal) at a time when global economic growth was about 3%). There was a caveat, in that the figure related only to fossil fuel burning, but the news was encouraging. In the same month, Donald Trump won the US Presidential election. During the televised debates between the prospective candidates Donald Trump and Hillary Clinton, there was no reference at all to climate change and its potential impact on the USA or the world. In Twitter feeds Donald Trump has noted that 'the concept of global warming was created by the Chinese in order to make US manufacturing non-competitive' (2012) and 'global warming is a total, and very expensive, hoax' (2013) – though, interestingly, one of his companies, a golf resort in Ireland, has applied for planning permission to build a sea wall for the specific purpose of protecting the site from the effects of global warming – so the hopes generated by Obama's position seem likely to be dashed. On the difference between a sceptic and cynic suggested above, Trump would be considered a climate cynic rather than a climate denier. In November 2017 it was reported that $CO_2$ emissions for the year were likely to be 2%, a stark change from the 0.2% annual increase which had raised so many hopes.

The Paris agreement is based almost entirely on trust and goodwill. It would appear Donald Trump – a supporter of the coal industry who has called coal 'clean energy' – will undermine both, a concern heightened when President Trump appointed Scott Pruitt as head of the US Environmental Protection Agency (EPA). Pruitt has stated that he disagrees with the premise that $CO_2$ is the primary contributor to global warming and that there was 'tremendous disagreement' about its impact, when among climate scientists 'tremendous agreement' would be a more accurate description. One of Pruitt's first decisions as EPA head was to decline renewal of contracts for half the scientists on the Agency's Board of Scientific Counsellors. Another was the decision not to ban Vulcan, a chloropyrifos pesticide which the Obama administration had earmarked for banning because it was known to be harmful to humans, particularly children: the pesticide is also doubtlessly toxic to wildlife as well.

As well as appointing Pruitt, President Trump also cut the Agency's budget by one-third. One leading US climate expert noted that because Donald Trump has said he would rip up the US commitments in the fight against climate change, his Presidency would mean 'game over' for the Earth's climate. At first Trump appeared to have softened his attitude towards the Paris agreement, but in June 2017 he announced US withdrawal. Though the US cannot actually withdraw until 2020, the announcement was deeply symbolic: with one of the world's largest climate polluters

declaring the agreement meaningless, it was likely that other countries would take a similar line. However, in practice many US States and corporations stated that they would continue to abide by the principles laid down in Paris, and many countries said they would also continue as before. The latter group included China, which as it is a major climate polluter, was extremely important. It was also highly symbolic as it appeared to represent a change in world order, with China assuming the mantle of world leader, something which Trump had almost certainly not intended.

It is also worth noting that although the USA's attitude to global warming is very important, the scepticism at the highest levels of government there are not unique. In February 2017 an Australian government minister brought a lump of coal into parliament, telling everyone they should not be afraid of it when the opposition had suggested the phasing out of coal-fired power stations in favour of renewable energy sources. At the time the temperature in the Australian capital was over 45°C.

To consider the plight of the Arctic and its wildlife in the face of the global problems associated with climate change might seem indulgent. Yet the fact that the temperature rise in the Arctic is more rapid than across the rest of the Earth means the region acts as a litmus test, an indicator of what the future holds. The stress of environmental change will affect all Arctic species, but particularly the marine mammals for which sea ice is the habitat of choice. It will also affect the Inuit and other native northern dwellers, whose way of life depends on an increasingly scarce resource.

Recent studies have shown that in the absence of sea ice Polar Bears have been forced to eat eggs at goose breeding colonies. In the longer term this will be disastrous for the goose populations and then for the bears as colonies disappear. In the Polar Bear, the Arctic has one of Earth's most iconic animals. Some populations of bear are already under threat and the species may perhaps become extinct in the wild even before the ice disappears. Male Polar Bears coming ashore in southern Hudson Bay, then moving to Churchill to await winter's freeze may lose up to 30% of their body weight as they wait. Pregnant females giving birth in maternity dens around Hudson and James bays may lose 55%. As summer ice has thawed earlier and

## Alternative solutions, alternative energy sources

As concerns have mounted over climate change, various technological options – which allow continued use of fossil fuels, but 'fix' the problem of their use – have been explored. The most obvious is carbon capture and storage which collects the carbon by-product of combustion and sequesters it. Other suggested schemes have involved the science-fictional, such as placing mirrors in space to reflect sunlight; the surreal, such as injecting sulphur into the atmosphere for the same end, with the risk of halting the monsoon season and so endangering the lives of 1.4 billion people in the Indian sub-continent; and some which could possibly work, e.g. seaweed farming (though the idea is based on dubious sums), but which require work on such a massive scale that it is reasonable to ask whether they are viable in a divided world.

Alternative energy sources bring their own problems. Fracking has been blamed in the USA for polluting groundwater used as public water supplies, and has caused an increase in seismic activity in previously stable Netherlands with building collapses and instabilities which may cost the government millions in compensation claims. Tar shales are more polluting than coal and oil, biofuels use land which could otherwise grow food, or decimate important forest and environmental areas (as well as creating $CO_2$ when burned). Uranium mining and power station construction produce $CO_2$ but once operating, nuclear power stations are carbon neutral, though they do bring other risks.

Renewable energy sources bring problems, though usually of a different magnitude. Dam building for hydropower floods land and may affect irrigation downstream – the combination of climate change and dam building has caused havoc in Indo-China, for instance to the 70 million people living on the Mekong delta. Reservoir construction in northern Scandinavia (for both power production and water supplies) has occasionally caused traditional Reindeer migration routes to be flooded, with animals drowning as they attempted to cross newly formed lakes. The James Bay project in Canada involved the construction of dams on the La Grande River on the bay's eastern shore, and the alteration of local watersheds to provide enhanced river flow. The local Cree and Inuit people were not consulted before the work began. The scheme resulted in the flooding of more than 10,000km² of local forest, and with the watershed alteration involved an area of over 350,000km² in total – almost 20% of Quebec's land area. The flooding resulted

in the release of large amounts of methyl mercury into the water, which poisoned the fish that were a mainstay of local communities. In Iceland, which has both geothermal energy and abundant possibilities for hydro power, and has promoted itself as a source of clean energy, experience has shown that even clean energy is not problem-free. In 2009 the Kárahnjúkar project in east Iceland provided power for the Fjarðaál aluminium smelter built by Alcoa, but the environmental cost was significant. The Icelandic government granted Alcoa a licence to emit more sulphur dioxide than the internationally agreed limit. The project required three huge dams and put an area of outstanding natural beauty under water permanently while infrastructure buried an additional area of pristine wilderness; the effect on local wildlife was significant; and the suggested employment opportunities were both limited and, in general, filled by immigrant labour. Now a study is underway on the feasibility of creating an undersea cable to export 'green' energy to Britain. While the cable would carry only some 1.5% of British power needs, many fear the effect on Iceland's environment would be severe, with more wilderness areas being destroyed or degraded.

Wind farms kill birds and onshore farms blight landscapes: they are also no value if there is no wind or, ironically, if the wind blows too hard, and so require conventional power stations as back up. Tidal power stations damage estuarine environments. The sun, of course, is the ultimate provider of free power, and it is certainly the case that covering the desert with solar panels would solve the world's energy problems. But it would also require international co-operation on a (currently) unimaginable scale, as well as the problem of many suitable sites being found in what is at present a highly volatile part of the world. Renewables also suffer from a cost problem. While costs are falling so that solar farms and offshore wind installations might reach parity with gas-firing in the 2020s, the financing of plants is different with high upfront costs being followed by low operating costs (e.g., a solar panel is expensive, but will give 25–30 years of cost free power), a complete reversal of the conventional power station funding where low upfront cost (measured as £/kW) is followed by high running costs (even if environmental costs are ignored). There is also the odd curio – recently the inhabitants of one US town vetoed the building of a solar power station, claiming it would suck up all the sun's energy and so stop photosynthesis, meaning all crops would die.

winter's freeze has happened later, the condition of the bears has declined over the past two decades, with a 15% reduction in average weight, and a reduction in the number of cubs reaching adulthood. There has also been a reduction in the number of cubs that become independent in their first year, a figure that was once around 40% in Hudson Bay, but which has dropped to around 5%. An ice-free ocean would probably mean the extinction of the Polar Bear, and this may well happen within the lifetime of readers of this book. Loss of the Polar Bear would be the tragedy touching the soul of millions.